BIOLOGICAL AND SYNTHETIC POLYMER NETWORKS

Selected papers from NETWORKS 86, being the 8th Polymer Networks Group Meeting held in Elsinore, Denmark, 31 August–5 September 1986.

BIOLOGICAL AND SYNTHETIC POLYMER NETWORKS

Edited by

O. KRAMER

Department of Chemistry,
University of Copenhagen,
Denmark

ELSEVIER APPLIED SCIENCE
LONDON and NEW YORK

ELSEVIER APPLIED SCIENCE PUBLISHERS LTD
Crown House, Linton Road, Barking, Essex IG11 8JU, England

Sole Distributor in the USA and Canada
ELSEVIER SCIENCE PUBLISHING CO., INC.
52 Vanderbilt Avenue, New York, NY 10017, USA

WITH 44 TABLES AND 254 ILLUSTRATIONS

© 1988 ELSEVIER APPLIED SCIENCE PUBLISHERS LTD

British Library Cataloguing in Publication Data

Biological and synthetic polymer networks.
1. Graft copolymers
I. Kramer, O.
547.7   QD382.G7

Library of Congress Cataloging in Publication Data

Biological and synthetic polymer networks.

Bibliography: p.
Includes index.
1. Polymer networks—Congresses.   2. Biopolymers—
Congresses.   I. Kramer, O. (Ole)
QD382.P67B56   1988   547.7   87-27432

ISBN 1-85166-166-2

Printed in Great Britain by Galliard Printers Ltd, Great Yarmouth.

PREFACE

Biological and Synthetic Polymer Networks contains 36 papers selected from the papers presented at NETWORKS 86, the 8th Polymer Networks Group Meeting. NETWORKS 86 was held in Elsinore, Denmark, on 31 August–5 September 1986. A total of nine invited main lectures and 68 contributed papers were presented at the meeting.

A wide range of important biological and synthetic materials consist of three-dimensional polymer networks. The properties range from very stiff structural materials to extremely flexible rubbery materials and gels. Most polymer networks are permanent networks held together by covalent bonds. Such networks are insoluble but they may swell considerably in good solvents. Polymer networks held together by ionic bonds, hydrogen bonds or so-called entanglements are of a more temporary nature. At long times they exhibit a tendency to flow, and they are soluble in good solvents. The paper by Professor Walther Burchard and his co-workers, 'Covalent, Thermoreversible and Entangled Networks: An Attempt at Comparison', serves as a general introduction to polymer networks.

The book contains both theoretical and experimental papers on the formation, characterisation and properties of polymer networks. Two topics were given special sessions at the meeting, namely Biological Networks and Swelling of Polymer Networks.

It was decided to bring scientists who study biological polymer networks together with scientists who study synthetic polymer networks. The two areas have developed different concepts and methods over the years, yet they are related enough for some of the ideas and methods of one area to be useful in the other. Synthetic polymer networks have been studied in great detail for many years and a considerable amount of basic knowledge has been obtained, both regarding the formation and characterisation of

v

polymer networks as well as the relationship between network structure and properties. The synthesis of model networks with simplified network structures has contributed considerably to this knowledge. Most biological polymer networks are found to exhibit much more complex structures than amorphous synthetic polymer networks. On the other hand, biological synthesis often leads to more perfect and regular structures than can be obtained by chemical methods.

Swelling of polymer networks is an old subject which has given rise to much confusion in the literature. However, both theory and experiment have progressed considerably during the last few years. Professor Paul J. Flory had agreed to give a main lecture on his latest theoretical developments. This was, however, prevented by the much too early death of Professor Flory in the early autumn of 1985. The two main lectures on swelling were given by Dr Moshe Gottlieb and Dr Bruce E. Eichinger.

The book has been divided into five sections: Biological Networks, Formation of Networks, Characterisation of Polymer Networks, Swelling of Polymer Networks and Rubber Elasticity.

Finally, I should like to thank Dr Søren Hvilsted for his valuable assistance in the editorial work.

OLE KRAMER

CONTENTS

SECTION 2: FORMATION OF NETWORKS

SECTION 3: CHARACTERISATION OF POLYMER NETWORKS

SECTION 4: SWELLING OF POLYMER NETWORKS

INTRODUCTORY PAPER

1

COVALENT, THERMOREVERSIBLE AND ENTANGLED NETWORKS: AN ATTEMPT AT COMPARISON

W. Burchard, R. Stadler, L. L. Freitas, M. Möller,
J. Omeis and E. Mühleisen

*Institute of Macromolecular Chemistry, University of Freiburg,
Stefan-Meier Str. 31, 7800 Freiburg, FRG*

ABSTRACT

The main features of reversibly gelling systems are reviewed and compared with permanent and entangled networks. Several techniques of characterizing the structure of junction zones are discussed. The necessity of combining the common thermal methods with viscoelastic and light scattering techniques is emphasized, since each technique alone gives only a limited answer.

In a second section, two examples of reversible networks with point-like network junctions are described. These are (i) linear polybutadiene chains which were modified by urazole derivatives and (ii) end-tagged ionomers. Analysis of the viscoelastic moduli in terms of a relaxation spectrum revealed that these gels have no equilibrium shear modulus. Dynamic light scattering exhibited a pronounced slow motion, which is not present in permanent gels, in addition to the common fast motion which is related to the correlation length. The ionomers of one-end tagged polystyrene chains show in cyclohexane inverse micelle formation with an aggregation number of 11–12. The both-end tagged ionomers form clear gels which become liquefied on sonification but are reformed on standing.

Finally, the properties of entangled networks are discussed and compared with reversibly gelling systems. Examination of various polymer systems in semidilute solutions shows that pure entanglement, with no fixed junctions, is very rare. Mostly crosslinks are formed for a short time when a certain concentration is exceeded, and the systems start to display behaviour of association and eventually of reversible gelation.

3

1 INTRODUCTION

The main intention of this contribution is to give an overview on what is known of three types of network: covalent, thermoreversible and entangled.[1] A comprehensive review is not intended, but an attempt is made to find a conceptual line which would link these various types of network. This contribution mainly deals with reversible networks in the presence of a solvent, i.e. with reversible gels.

Theoreticians have dealt in the past mainly with idealized networks.[2a−f] These are at one extreme the model networks, with permanently localized crosslinks which are point-like and hold together f flexible chains, where f is the functionality of the crosslink units. At the other extreme stand the transient networks, in which entanglements form point-like, but non-localized, junctions.[3] In between these two limits of idealized models we find the vast field of physical networks (see Fig. 1). These are often thermoreversible, i.e. a gel at low temperature can 'melt' and form a liquid on heating, and this process can be reversed on cooling.

Remarkable progress has been achieved in the theories of both fields of idealized networks, and with regard to the pure entanglement model, polystyrene chains were found to agree with theory very satisfactorily.[4,5] However, for solvent-dependent characteristics, significant deviations are

Type	Covalent networks	Reversible networks	Pure entanglement
Crosslinks	Localized Permanent	Localized Finite lifetime	Not localized Freely fluctuating
Structure of junctions	Point crosslinks	(a) Point-like (b) Bundles of chains (one-dimensional) (c) Extended domains (three-dimensional)	Topologic constraints
Examples	Copolymers of mono- with di-functional monomers	(a) Ionomers H-bridged networks	Polymer melts
	Epoxies Radically crosslinked chains	(b) Gelatin/collagen Fibrin Polysaccharides	PS in toluene
	End-linked 'Model' networks	(c) PVC Isotactic polymers Block copolymers Heat-denatured globular proteins	

Fig. 1. Overview of network types.

observed with the same samples.[5] These appear to arise from crosslinks having a certain lifetime. The observations make clear that the entanglement network is only an idealization.

Crosslinks in reversible gels are localized, but not permanent.[6] The interacting groups dissociate and associate according to the thermodynamic equilibrium conditions. We nevertheless may ask the following:

(1) Are such junction zones a necessary requisite of reversible gels, or can networks also be prepared with point-like reversible crosslinks?

(2) Supposing that such networks can be chemically prepared, what will the equilibrium and dynamic properties be like? Point-like reversible crosslinks are localized at a certain position of the chain only for a short time. How will such networks differ in behaviour from those of pure entanglement?

To give a comprehensive description of the characteristics of reversible network formation, structure and properties must be considered. The first part of this report (Section 2) deals with various techniques which have been used to evaluate structural information on reversible gels, i.e. helices or crystalline bundles. It is intended to show that, besides the valuable information obtained from any technique, none of these, applied alone, give a comprehensive picture.

In Section 3, two examples of reversible gels with point crosslinks will be represented, and initial results from an uncompleted project will be discussed.

Finally, in Section 4 these two systems are compared with others, where only entanglement was expected.

2 STRUCTURE OF JUNCTION ZONES

2.1 Some General Remarks

As already mentioned, there is much evidence in many systems for the presence of extended junction zones. The observations are probably familiar, but it may nevertheless be useful to recall the facts for some of the products.

Well-known examples of reversibly gelling synthetic polymers[7,8] are polyvinyl chloride (PVC) of a certain syndiotacticity[9a] and atactic poly(butylmethacrylate) (PBMA) in isopropanol.[9b] Takahashi et al.[9a] carried out X-ray diffraction measurements from gels of such PVC samples with a syndiotacticity of $s = 0.52$. They observed two very weak rings embedded in a diffuse background, which evidently results from crystallites. The authors did not determine the amount of crystalization but took these diffraction rings as enough evidence for a gel that is based on crystallites. In view of the very weak crystalline rings one may wonder, however, whether this small amount of crystallites do form a network. Figure 2 gives a schematic picture of a network formed by crystallites.[10]

As another example, Fig. 3 shows a network of collagen fibres which was obtained by precipitation from an acid solution of skin collagen.[11] Electron micrographs from coarse fibrin gels show characteristic fibril structures with bifurcation.[55] With special techniques fine fibrin gels can be prepared. It is useful to have a closer look at the electron micrographs. Figure 4 shows a micrograph from a strained fine fibrin network that was taken by Müller et al.[12] This picture shows clearly that in the initial stage of fibrin network formation only two strands come together. This dimerization process is even better seen in a further figure (Fig. 5(a) and (b)).[13]

Another argument for such a bimolecular aggregation process comes from X-ray diffraction measurements of stretched ι-carrageenan gels which revealed a double-helix structure.[14] Furthermore, the gel melting temperature was found to coincide with a helix–coil transition temperature.

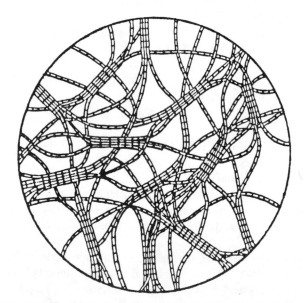

Fig. 2. Schematic diagram of network formed by crystallites (from Herrmann and Gerngross[10]).

Fig. 3. Reconstituted collagen fibril obtained when acid solution of native collagen is neutralized.[11]

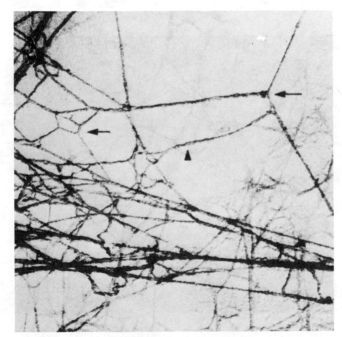

Fig. 4. Details from a stretched fine fibrin network.[12] ⟶, branching points; ▼, untwined double helix section.

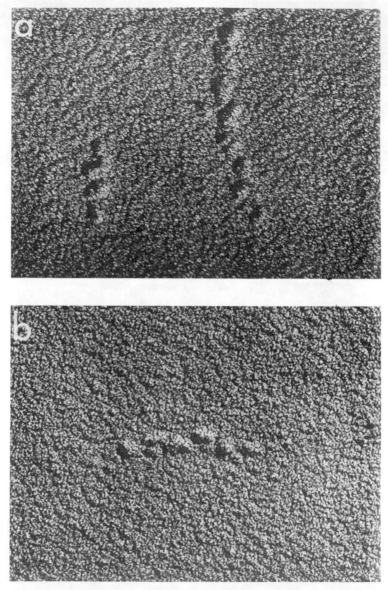

Fig. 5. Electron micrograph from (a) a fibrin dimer (left) and part of higher polymer (right) and (b) from a fibrin tetramer.[13]

Thus, it is persuasive to assume in these cases that the gel melting process is based on the denaturation of double helices, i.e. the separation of the helix into two independent coils. In gelatin at least three chains are involved in gel setting, but here one could think of the triple helix formation as being caused by three essentially independent pair formations. Also, if bundles are formed one may consider this process as a series of bimolecular processes.

These few examples may be sufficient to pose the following questions:

(1) How many chains are in a bundle or a high density domain, and how many strands are involved in the melting or gelation process? Not necessarily all domains have to melt before the gel becomes a solution.

(2) How long are the junction zones, or, more generally, what is the size of the dense domains?

(3) Have these bundles or domains crystalline order or are they only liquid crystalline or even amorphous?

In the following, different techniques will be discussed which all have their disadvantages but give partial answers to these questions. Two of these techniques are based on thermoanalysis measurements, two others on the concentration dependence of the elastic shear modulus and the last one on light scattering measurements.

2.2 Evaluation of DSC Curves

The discussion may be started with the DSC measurements from *ι*-carrageenan. This type of carrageenan forms clear gels and for this reason it may be justified to assume that only double-helix formation is involved in the gelation process.[14] Figure 6 shows the chemical structure. The chain consists of a disaccharide repeating unit: the bond in the (1,3)-galactose-(1,4)-3,6-anhydrogalactose structure favours a stretched chain conformation. However, approximately every 35th repeating unit contains a sugar, in which the 3,6-anhydro ring has been cleaved. As a consequence, the chain conformation is flipped into another, more stable, conformation and forms a kink in the stretched chain conformation which prevents a macroscopic crystallization. Furthermore, the chain can be segmented at this position because this kink unit can be oxidized, and thus the chain can be split at this position by a weak acid degradation (Smith degradation). Segments alone do not form gels.

Figure 7 shows a typical DSC diagram after correction for the baseline.[15] Stepwise integration yields a transition curve as a function of temperature.

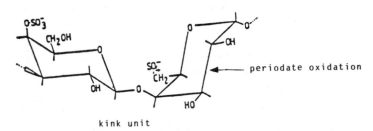

Iota-Carrageenan

repeating unit

kink unit

Fig. 6. Repeating unit and kink unit in ι-carrageenan. The arrow indicates the position of periodate oxidation which leads to a cleavage of the sugar ring. Subsequent weak acid degradation (Smith degradation) leads to chain scission at the C1-position.

In Fig. 8(a) and (b) this melting curve is compared with that from the segments and with those obtained from optical rotation measurements. The transition curves demonstrate that it is the denaturation of the helix structures of the segments which governs the melting process of the gel.

To describe the melting process more quantitatively one can try to apply a simple association equilibrium between *two chains*:

$$2\,A \underset{}{\overset{K}{\rightleftharpoons}} A_2 \qquad (1)$$

With the law of mass action one then finds

$$K = [A]/[A_2] = c_o(1 - \theta)^2/\theta = \exp\left[-\Delta H_m/RT\right] + \text{const} \qquad (2)$$

where $c_o = [A] + 2[A_2]$ is the overall concentration and $\theta \equiv 2[A_2]/c_o$ is the degree of association. A plot of $\ln K$ versus $1/T$ should give a straight line with a slope that is proportional to the heat of melting ΔH_m. Such straight lines are indeed observed (see Fig. 9). A value of $\Delta H_m = 295 \pm 60\,\text{kJ/mol}$ is found on average in the concentration range of $2 \cdot 77 - 16 \cdot 58\,\mu\text{mol/ml}$, and

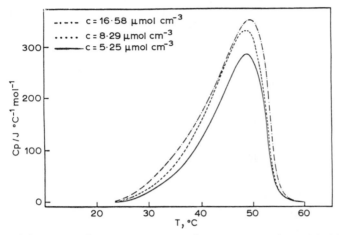

Fig. 7. DSC curves of ι-carrageenan at three concentrations: (a) 5·25 μmol; (b) 8·29 μmol; (c) 16·58 μmol.[15]

this rather large value cannot be assigned to the breaking of only one physical bond.

From the area of the total melting curve a heat of transformation per residue of $\Delta H_{res} = 4·73 \pm 12\ kJ/mol$ is obtained, which is reasonably in the order of the breaking of one H-bond in an aqueous medium. Comparison of the two heats of melting shows that about 58 ± 20 units are involved in a

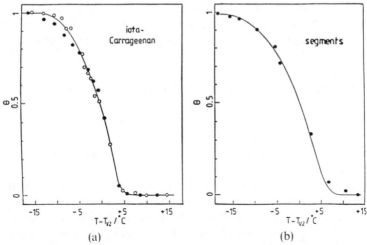

Fig. 8. Normalized transition curves of (a) ι-carrageenan and (b) its segments.[15] ●, from DSC; ○, from optical activity measurements.

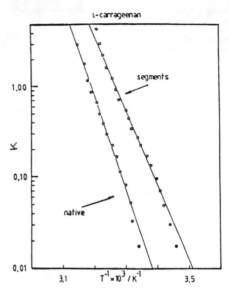

Fig. 9. Arrhenius plot of the helix–coil equilibrium constant from native and segmented ι-carrageenan.[15]

co-operative melting process. If only a double helix formation is assumed then this number corresponds to the helix which melts simultaneously, and $\Delta H_m/\Delta H_{res} = L_{coop}$ is therefore often called a co-operativity length (expressed in number of repeating units).[16] In the present case this length should be approximately the segment degree of polymerization, which is indeed observed.

2.3 Concentration Dependence of the Melting Temperature

The detailed analysis of a DSC melting curve is difficult if more complicated processes than a simple helix–coil transition are involved in the gel melting. Also, a very sensitive calorimeter is needed. Much easier is the determination of the melting temperature where a number of different techniques can be used.

Strictly speaking the melting temperature is defined as the midpoint temperature in the transition curve. In all practical applications, however, simply the temperature where flow starts is taken. Eldridge and Ferry,[17] for instance, filled a small test-tube with the solution at high temperatures, quenched the material to temperatures where gelation occurs and turned the test-tube upside down. This test-tube was then slowly heated, and the

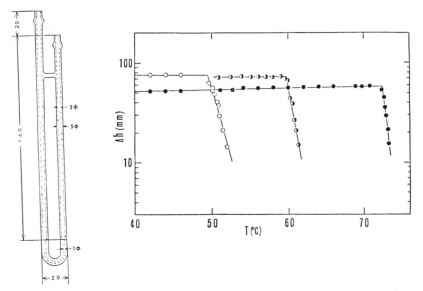

Fig. 10. Gel melting temperature apparatus (left) and examples for determining T_m[9a] (right).

temperature at which the material fell down was taken as T_m. Takahashi *et al.*[9a] filled a U-formed capillary tube with the solution at elevated temperature and let the gel form such that there remained a certain difference between the two levels of the U-tube. The upper level was then observed by a cathetometer while the tube was slowly heated. The melting temperature was taken then as the kinkpoint in Fig. 10. Baer and his coworkers[18] applied a similar technique, but followed the sinking of a small ball as a function of temperature.

From eqn (2) it follows that[9a]

$$-\ln c_o = \ln \phi_2/\bar{v}_2 = -\Delta H_m/RT_m + \text{const}_1 \qquad (3)$$

Thus, the melting temperature for a bimolecular association process must depend on the logarithm of the polymer concentration or on the volume fraction ϕ_2, since $\phi_2 = \bar{v}_2 c_o$, where $\bar{v}_2$ is the partial specific volume of the polymer.

In their now famous equation, Eldridge and Ferry[17] went one step further and took into account a molecular weight dependence of the melting temperature. This is a reasonable assumption for longer chains where several sections can form junction zones. The reactions of these

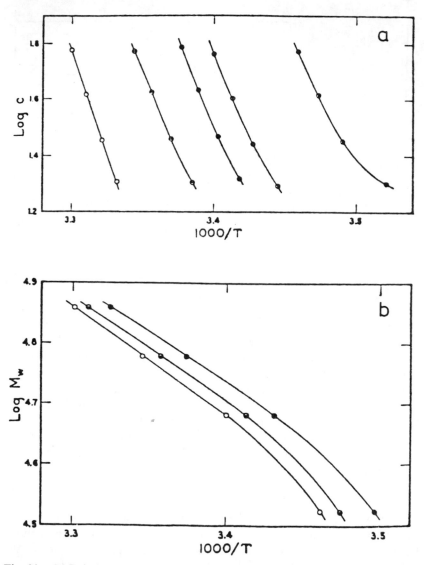

Fig. 11. (a) Relation between gelation concentration and melting temperature for gelatin of $M_w = 72\,000$, $60\,000$, $52\,700$, $48\,000$ and $33\,400$ (from left to right). (b) Relation between weight-average molecular weight and melting temperature for gelatin,[17] $c = 55$, 40 and 25 mg/ml (from left to right).

zones will in general not be independent of each other. Since it is difficult to make an estimation of the interdependence purely on theoretical grounds, Eldridge and Ferry[17] made a semi-empirical estimation. They measured the shear modulus as a function of molecular weight. According to the theory of elasticity, this modulus is proportional to the number of junctions n per unit volume. Thus, an empirical relationship between n and the molecular weight M_w can be established, which for gelatin was found to be

$$n \sim M_w^{7 \cdot 0} \tag{4}$$

Using this relationship the authors obtained

$$\ln M_w = (1/7)\Delta H_m / RT_m + \text{const}_2 \tag{5}$$

The two equations (3) and (5) can be combined, which yields[9a]

$$\ln c M_w = (8/7)\Delta H_m / RT_m + \text{const}_3 \tag{6}$$

Thus a common line should be obtained from graphs of the concentration *and* molecular weight against the melting temperature. Such common lines are obtained for gelatin[17] but also for the PVC gels[9a] (see Fig. 11(a) and (b)). Linearity was found with gelatin, however, only if $x\phi_2 \sim cM_w$ is sufficiently large, where x is the degree of polymerization. The deviations for shorter chains were attributed to network imperfections, e.g. dangling chain ends and ring formation. The validity of eqn (6) for PVC with the same prefactor of 8/7 is rather unexpected, since the scaling behaviour of eqn (4) cannot be assumed to be universal.

Takahashi *et al.*[9a] refined the Eldridge–Ferry theory by:

(1) assuming copolymeric block sequences which can crystallize;
(2) taking into account the segment or junction interface free energy σ_{ec}, i.e. the interface of the junction ends towards the amorphic chain section part;
(3) taking into account the melting temperature depression of copolymers.

Their equation, which is not produced here, represents many observed features more satisfactorily than the Eldridge–Ferry theory does, but the final result for PVC still remains a little disappointing. For the crystallizable segments the authors found lengths of 10–12 repeating units and interface free energies of $\sigma_{ec} = 250$ to -1827 J per junction, depending on the solvent used. A similar length would be estimated by the naive comparison of the co-operative heat of melting with the heat of fusion per repeating unit. In fact the low value of σ_{ec} is not informative for an estimation of the *thickness* of the crystallites. Attention may be also drawn to the very short length of the isotactic blocks in the PVC sample studied, which makes the conception of crystalline junctions questionable.

2.4 Concentration Dependence of the Shear Modulus: The Theory of Oakenfull

Oakenfull[19] uses the increase of the shear modulus with concentration to estimate length *and* number of chain segments involved in the melting process. He starts with the relationship for the shear modulus of an *ideal rubbery elastic* gel[20] with flexible network chains

$$G = cRT/M_c \tag{7}$$

where c is the weight concentration and M_c the number-average molecular weight of a chain joining two adjacent crosslinks, which in these cases are the extended junction zones. Network imperfections, i.e. dangling chains and ring formation, are neglected. M_c is then expressed in terms of: (1) M_n, the number-average molecular weight of the non-crosslinked chains; (2) J, the number of junction zones per volume; and (3) M_j, the molecular weight of a junction. This leads to

$$M_c = [M_n - (M_j M_c/cN_A)]/(JM_n cN_A - 1) \tag{8}$$

The denominator can be recognized as the number of active chains, and the second term in the numerator represents the fraction of chains involved in junction zones. Passing from number concentrations of the junction zones to *molar* concentrations, $[J] = J/N_A$, and using eqn (8) he finds for the modulus

$$G = (RTc/M_n)(M_n[J] - c)/(M_j[J] - 1) \tag{9}$$

Next an n-fold co-operative lateral association of strands is assumed, instead of the bimolecular association assumed in older theories. With the law of mass action the equilibrium constant results:

$$K_j = [J]M_j^n/[n(c - M_j[J])]^n \tag{10}$$

If $[J]$ is eliminated from eqns (9) and (10), one obtains a relationship for G as a function of the concentration c.

In general there are four unknown quantities, i.e. (1) the molecular weight of the chains, M_n; (2) the molecular weight of a junction, M_j; (3) the number of strands involved in a junction n; and (4) the equilibrium constant, K_j. The fit of an experimental curve is not a simple procedure since eqn (10) cannot be solved analytically for $[J]$.

Oakenfull[21] applied his technique to gelatin and several pectins which differ in the degree of esterification. He found for gelatin $n = 3$ and for the various pectins $n = 2$, i.e. in the first case only three strands and in the others only two strands are involved in the gelation process. It would be interesting to also apply this technique to PVC gels, to get information on the number of strands involved in the crystallites of the junctions; unfortunately this was not done.

The length of the junction zones was found in gelatin to be 47 amino acid residues per strand, which corresponds to 16 turns in a helix. This value is about three times larger than was estimated by Eldridge and Ferry.[17] However, in their original paper the authors estimated an enthalpy of fusion per residue of $\Delta H_m^{res} = 20 \, kJ/mol$, which is probably about four times too large for the breaking of one H-bond in the aqueous medium. With the more realistic value of about $5 \, kJ/mol$, the length of a junction is also in these older experiments about 20 turns in the single helix, in good agreement with 16 turns found by Oakenfull.

For the pectins a pronounced increase of the junction length with increasing degree of esterification was obtained. The lengths varied from 34 repeating units to 460 units if the degree of methylation was changed from 0·63 to 0·93. Evidently the junction zones become stabilized by hydrophobic interaction.

2.5 Concentration Dependence of the Shear Modulus: The Technique of Clark and Ross-Murphy

The results obtained by Oakenfull for gelatin appear to be reasonable and are in fair agreement with the DSC and T_m measurements as a function of concentration. In view of the drastic oversimplifying assumptions, which neglect the remarkable rigidity of the junction zones and all network imperfections, this good agreement is surprising. It appears questionable that the simple equation of an ideal rubber can be used in this unmodified version.

These doubts become strengthened by the recent study of Clark and Ross-Murphy[22] on biological gels. The authors calculated the number of active network chains with the aid of Gordon's branching and gelation theory.[23] They incorporated in this theory the basic assumption, first introduced by Hermans,[24] that the equilibrium of association is established only between *two functional groups* and not between strands of chain sections. The equation for the shear modulus, derived in this manner, contains three unknown parameters, i.e. (1) f, the number of functional groups per chain; (2) c_{cr}, the critical concentration below which no gelation

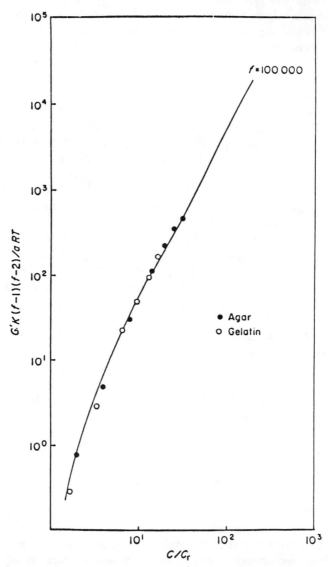

Fig. 12. Shear modulus data reduced to the high f-cascade limit.[22]

occurs; and (3) a, a generalized front factor. The value of c_{cr} can be obtained from experiments by extrapolating the shear modulus $\lim G(c) \to 0$. This concentration is related to the equilibrium constant K and the number of functional groups f as

$$c_{cr} = M(f-1)/Kf(f-2)^2 \qquad (11)$$

Figure 12 shows the result of measurements of $G(c)$ represented in a scaled master curve (for details see the original paper[22]). For all biopolymers examined by the authors, the best fit was obtained with a very large number of functional groups in the order of $f = 10^5$, which appears to be reasonable since f can be assumed to be equivalent to the degree of polymerization. The front factors differ widely for the three types of gel and were found (for $f = 10^5$) to have the following values:

$a \simeq 0.3$ for gelatin
$a \simeq 15$ for agarose
$a \simeq 0.6{-}30$ for bovine serum albumin gels prepared at different pH values

In all cases this empirical front factor appeared to increase with the rigidity of the extended junction domains. These results suggest that the rigidity has a significant effect on the front factor and thus on the validity of the common theory of rubber elasticity. The high front factors possibly reflect an enthalpic part of the modulus. The enthalpic part of the modulus, which in permanent gels is determined from the temperature dependence of the modulus, no longer will show in such cases an increase of the modulus proportional to the temperature. Corresponding experiments are, however, not conclusive with thermoreversible gels, since the extent of crosslinking usually decreases with increasing temperature.

The Clark and Ross-Murphy approach does not give information on the extension of the domains. It should be mentioned, however, that values for heterogenic structures can to some extent also be calculated by means of Gordon's branching and gelation theory,[25] and this may allow a theoretical prediction of the high front factors.

2.6 Direct Structure Determination by Scattering Experiments
Direct information on the structure of the junction zones can be obtained from scattering experiments, and it is of interest to compare these results with those obtained indirectly from thermodynamic and viscoelastic experiments. Only in a very few cases has such direct comparison been

made, and the following discussion is confined to the light scattering results obtained with ι-carrageenan.[15,26]

Figure 13 shows a typical light scattering result in the representation of a Holtzer plot,[26] where qR_θ/Kc is plotted against $q = (4\pi/\lambda)\sin\theta/2$, with θ being the scattering angle. R_θ is the scattering intensity (Rayleigh ratio) at the scattering angle θ, K an optical contrast constant and c the concentration. From such a curve the following conclusions can be drawn:[26,27]

(1) The occurrence of a q-independent plateau indicates the presence of long rod-like filaments.

(2) The value of the non-normalized plateau gives the linear mass density $M_1 = (M/L)$, where M is the molecular weight of the filament and L its contour length.

(3) The ratio of the height of the maximum to that of the plateau gives the number of Kuhn segments N_K per filament.

Thus if the molecular weight can be estimated, by independent measurements, the contour length can be determined, and with the aid of Kuhn's equation for the contour length of a worm-like chain $L = N_K l_K$, the length of a statistically independent Kuhn segment l_K is obtained. Since the linear mass density of a single strand can be calculated from the bond length of a repeating unit and its molecular weight, the number of strands in

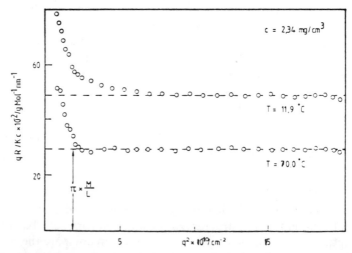

Fig. 13. Holtzer plot of ι-carrageenan light scattering data at $c = 2.34$ mg/ml and $T = 11.9$ and 70°C.[15]

TABLE 1

Contour Length L, Kuhn Segment Length l_K and Number n of Associated Strands of Carrageenan Filaments in Gels of $0.1\,N$ NaCl solutions of Low and High Carrageenan Concentrations at 20°C

Low concentration	High concentration
$c = 1.61\,\mu\mathrm{mol}$	$c = 8.86\,\mu\mathrm{mol}$
$L = 1\,700\,\mathrm{nm}$	$L = 3\,600\,\mathrm{nm}$
$l_K = 470\,\mathrm{nm}$	$l_K = 300\,\mathrm{nm}$
$n = 5.6$	$n = 3.35$
$L_{\mathrm{coop}} \cong 30\,\mathrm{nm}$	

a filament can be estimated from the ratio of the experimentally determined to the theoretically expected values for a single strand. The whole curve can be calculated by Koyama's[28] theory for semi-flexible chains with the three parameters (i) n, the number of strands; (ii) L, the contour length of the filament; and (iii) N_K, the number of Kuhn segments. The result is schematically listed in Table 1. These results may be summarized as follows. The filaments are about 60–120 times longer than the co-operative melting length; the Kuhn segments are at the low concentration about 16 times larger than the co-operative length and about 10 times larger at the high concentration. The number of laterally aggregated strands is about eight in dilute solution and decreases to a little more than two at the highest concentrations. A similar decrease to about two strands is observed on heating the solutions. The apparent decrease of the Kuhn segment length with increasing concentration, which is accompanied by an increase of the filament length, can be interpreted as the correlation length in a transient network.[3] The structures occurring at different concentrations are schematically shown in Fig. 14. These findings demonstrate that the

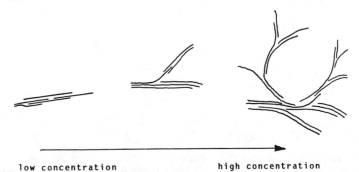

low concentration   high concentration

Fig. 14. Model suggestion for aggregation of the K-form of ι-carrageenan.[15]

structures actually existing in a gel can be much more extended than would be estimated from thermodynamic measurements.

3 ARE JUNCTION ZONES AN IMPERATIVE REQUISITE OF REVERSIBLE GELS?

So far we have tacitly assumed the presence of extended junction zones in thermoreversible gels. We may ask now whether such junction zones are really needed in reversible gels or whether reversible gels can also be prepared with point-like reversible crosslinks. In the following, two examples will be discussed where point-like bonds are evidently involved. These are modified polybutadiene chains in one case and end-tagged ionomer chains in the other.

3.1 H-Bond Modified Polybutadiene Networks
By addition of 4-phenyl-1,2,4-triazoline-3,5-dione (PTD) to polybutadiene, either in solution or in the melt, the polybutadiene chain is modified by substituents which are capable of H-bond formation.[29,30a−c] The reaction with the double bond of the polybutadiene repeating unit is shown in Fig. 15. The reaction takes place readily, which can be easily followed by the disappearance of the deep red colour of the reactant. The substituents were

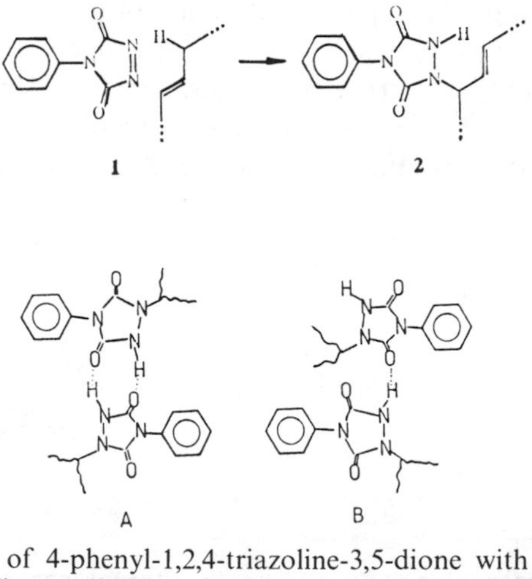

Fig. 15. Reaction of 4-phenyl-1,2,4-triazoline-3,5-dione with double bonds of polybutadiene.[30a] The possibilities of H-bond formation between two groups of (2) are demonstrated by A (two H-bonds) and B (one H-bond).

shown being randomly distributed along the polymeric chain.[30b] For large chain lengths gelation is observed already with a very low degree of substitution, but if lower molecular weight polybutadienes are used the degree of substitution can be increased up to about 2% before gelation is observed.

The fraction of free and H-bonded urazole groups could be measured by IR spectroscopy evaluating the carbonyl stretching vibration bond which is shifted from $1723 \, cm^{-1}$ for the free groups to $1701 \, cm^{-1}$ for the H-bonded structures. The change of the extinction coefficient with temperature is shown in Fig. 16. From these data the equilibrium constant can be calculated and plotted according to Arrhenius. The slope in this curve gives a heat of complex formation of $\Delta H_m = 24 \, kJ/mol$. This value is in the expected order for two H-bonds in a non-polar medium, and the melting process thus indicates no co-operativity. This does not, however, mean that the H-bond formation of the various substituents occurs fully at random. Preliminary model calculations[31] show that such an independence is not possible for substituents which are localized on a chain, even if this chain is ideally flexible and the substituents are placed at random along the chain.

The effect of the modification on the viscoelastic properties of the

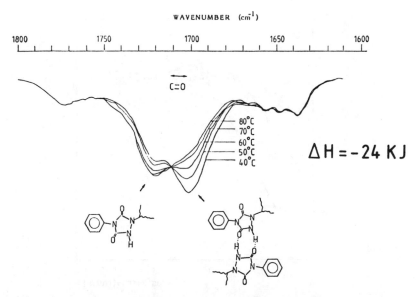

Fig. 16.   Infra-red spectra of the H-bonded complex of modified polybutadiene at various temperatures.[30b]

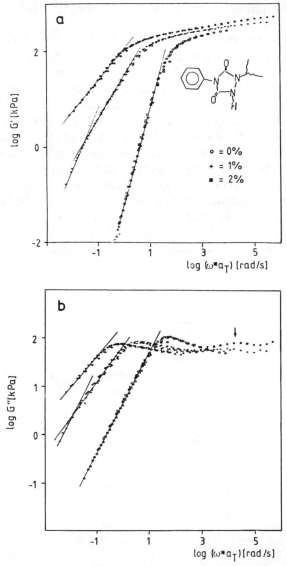

Fig. 17. Isothermal viscoelastic mastercurves for a polybutadiene with $M_w = 49\,500$ and modified polybutadienes with 1% and 2% urazole.[30c]

corresponding bulk system is demonstrated by Fig. 17, where the storage modulus $G'(\omega)$ and the loss modulus $G''(\omega)$ are given for a polybutadiene with different groups. These mastercurves (reduced to 273 K) were obtained from isothermal data in the temperature region between 220 K and 340 K. From these data the relaxation time spectra $H(\tau)$ were obtained according to the approximate formula derived by Tschoegl.[32] Starting from G'

$$H(\tau) = G'\left[\frac{d \log G'}{d \log \omega} - \frac{1}{2}\left(\frac{d \log G'}{d \log \omega}\right)^2 - \left(\frac{1}{4 \cdot 606}\right)\frac{d^2 \log G'}{d (\log \omega)^2}\right]_{\omega = \sqrt{2}/\tau} \quad (12)$$

Starting from G''

$$H(\tau) = \frac{2G''}{\pi}\left[1 - \frac{4}{3}\left(\frac{d \log G''}{d \log \omega}\right) + \frac{1}{3}\left(\frac{d \log G''}{d \log \omega}\right)^2 + \frac{1}{6 \cdot 909}\frac{d^2 \log G''}{d (\log \omega)^2}\right]_{\omega = \sqrt{5}/\tau} \quad (13)$$

The approximate spectra were optimized in an iterative procedure[33] similar to a method described by Graessley[34] (Fig. 18(a)). While the modified polybutadienes at 273 K and the experimental frequency of 0·0443 Hz apparently show the typical behaviour of a network, the calculated frequency dependent storage modulus (eqn (12) and solid curve in Fig. 18(b)) is identical to the measured storage modulus, i.e. the equilibrium network modulus G_e is 0. In other words, these reversible gels have no real equilibrium elastic shear modulus, and the network starts to flow under the influence of an external force, though at a very low creeping rate. This flow may arise from the breaking of bonds at a certain time and the reformation of other bonds shortly after.

A similar conclusion was drawn from dynamic light scattering behaviour of ι-carrageenan in the gel state.[15,35] Figure 23 (see later) shows the presence of very slow motion besides the expected fast motion. Such slow motion is not observed with permanent gels[36] and results in the present case from an apparent translational motion of clusters. These clusters evidently decompose at one side and grow at random at another side which causes a random displacement of the centre of mass.

3.2 End-linked Ionomers
Another example of a physically crosslinked network is given by the so called ionomers. Typically, ionomers are non-polar polymers to which small amounts of ionic groups are attached. Several extensive reviews describe their synthesis and properties.[37,38] In bulk or dissolved in a non-polar solvent, the interaction of the ion pairs results in the formation of

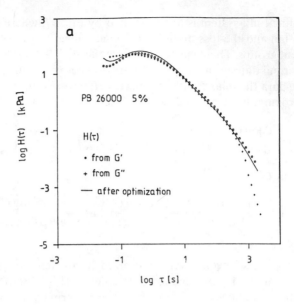

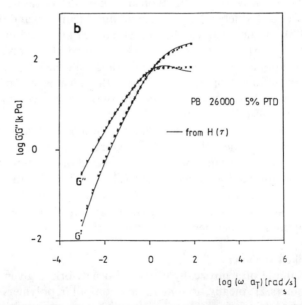

Fig. 18. (a) Relaxation spectra $H(\tau)$ for a 2% modified polybutadiene derived from G' and G''. (b) From $H(\tau)$ calculated moduli G' and G'' in comparison to experiments.[30c]

multiplets or clusters. The studies demonstrate how ionic interactions alter the solution behaviour as a function of polymer concentration fraction, nature of the ionic groups and polarity of the solvent. Some examples have viscoelastic properties similar to higher molecular weight analogues, while other ionomers can act even at low frequency as effectively crosslinked systems.[39] While in most examples studied the ionic groups were introduced statistically along the chain, polymers can be synthesized in which the ionic groups are attached solely at the chain ends.[40,41] In this case the number of interacting groups per molecule is two and the average distance between the interacting groups is related to the molecular weight. In the case where only one of the chain ends is functionalized by an ionic group, soluble well-defined multimers can be formed. Such systems offer a unique opportunity for the investigation of the correlation between strength of ionic pair interactions, multiplet size, dynamics and gelation behaviour.

We now discuss an ionomer where the ionic sulphonate group is placed either at only one chain end or at both ends. Such ionomers have been prepared by living anionic polymerization of polystyrene, either with butyl lithium or with lithium naphthalene as initiator, and where the living endgroups are terminated by addition of propansulton.[42,43] The reaction is schematically shown in Fig. 19. In the first case, where only one chain end is tagged, soluble clusters are expected, while in the second case, with ionic groups at both ends, network formation can be expected.

To get information on the strength of the interaction, the association mechanism was studied first with the one-end functionalized ionomers.[42,43] Two samples with molecular weights of 40 000 and 80 000 were studied. Light scattering measurements were carried out in cyclohexane at 27°C which proved to be a theta solvent for these modified polystyrene chains. Figure 20 shows the result of the normalized inverse osmotic compressibility $(M/RT)(\partial \pi/\partial c)$ as a function of the concentration, where c has been varied from the typical dilute regime, i.e. 0·01%, up to about 10%. The osmotic compressibility corresponds to the reciprocal zero angle scattering intensity and is defined as

$$Kc/R_{\theta = 0} = (1/RT)(\partial \pi/\partial c) \tag{14}$$

In Fig. 20 this inverse osmotic compressibility has been multiplied with the molecular weight which is 11–12 times the single chain molecular weight and corresponds to the plateau value in the concentration region of 0·1–1%. Below 0·1% a strong increase of the inverse osmotic compressibility occurs which indicates dissociation into single chains. The plateau value,

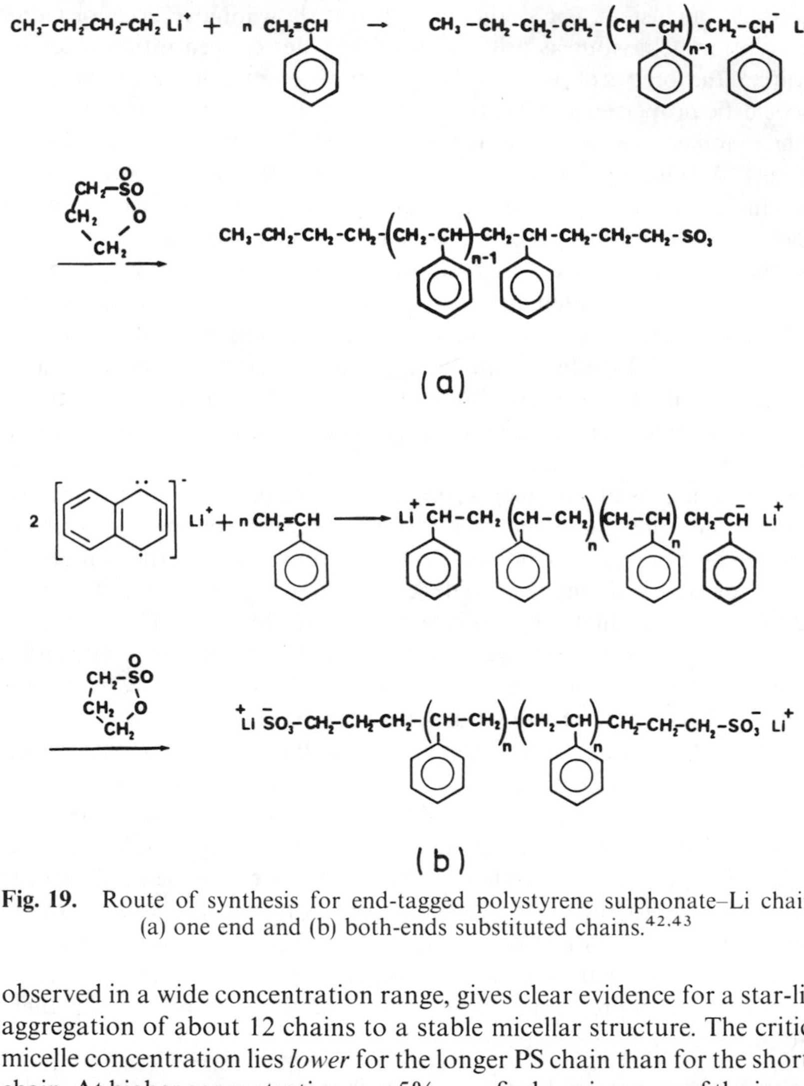

(a)

(b)

Fig. 19. Route of synthesis for end-tagged polystyrene sulphonate–Li chains: (a) one end and (b) both-ends substituted chains.[42,43]

observed in a wide concentration range, gives clear evidence for a star-like aggregation of about 12 chains to a stable micellar structure. The critical micelle concentration lies *lower* for the longer PS chain than for the shorter chain. At higher concentration, $c < 5\%$, one finds an increase of the inverse osmotic compressibility for the shorter chain, whereas for the longer chain the opposite behaviour is observed. In Fig. 20(b) the same data as shown in Fig. 20(a) are plotted against c/c^*, where the c^* is the coil overlap concentration defined through the hydrodynamic volume as

$$c^* = M/(N_A(4\pi/3)R_h^3) \tag{15}$$

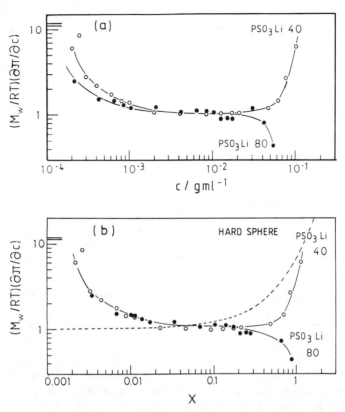

Fig. 20. Double logarithmic plot of the inverse osmotic compressibility $M_w/RT \times (\partial\pi/\partial c) = M_w Kc/R_{\theta=0}$ versus (a) concentration and (b) $X = c/c^*$, where $c^* = M_w/(N_a 4\pi/3 R_h^3)$.

The hydrodynamic radius R_h was obtained from the translational diffusion coefficient D measured by means of dynamic light scattering, using the Stokes–Einstein relationship

$$R_h = kT/(6\pi\eta_0 D) \qquad (16)$$

with η_0 the solvent viscosity.

In this plot the results from the two different chain lengths now form a common curve up to $c/c^* = 0.2$. The dotted line describes the predicted behaviour of hard, impenetrable spheres. Clearly the osmotic compressibility must diverge at a concentration of closed sphere packing. The steep increase of the osmotic compressibility occurs for the micelles at larger c/c^*

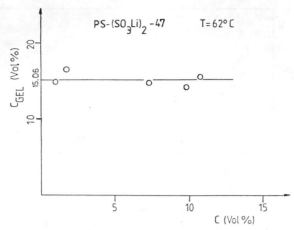

Fig. 21. Gel concentration of a both-ends tagged polystyrene sulphonate–Li chain for various initial overall concentrations as shown on the abscissa.[43]

values since these molecules can partly interpenetrate (note, however, the ambiguity of c^*).†

The decrease of the inverse osmotic compressibility for the larger molecular weight results from an association of micelles. Here the segment density at the periphery is lower than for the short chain PS, therefore the micelles can interpenetrate more deeply, and this gives rise to a marked segment–segment association of different micelles. This part will be discussed in greater detail in the next chapter.

The investigation of the polystyrene with the sulphonate groups at both chain ends is not complete yet. In non-polar solvents fully functionalized samples give clear gels, even at very low concentrations.[43] The samples could be 'dissolved' at elevated temperature in cyclohexane only under the action of an ultrasonic source. The solutions remained clear at 62°C for a short time; then a clear gel was obtained at the bottom of the cell. The gel fraction was found to increase with the overall polymer concentration, but the concentration in the gel was found to remain constant. This gel concentration of 15% is about two times larger than the overlap concentration c^* of the corresponding star aggregate (see Fig. 21).

Addition of polystyrene in which the chains were functionalized by

† The hard sphere curve was calculated for low c/c^* values from a virial expansion including the fourth virial coefficient,[6,44,45] and for higher c/c^* the free volume approximation[38] was used, where $1/c^* = 4(4\pi r^3/3M)0.74$, with r being the sphere radius.

sulphonate groups only at one chain end changes the gelation behaviour significantly. In the case when only about 50% of the chains have the sulphonate groups at both chain ends while the rest of the chains have only one sulphonate group, gelation becomes concentration dependent. At low concentrations solutions are formed which have only slightly higher viscosities than the non-functionalized polystyrenes of equivalent molecular weight. At a concentration c^*, which corresponds to the overlap concentration, the viscosity is increased steeply up to 6 orders of magnitude. Clearly the gelation process depends on the number of crosslinking chains per ionic multiplet.

4 ENTANGLEMENT AND REVERSIBLE GELATION

4.1 Entangled Systems

It is now about 10 years since a rather detailed picture was developed for interpenetrating linear and flexible chains in a concentration region where the chains become highly entangled.[3,46] In the limit of very good solvents the solvated segments exert repulsion on each other. Because of the interaction with the solvent molecules, the interaction between bare segments is highly screened. As a result, no polymer–polymer interactions have to be considered. The points of entanglement are neither localized nor fixed. The entanglement may be regarded as a topological constraint which can fluctuate and disentangle provided there is enough time.

It was mainly de Gennes[3,46] who derived, by means of simple scaling arguments, an essential relationship for the inverse osmotic compressibility which is given by

$$(M/RT)(\partial \pi / \partial c) \sim (c/c^*)^{1 \cdot 25}$$

Scaling laws can be applied to derive asymptotic relationships, which in the present case means $c/c^* \gg 1$, but the crossover from dilute to semi-dilute behaviour cannot be derived by this technique. This limitation has been overcome in the last few years by employing the space group renormaliz- ation theory, and a most useful relationship has recently been derived by Ohta and Oono.[47]

Figure 22 shows the prediction of this theory. It gives a satisfactory agreement with the experimental data for polystyrene (PS) in toluene, which is a very good solvent for PS. If on the other hand the solvent is changed, for instance to methylethylketone (MEK), the experimental points follow the theory only up to about five times the overlap

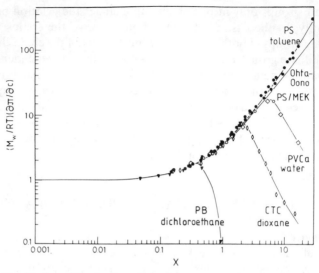

Fig. 22. Double logarithmic plot of the inverse osmotic compressibility against $X = A_2 M_w c \sim c/c^*$ for various systems (PS = polystyrene, PVCa = polyvinyl caprolactam, CTC = cellulose-tri-carbanilate, PB = polybutadiene). The turnover indicates the onset of reversible association. The curve labelled 'Ohta–Oono' presents results from renormalization group theory.[48]

concentration. Then a strong excess scattering occurs at small angles, and the inverse osmotic compressibility shows a turnover to lower values. All details of static and dynamic light scattering indicate the formation of clusters which grow in size when the concentration is increased.[49] This cluster formation is reversible, i.e. the clusters disappear on dilution, and consequently the crosslinks are no longer pure entanglements but are fixed with a certain lifetime at a special chain position.

This behaviour is not a unique property of PS in marginal solvents but is found with almost all other systems. Only a few examples are shown in Fig. 22; linear polybutadiene (PB) in dichloroethane also displays this behaviour.

Dynamic light scattering revealed in the same concentration region a typical bimodal relaxation spectrum (see Fig. 23(a)). The first relaxation can be assigned to the predicted co-operative diffusion mode[3] in an entangled system and is described by the topologically defined correlation length ξ which is the average length between two points of entanglement. The slow motion must be attributed to the translational motion of clusters.[50]

It is now of interest that the very same type of slow motion is also

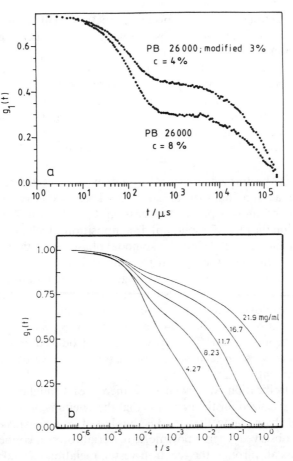

Fig. 23. (a) Time correlation function obtained by dynamic light scattering for a modified PB (above) and non-modified PB (below) in dichloroethane. (b) Time correlation functions from a K-ι-carrageenan gel at various concentrations.[15]

observed in the ι-carrageenan gels,[15,40] whereas in permanently cross-linked gels the slow motion suddenly disappears when the gel point is passed.[41]

4.2 Reversible Gelation and Phase Separation

All the results presented in the preceding section indicate that pure entanglement is observed only in systems where the interaction between monomers is sufficiently well screened. Otherwise, non-specific van der

Waals interactions or specific polar interactions will cause a reversible association which becomes observable and will increase at higher concentrations.

This interpretation of the various results obtained in the semi-dilute regime becomes confirmed by reconsidering the results by Tan *et al.*[18] These authors noticed a thermoreversible gel setting of PS at low temperatures in almost all solvents when the overlap concentration was exceeded. The low angle excess scattering observed at room temperature turned out to be directly related to the capability of gelation for these systems.[51]

Tan *et al.*[18] performed an extensive study with PS of different molecular weights in a large number of solvents. These experiments allow the clarification of another point of discussion. In 1974 Prins[52] brought into consideration the idea of a spinodal decomposition which takes place if the system is quenched well below the spinodal phase separation temperature. Cahn[53] predicted in a theory that the diffusion-controlled onset of phase separation introduces a special heterogeneity with periodic domains of high and low segment densities. Such periodicity has indeed been observed with metallic alloys. Prins[52] speculated that the bundle formation observed with agar-agar on gel setting may be caused by this periodicity during the spinodal decomposition. This speculation introduced much confusion since thermoreversible gelation and phase separation were tacitly assumed to be equivalent.

A first clarification was given by a theory of Coniglio *et al.*[54] These authors emphasized for the first time that the phenomenon of gelation is a question of chain connectivity and is thus based on *topological* arguments, i.e. a gel is observed when a polymeric material forms a connected cluster which spans all through the reaction vessel. Gelation can take place in a good and in a poor solvent and is not neccessarily related to thermodynamics. Phase separation, however, is defined by the *thermodynamic function* $\Delta\mu_1$, i.e. by the instability condition $\partial\Delta\mu_1/\partial\phi_2 = 0$, where $\Delta\mu_1$ is the chemical potential of the solvent in the solution at concentration c.

Coniglio[54] now performed a bond percolation calculation on a Bethe lattice where he introduced thermodynamic interactions of the solvent molecules with polymer segments, in addition to the capability of the segments to form a chemical bond when they come to lie on adjacent lattice sites. Figure 24 shows the theoretical results for different types of solvent interaction with the polymer. In every case gelation occurs at high concentrations prior to phase separation. In most cases the critical point of

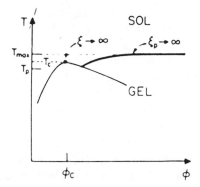

Fig. 24. Phase diagrams predicted from percolation theory taking into account special solvent–polymer interactions. ξ and ξ_p indicate correlation lengths of thermodynamic fluctuations and extension of polymer clusters. $\xi \to \infty$ describes the critical point of mixing and $\xi_p \to \infty$ indicates infinite clusters or gelation.[54]

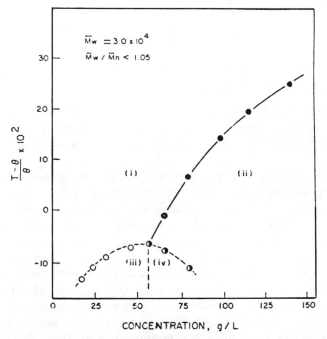

Fig. 25. Typical phase diagrams for PS in solution at low temperatures: (i) one phase solution; (ii) gel; (iii) liquid–liquid immiscibility area; and (iv) gel–gel immiscibility area. θ represents the theta temperature. The solvent in the present system was CS_2.

phase separation appears at lower concentrations than the intersection point of the gel curve with the spinodal. The theory shows clearly that spinodal decomposition will be a very rare process with common gels, and can be observed only if the solution is quenched below the spinodal decomposition temperature. The development of periodic density heterogeneities will be not observed if a homogeneous gel has already been formed, because then the already formed crosslinks prevent the structure formation as is predicted by Cahn.

The theoretical prediction was very nicely proved experimentally by Tan et al.,[18] although these authors apparently did not know of Coniglio's theory. Figure 25 represents a selection of their measurements in various solvents. The measurements were certainly difficult to perform at those low temperatures, and traces of moisture may be responsible for the observation of phase separation. The principal feature of the gel setting and the spinodal curves remains, however, unaffected.

5 CONCLUSION

Summarizing the various details of this comparison we come to the following conclusions for gels.

(1) Reversible gels fall in between two idealized limits: (a) where permanent crosslinks are fixed at defined positions of the chain; and (b) where entangled chains form transient networks and crosslinks cannot be localized.

(2) The junctions in a reversible network are often extended domains consisting of bundles of chains (one-dimensional order) or sometimes of small crystallites (three-dimensional order), but these junctions can also be point-like and positioned either at random along a chain (modified polybutadiens) or fixed at the two chain ends (end-tagged ionomers).

(3) The junctions have often only a short lifetime, and this fact gives rise to (a) an apparent diffusion of clusters caused by random dissociation and association of chains and (b) a slow creep. The materials show elastic moduli only at a finite frequency, but have no equilibrium modulus G_e.

(4) DSC and viscoelastic measurements give only limited information on structures in gels. To receive a more complete picture these methods have to be combined with results from scattering experiments.

(5) Pure entanglement with no fixed junctions exists only in extremely good solvents where all interactions among polymer segments are fully

screened. Special segment–segment interactions may cause association and finally reversible gelation.

(6) Gel formation is a question of connectivity, and a gel is defined through topological constraints. The gel-setting curve has clearly to be distinguished from a spinodal decomposition, which is defined by the laws of thermodynamics. Thermodynamic interactions are certainly the reason for reversible gelation, but a reversible gel is mostly far away from the spinodal decomposition curve where a collapse of the gel accompanied with syneresis is observed.

REFERENCES

1. Flory, P. J., *Trans. Farad. Soc.*, 1974, **57**, 7.
2. (a) Flory, P. J., *Polymer*, 1979, **20**, 1317; (b) Eichinger, B. E., *Ann. Rev. Phys. Chem.*, 1983, **34**, 359; (c) Mark, J. E., *Adv. Polym. Sci.*, 1982, **44**, 1; (d) Candau, S. J., Bastide, J. and Delsanti, M., *Adv. Polym. Sci.*, 1982, **44**, 27; (e) Heinrich, G., Straube, E. and Helmis, G., *Acta Polym.*, 1980, **31**, 275; (f) Helmis, G., Heinrich, G. and Straube, E., *Wiss. Zeitschr. Leuna. Merseburg*, 1984, **26**, 3.
3. De Gennes, P.-G., *Scaling Concepts in Polymer Physics*, Cornell University Press, Ithaca, New York, 1979.
4. Wiltzius, P., Haller, H. R., Cannell, D. S. and Schaefer, D. W., *Phys. Rev. Lett.*, 1983, **51**, 1183.
5. Burchard, W., unpublished work.
6. Ferry, J. D., *Adv. Protein Chem.*, 1948, **4**, 1.
7. (a) Atkins, E. D. T., Isaac, D. H. and Keller, A., *J. Polym. Sci., Physics Ed.*, 1980, **18**, 71; (b) Atkins, E. D. T., Hill, M. J., Jarris, D. A., Keller, A., Sarhene, E. and Shapiro, J. S., *Coll. & Polym. Sci.*, 1984, **262**, 22.
8. Tuzar, Z. and Kratochvil, P., *Adv. Coll. Interface Sci.*, 1976, **6**, 201.
9. (a) Takahashi, A., Nakamura, T. and Kagawa, J., *Polymer J.*, 1972, **3**, 207; (b) Wolf, B. A., *Macromolecules*, 1987, **20**.
10. Herrmann, K. and Gerngross, O., *Kautschuk*, 1932, **8**, 565.
11. Gross, J., in *The Chemical Basis of Life, Readings from Scientific American*, Hannawalt, P. C. and Haynes, R. H., (Eds), Freeman, San Francisco, 1973, p. 65.
12. Müller, M. F. Ries, H. and Ferry, J. D., *J. Mol. Biol.*, 1984, **174**, 369.
13. Scheraga, H. A., *Ann. N.Y. Acad. Sci.*, 1983, **408**, 330.
14. Andersen, N. S., Dolan, J. W., Harshing, M. H., Rees, D. A. and Samuel, J. W., *J. Mol. Biol.*, 1969, **45**, 85.
15. ter Meer, H.-U., PhD Thesis, Freiburg, 1984.
16. Applequist, J. and Damle, V., *J. Am. Chem. Soc.*, 1965, **87**, 1450.
17. Eldridge, J. E. and Ferry, J. D., *J. Phys. Chem.*, 1954, **58**, 992.
18. Tan, H.-M., Moet, A., Hiltner, A. and Baer, E., *Macromolecules*, 1983, **16**, 28.
19. Oakenfull, D., *J. Food Sci.*, 1984, **49**, 1103.
20. Treloar, L. R. G., *The Physics of Rubber Elasticity*, Clarendon Press, Oxford, 1975.

21. Oakenfull, D., *J. Food Sci.*, 1984, **49**, 1093.
22. Clark, A. H. and Ross-Murphy, S. B., *Br. Polym. J.*, 1985, **17**, 164.
23. Dobson, G. and Gordon, M., *J. Chem. Phys.*, 1965, **43**, 705.
24. Hermans, J., *J. Polym. Sci.*, 1965, **A3**, 1859.
25. Burchard, W., *Adv. Polym. Sci.*, 1983, **48**, 1.
26. ter Meer, H.-U. and Burchard, W., *Polymer*, 1985, **26**, 273.
27. Schmidt, M., Paradossi, G. and Burchard, W., *Makromol. Chem., Rapid Commun.*, 1985, **6**, 767.
28. Koyama, R., *J. Phys. Soc., Japan*, 1973, **34**, 1029.
29. Leong, K. W. and Butler, G. B., *J. Macromol. Sci.*, 1980, **A-14**, 287.
30. (a) Stadler, R. and Freitas, L. L., *Polym. Bull.*, 1986, **15**, 173; (b) Stadler, R. and Burgert, J., *Makromol. Chem.*, 1986, **187**, 1681; (c) Stadler, R. and Freitas, L. L., *Coll. & Polym. Sci.*, 1986, **264**, 773.
31. Stadler, R., *Macromolecules*, in press.
32. Tschoegl, N. W., *Rheol. Acta*, 1971, **10**, 582.
33. Freitas, L. L. and Stadler, R., *Macromolecules*, 1987, **20**.
34. Graessley, W. W., *Rheol. Acta*, 1977, **16**, 291.
35. Burchard, W., *Br. Polym. J.*, 1985, **17**, 154.
36. Candau, S. J., Ankrim, M., Munch, J. P. and Hild, G., *Br. Polym. J.*, 1985, **17**, 210.
37. Eisenberg, A. (Ed.), Ionsin polymers, *Adv. in Chem. Series, Am. Chem. Soc.*, 1980, **187**.
38. Eisenberg, A. and Bailey, F. E. (Eds), Coulombic interactions in macromolecular systems, *ACS Symposium Series, Am. Chem. Soc.*, 1986, **302**.
39. Agarwal, P. K. and Lundberg, R. D., *Macromolecules*, 1984, **17**, 1918, 1928.
40. Jerome, D., Horrion, J., Fayt, R. and Teyssie, P., *Macromolecules*, 1986, **17**, 2447.
41. Galland, D., Belakhavsky, M., Medrignac, F., Pineri, M., Vlaic, C. and Jerome, R., *Polymer*, 1986, **27**, 883.
42. Omeis, J., PhD Thesis, Freiburg, 1986.
43. Möller, M., Omeis, J. and Mühleisen, E., in *Reversible Polymeric Gels and Related Systems*, Russo, P. S. (Ed.), Symposium Series, J. Am. Chem. Soc., **350**, 1987.
44. Kirkwood, J. G., Maun, E. and Alder, B., *J. Chem. Phys.*, 1950, **18**, 380, 1040.
45. Eyring, H. and Hirschfelder, J. O., *J. Chem. Phys.*, 1937, **41**, 249.
46. Des Cloiseaux, J., *J. Phys. (Paris)*, 1982, **36**, 281.
47. Ohta, T. and Oono, Y., *Physics Lett.*, 1982, **89A**, 460.
48. Burchard, W., manuscript in preparation.
49. Eisele, M. and Burchard, W., *Macromolecules*, 1984, **17**, 1636.
50. Wenzel, M., Burchard, W. and Schätzel, L., *Polymer*, 1986, **27**, 195.
51. Gan, J. Y. S., Francoise, J. and Guenet, J.-M., *Macromolecules*, 1986, **19**, 173.
52. Feke, G. T. and Prins, W., *Macromolecules*, 1974, **7**, 527.
53. Cahn, J. W., *J. Chem. Phys.*, 1965, **42**, 93.
54. Coniglio, A., Stanley, H. E. and Klein, W., *Phys. Rev.*, 1982, **25**, 6805.
55. Schmitt, F. O., *The Molecular Basis of Life*, Haynes, R. H. and Hanawalt, P. C. (Eds), Freeman, San Francisco, 1968, p. 18.

SECTION 1

BIOLOGICAL NETWORKS

2

STRUCTURE AND RHEOLOGY OF FIBRIN NETWORKS

John D. Ferry

Department of Chemistry, University of Wisconsin,
Madison, Wisconsin 53706, USA

ABSTRACT

Fibrin, the structural component of a blood clot, is formed by polymerization of fibrinogen, a megamonomer (molecular weight 340 000) present in blood plasma at a concentration of about 0·25%. The enzyme thrombin splits off certain polypeptides from fibrinogen to expose association sites (A and B), and the resulting rod-like fibrin monomers assemble with strong non-covalent bonding in a staggered overlapping pattern to form long straight protofibrils. At high pH and ionic strength, the protofibrils associate laterally to some extent and a network is formed in which branching appears to occur by intermittent twisting of the protofibrils around each other. This gel-like 'fine' clot formed from pure fibrinogen is remarkably close to perfectly elastic, obeying Hooke's Law in small shearing deformations and having very little viscoelastic loss over many decades of time scale. The mechanism for elasticity is not rubber-like and is tentatively ascribed to bending of the fibrils. At very long times, the clot undergoes creep and creep recovery with some irrecoverable deformation; the Boltzmann superposition principle is obeyed and there is no net structural damage, as evidenced by constant differential modulus measured with intermittent application of additional stress. At large strains, there is pronounced strain-hardening of the structure followed by some structural damage that appears to be largely reversible. Introduction of a certain tetrapeptide (glycine–proline–arginine–proline, GPRP), that competes with A association sites for binding, lowers the shear modulus and increases enormously the creep rate and irrecoverable deformation; the tetrapeptide apparently catalyzes interchange of intrafibril junctions so that the clot flows with a high but finite viscosity, again with no net structural damage. Under physiological conditions, another enzyme (fibrinoligase) introduces covalent bonds by a chemical reaction after the protofibrils have assembled. The resulting ligated clot experiences practically no irrecoverable deformation in creep nor susceptibility to GPRP.

INTRODUCTION

The clotting of blood is a polymerization process in which the monomer is fibrinogen, a protein with molecular weight 340 000 that is present in blood plasma at a concentration of about 0·25%. Following injury or certain other conditions, a complicated cascade of biochemical reactions culminates in the appearance of the enzyme thrombin, which splits off two pairs of small polypeptide fragments from the fibrinogen molecule; the product is fibrin monomer, which polymerizes spontaneously to form a three-dimensional network, and thus produces a solid gel. We are concerned here only with the structure of the gel and its rheological properties when formed from highly purified fibrinogen in the absence of the other elements of blood.

THE FORMATION AND STRUCTURE OF FIBRIN

The structures of fibrinogen and its polymerization product have been elucidated by a large volume of work in many laboratories.[1,2] A model of fibrinogen which portrays many of its essential features is shown at the top of Fig. 1. It contains three pairs of peptide chains (α, β, γ) that are tied together in a relatively compact central nodule. The central region contains the four fibrinopeptides, two As and two Bs, that are split off by thrombin to uncover combining sites. From the center region extend, in almost opposite directions, two rod-like structures, each of which contains an α, β and γ chain coiled around each other in a helical pattern. At each end is a compact globular structure consisting of the remainders of the β and γ chains in a highly specialized and not yet fully elucidated configuration, while the remainder of each α chain forms a flexible random coil. At the bottom of Fig. 1 is a simplified abstraction of this model, showing only two rods that connect a central nodule E and two terminal nodules D, labeled in accordance with current fibrin nomenclature. This overall trinodular rod shape was clearly identified by electron microscopy[3] in 1959 and has been confirmed by several recent studies, while the more detailed structure shown at the top of Fig. 1 has been deduced from a variety of biochemical investigations.[1,2]

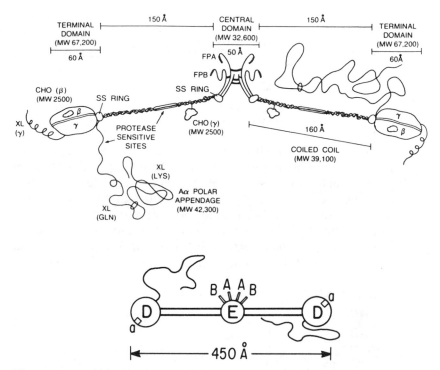

Fig. 1. Schematic sketches of the fibrinogen molecule. Top: detailed structure showing the three pairs of polypeptide chains and the locations of binding sites, with two coil–coil rod-like regions connecting three regions of more compact and complicated structure (reproduced with permission from Ref. 2, © 1984 by Annual Reviews Inc.). Bottom: abstraction of the above as two end nodules (D) and a central nodule (E) connected by rods. A, B (exact location not implied) denote fibrinopeptides split off by thrombin to uncover binding sites.

The letters A and B in Fig. 1 refer to (but do not imply the exact location of) the central ends of the α and β chains. Short lengths of these (the fibrinopeptides A and B) are split off by thrombin, uncovering sites A and B which are able to associate by strong non-covalent bonding with complementary sites a and b on the D nodules of other monomers, located somewhere on the γ and β lobes, respectively. Since the Aa associations are formed first and are sufficient to cause polymerization, we focus attention on these and show in Fig. 2 how an oligomer can be built up by pairs of Aa associations with half-staggered overlap of the fibrin monomer units. It is well established that the oligomers have twice the mass per unit length of a monomer unit.

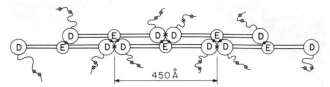

Fig. 2. Fibrin pentamer formed by staggered overlapping of monomer units non-covalently bonded by association of A sites on E nodules with a sites on D nodules.

Under physiological conditions (pH 7·4, ionic strength 0·15), oligomer formation is followed by lateral association of these rod-like fibrils to form thick bundles that branch by anastomosis to make up a coarse network that is optically opaque. However, at pH 8·5, ionic strength 0·45, there is very little lateral association and the so-called 'fine' clot is transparent. Electron micrographs of fine clots (Fig. 3) show a network of very thin strands that run straight for long distances and appear to branch at sharp angles from segments where two or three strands are twisted together.[4] The contrast

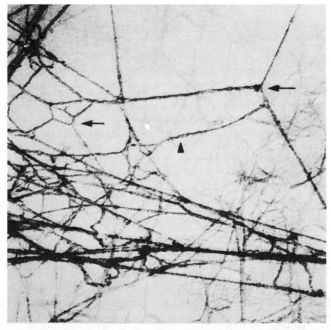

Fig. 3. Electron micrograph of 'fine' clot network formed by extended polymerization in the manner of Fig. 2, possibly with a small degree of lateral association, and branching through twisted segments.

between this picture and the accepted concept of a rubbery polymer network should be emphasized. This review is concerned *only* with fine clots because of their simpler structure.

Since the Aa associations (and also the Bb associations, not yet considered) are non-covalent, they are in principle reversible, and in fact the clot we have described can be depolymerized back to monomeric fibrin by changing the pH to 5 or by use of a suitable solvent such as 1M lithium bromide. However, in shed blood, another enzyme is activated, fibrinoligase or Factor XIIIa, which forms *covalent* amide bonds between adjacent D nodules in the half-staggered pattern, shown by crosses in Fig. 2. The reactive sites are on the γ chains and this process is called $\gamma-\gamma$ ligation. (Biochemists call it crosslinking, but it is not crosslinking in a polymer chemist's sense because it does not introduce branching.) In coarse clots, there is also $\alpha-\alpha$ ligation between the dangling α chains, at locations indicated by crossed circles in Fig. 2, but this is absent, or very slight, in fine clots. Ligated clots, which can be formed by adding purified Factor XIIIa and calcium ions to a fibrinogen–thrombin mixture, are not soluble by pH change or lithium bromide. They look exactly the same as unligated clots macroscopically or in the electron microscope, but have quite different rheological properties.

RHEOLOGICAL BEHAVIOR AT SHORT TIMES

During the polymerization process, a finite shear modulus appears at the gel point (termed by coagulationists and clinicians 'the clotting time'), just as in synthetic network polymers, and it increases to reach a constant value when the polymerization is complete. In all the rheological experiments described here, clots are first aged for many hours to make sure that the structure is fully developed. In small strains, over a time scale of 1–100 s, there are no relaxation processes, whether the clot is ligated or not. Hooke's law is strictly obeyed up to shear strains of at least 0·03, as illustrated[5] in Fig. 4. The linear storage and loss shear moduli, G' and G'', of an unligated clot are plotted[6,7] over a very wide frequency range in Fig. 5. (The values below 0·1 rad/s are converted from creep measurements.) The constancy of G' over many logarithmic decades of frequency is remarkable and is probably unmatched by any other material with such a low modulus (order of 10^3 dyn/cm^2 or 10^2 Pa). At frequencies above 1 rad/s, loss appears and its frequency dependence is consistent with a retardation time of about 2×10^{-3} s; the mechanism for this has not been elucidated.

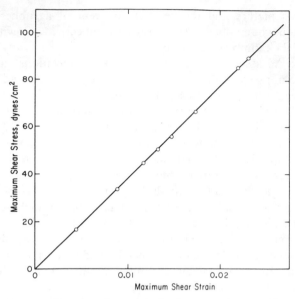

Fig. 4. Shear stress plotted against strain for fine ligated clot, fibrin concentration 8·0 g/liter. (Stress and strain for circular disc in torsion vary with radial distance: 'maximum' refers to value at disc edge. Measurements made 25 s after imposition of stress. Reproduced with permission from Ref. 5.)

The slight drop in G' at very low frequencies is associated with time-dependent processes discussed in the following section. This is eliminated by ligation, as illustrated in Fig. 6 where curves of shear creep compliance $J(t)$ for ligated and unligated clots are compared. (For such a very low-loss material, $G'(\omega) = [J(t)]^{-1}$, with $t = \omega^{-1}$.) The ligation is slow to complete and it eventually decreases the compliance (increases the modulus) somewhat.

Superficially, Fig. 5 resembles corresponding plots for a lightly cross-linked gum rubber, except for the magnitudes and time scales. Thus, a plot of log G' and G'' for styrene–butadiene gum rubber crosslinked with dicumyl peroxide looks very similar, except that log G' levels at low frequencies at a much higher value of 6·9 and the curves for log G' and log G'' meet at log $\omega = 5·2$ (units dyn/cm², at 25°C).[8] However, the clot elasticity cannot be rubber-like; Fig. 3 does not show a random coil network with highly flexible strands. Unfortunately, there have been naïve discussions of rubber-like elasticity. Their absurdity can be illustrated by use of the familiar equation $M_c = cRT/G$ where M_c is the average molecular

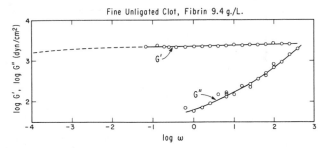

Fig. 5. Storage and loss shear moduli, G' and G'', plotted logarithmically against radian frequency for fine unligated clot, fibrin concentration 9·4 g/liter. The dashed line is converted from creep measurements (data from Refs 6 and 7).

weight between crosslinks in a network formed in the presence of solvent and c is the concentration of polymer in g/ml. For a fine unligated clot with $c = 0.008$, $G = 2.5 \times 10^3$ dyn/cm^2 or 2.5×10^2 Pa; then $M_c = 8 \times 10^4$, or one-quarter of a monomer unit, 55 Å in the sketch in Fig. 2.

At large shear strains, both ligated and unligated clots show strain hardening (opposite to the behavior of random coil rubber networks). The tangent modulus $G_i = \sigma/\gamma$, where σ is stress and γ strain, is independent of γ up to about $\gamma = 0.1$, but then increases, as illustrated in Fig. 7 for an unligated clot.[9] Moreover, the differential modulus $G'(\omega, \gamma)$, measured in small oscillating deformations at about 1 rad/s superposed on the static strain, increases enormously. At very high strains, it can increase by a factor of 20. If there were no time-dependent effects, $G'(\omega, \gamma)$ is calculated to be $G_i(\gamma) + \gamma \, dG_i/d\gamma$, represented in the figure by a dashed curve. However, at

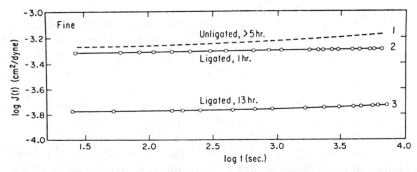

Fig. 6. Creep compliance $J(t)$ plotted logarithmically against time for unligated and ligated fine clots, fibrin concentrations 8.1 ± 0.1 g/liter. Times in hours refer to the interval between formation of clot and beginning of creep experiment. (Reproduced with permission from Ref. 5.)

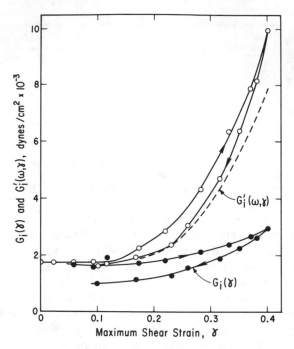

Fig. 7. Static (tangent) shear modulus $G_i(\gamma)$ measured 25 s after imposition of stress, and differential dynamic shear modulus $G_i'(\omega, \gamma)$ measured at about 1 Hz: data taken consecutively with increasing strain and then decreasing strain, with time intervals of about 40 s. Dashed curve is $G'(\omega, \gamma)$ calculated as $G_i(\gamma) + \gamma \, \mathrm{d}G_i/\mathrm{d}\gamma$ from upper G curve. Fibrin concentration 7·0 g/liter. (Reproduced with permission from Ref. 9.)

high strains in an unligated clot, there are time-dependent structural changes and the experiment cannot be completed fast enough to avoid some hysteresis in a sequence of ascending and descending strain, as seen in the figure, and the dashed curve does not agree exactly. Nevertheless, the marked strain hardening is evident. (Discussion of the structural changes is beyond the scope of this review; they indicate damage to the network which is gradually repaired after stress is released and is eliminated by ligation.)

The mechanism of energy storage in deformation is not established, but it has been tentatively ascribed to bending of relatively stiff network strands. The strain hardening could be qualitatively understood by a theory of Doi and Kuzuu[10] for a structure of randomly interpenetrating stiff rods, whose interparticle contacts increase in number with increasing deformation; but the observed dependence of G on strain and concentration does not agree quantitatively with this model. For example, G is proportional to the fibrin concentration to the power 1·6 (ligated) or 1·9 (unligated), while the Doi–Kuzuu model predicts an exponent of 5.

RHEOLOGICAL BEHAVIOR AT LONG TIMES

The time scale for creep in Fig. 6 goes up to 10^4 s. At longer times, creep of an unligated clot continues; when creep and creep recovery are plotted with linear scales as in Fig. 8 (in this case, up to 2×10^4 s), the behavior is remarkably similar to that of a high-viscosity viscoelastic liquid,[11] with a substantial irrecoverable deformation. For ligated clots, on the other hand, the creep is negligible and the recovery after removal of stress is complete. As long as the strains of unligated clots are small, linear viscoelastic behavior is followed. Tests with the Boltzmann superposition principle show that the creep recovery can be accurately predicted from creep to longer times in a parallel experiment.[5,11] Moreover, the differential modulus measured from a brief on–off step stress superposed on the constant applied stress remains *constant* throughout creep and creep recovery, even when the irrecoverable deformation is quite large. This shows that there are no net changes in structure during whatever rearrangements are taking place to permit permanent deformation.[11]

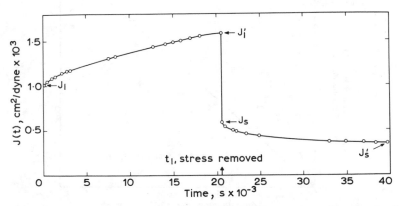

Fig. 8. Creep compliance and creep recovery plotted linearly against time for fine unligated clot, fibrin concentration 6·5 g/liter. For a ligated clot, the corresponding plot is a rectangle, with essentially no creep following initial deformation and immediate complete recovery following removal of stress. (Reproduced with permission from Ref. 11.)

The creep compliance has not quite attained a linear function of time at the end of the creep phase in Fig. 8, so one cannot calculate a viscosity as is done for a viscoelastic liquid. However, if the creep is terminated at different times in separate experiments, the irrecoverable deformation is directly proportional to the creep duration time over a considerable range. In Fig. 9, the ratio of irrecoverable compliance to the initial compliance (25 s after imposition of stress) is plotted against creep duration time for fine clots that have different fibrin concentrations.[11] (The clot must be at least 18 h old for complete development of structure before the creep is started.) For a fibrin concentration of 0·008 g/ml, this would correspond to a lower limit for effective viscosity of about 10^8 P and a lower limit for an average relaxation time of about 10^5 s (1 day).

The mechanism for irrecoverable deformation in unligated clots is uncertain; it has been postulated that the branch points, if represented by segments of strands twisted around each other, may occasionally slip to new positions. Alternatively, the strands may occasionally be severed by dissociation of a pair of Aa associations as illustrated in Fig. 10. In this case, the average relaxation time of 10^5 s would be qualitatively identified with reciprocal rate of dissociation as in the transient network theories of Tobolsky[12] and Lodge.[13] Note that if strands are severed they must form again in new positions since the differential modulus always remains

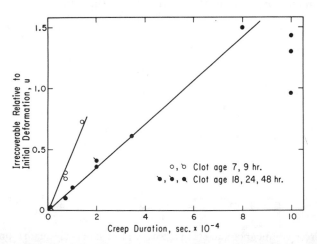

Fig. 9. Irrecoverable deformation in creep of fine fibrin clots, relative to initial deformation at 25 s (ratio of J'_s to J_1 as identified in Fig. 8) plotted against time of creep duration (t_1 in Fig. 8), for various fine unligated clots, fibrin concentrations 4·1–13·2 g/liter. (Reproduced, after corrections, with permission from Ref. 11.)

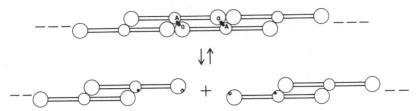

Fig. 10.   Sketch of possible dissociation of two Aa junctions in a protofibril to cause severance of a network strand.

constant. This may be somewhat implausible. However, the fact that there is no irrecoverable deformation in ligated clots is consistent with the concept of Fig. 10, since the severance would be impossible if pairs of D nodules are linked by covalent bonds.

DISSOCIATION OF OLIGOMERS AND NETWORK STRANDS BY COMPETING REAGENTS

There are circumstances under which a fine unligated clot behaves clearly like a viscoelastic liquid, and a finite viscosity and steady-state compliance can be calculated from creep and creep recovery measurements.

The A binding sites at the ends of the α chains in nodule E, after release of the A fibrinopeptides, have the amino acid sequence glycine–proline–arginine–valine. Laudano and Doolittle[14] synthesized tetrapeptides with similar sequences and found that they bound strongly to fibrinogen, presumably at the a site on the D nodule, and inhibited fibrin polymerization. Most work has been done with glycine–proline–arginine–proline (GPRP).

When GPRP is added at concentrations in the range 0·003–0·02M to a solution containing fibrin oligomers (the thrombin having been inhibited to stop the progress of polymerization), the oligomers are dissociated as would be expected if the tetrapeptide (P) competes with the A sites in binding to the a sites. The results are consistent with an equilibrium as depicted in Fig. 11, and an equilibrium constant can be estimated.[15]

When GPRP is introduced into a fine unligated clot by diffusion to reach a final concentration of 0·001–0·004M, dramatic changes in rheological properties are observed.[15,16] The modulus is lower and the creep rate and irrecoverable deformation are enormously increased (Fig. 12). The viscosity η and steady-state compliance J_e^0 are easily calculated. For 0·0035M GPRP, $\eta = 2\cdot7 \times 10^5$ P and $J_e^0 = 6 \times 10^{-3}$ cm^2/dyn (6×10^{-2} Pa^{-1}). The weight-average terminal relaxation time calculated as ηJ_e^0 is $1\cdot6 \times 10^3$ s. The

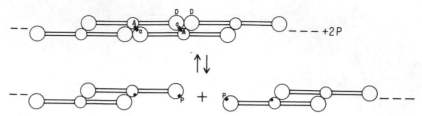

Fig. 11.   Sketch of possible reaction of two Aa junctions with two molecules of GPRP to cause severance of a network strand.

modulus is decreased by a factor of 0·2. If this decrease is attributed to severance of network strands, it is understandable that the effect is observed at lower GPRP concentrations than for oligomer dissociation, since only one break is presumably sufficient to remove the contribution of a strand to the modulus. Far more striking is the huge reduction in viscosity and relaxation time. Since the differential modulus again remains constant during creep and creep recovery, severed junctions must be forming with new configurations and it appears that the tetrapeptide is in effect catalyzing the interchange of fibril junctions. In *ligated* fine clots, GPRP has

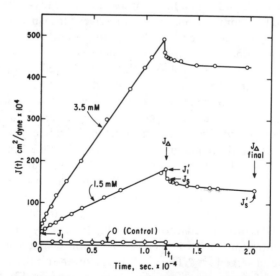

Fig. 12.   Creep compliance and creep recovery of fine unligated clots after introduction of GPRP by diffusion to the nominal concentrations shown. Fibrin concentrations 8·7 ± 0·4 g/liter. 'Control' is similar to Fig. 8: note the difference in ordinate scale. (Reproduced with permission from Ref. 16.)

no effect; again this is understandable since the severance depicted in Fig. 11 could not take place.

ROLE OF Bb ASSOCIATIONS AND OTHER FEATURES

Until this point, the polymerization has been discussed only in terms of Aa associations. However, the B sites on E nodules also form associations with b sites on the β lobes of the D nodules in a manner not clearly understood. When fibrinogen is treated with certain snake venom enzymes (ancrod, batroxobin), only the A sites are uncovered. The resulting monomers (α-fibrin or des-AA fibrinogen) polymerize to oligomers and then clot. However, the fine unligated clots have lower modulus and much lower effective viscosity, and are much more susceptible to the action of GPRP, than thrombin clots in which A and B sites participate in polymerization.[16,17] The Bb associations are usually considered to be important in lateral aggregation of protofibrils, but this may not be their only role. Another venom enzyme (venzyme) uncovers only the B sites, and the resulting β-fibrin or des-BB fibrinogen also polymerizes; the rheology of such clots is being studied. The non-covalent binding between D and E nodules may be stabilized by still other interactions not yet identified.[18]

The foregoing discussion has been concerned only with fine clots. The rheology of coarse clots, in which the fibers are made up of hundreds of protofibrils aligned laterally, has also been studied but is more difficult to interpret.[5,7,19] Generally, coarse unligated clots have higher moduli at the same fibrin concentration than fine clots; they creep more at short times but less at long times. In small deformations, the Boltzmann superposition principle is obeyed. The irrecoverable deformation cannot be attributed to strand severance in this case. Ligation of coarse clots increases the modulus and almost eliminates creep; recovery from deformation is essentially complete. In ligated coarse clots, the α–α covalent bonds between groups of dangling α chains are presumably an important addition to the structure. Such crosslinked random coil-like appendages might be thought to contribute some elasticity of a rubber-like nature, but it is very doubtful whether it would be significant compared to the bending of the coarse, massive fibers.

When blood clots under physiological conditions, there are other complications. Blood plasma contains large amounts of other proteins which may be responsible for minor modifications, and also platelets, formed elements at whose surfaces clotting is initiated and which may modify the mechanical properties at large strains somewhat as fillers do in synthetic polymer composites. Also, a large proportion of blood volume is of course occupied by red cells. However, studies of polymerized purified fibrinogen are revealing the basic structure and mechanical properties of fibrin, and showing the remarkable ingenuity and subtlety which nature has evolved to protect the organism from loss of its most vital fluid.

ACKNOWLEDGEMENTS

Investigations in our laboratory that have been mentioned here have been supported in part by the National Institutes of Health, mostly under Grant GM 21652, and have been conducted by Drs M. D. Bale, C. Gerth, P. A. Janmey, G. W. Kamykowski, O. Kramer, M. F. Müller, G. W. Nelb, W. W. Roberts, G. Schindlauer and A. Shimizu. Professors R. F. Doolittle, L. Lorand, D. F. Mosher, M. T. Record and H. Ris have aided us greatly by gifts of valuable materials, use of equipment and helpful advice.

REFERENCES

1. Doolittle, R. F., Structural aspects of the fibrinogen to fibrin conversion, *Adv. Protein Chem.*, 1973, **27**, 1–109.
2. Doolittle, R. F., Fibrinogen & fibrin, *Ann. Rev. Biochem.*, 1984, **53**, 195–229.
3. Hall, C. E. and Slayter, H. S., The fibrinogen molecule: its size, shape, and mode of polymerization, *J. Biophys. Biochem. Cytol.*, 1959, **5**, 11–15.
4. Müller, M. F., Ris, H. and Ferry, J. D., Electron microscopy of fine fibrin clots and fine and coarse fibrin films, *J. Mol. Biol.*, 1984, **174**, 369–84.
5. Nelb, G. W., Gerth, C. and Ferry, J. D., Rheology of fibrin clots. III. Shear creep and creep recovery of fine ligated and coarse unligated clots, *Biophys. Chem.*, 1976, **5**, 377–87.
6. Roberts, W. W., Kramer, O., Rosser, R. W., Nestler, F. H. M. and Ferry, J. D., Rheology of fibrin clots, I. Dynamic viscoelastic properties and fluid permeation, *Biophys. Chem.*, 1974, **1**, 152–60.
7. Gerth, C., Roberts, W. W. and Ferry, J. D., Rheology of fibrin clots. II. Linear viscoelastic behavior in shear creep, *Biophys. Chem.*, 1974, **2**, 208–17.
8. Kramer, O. and Ferry, J. D., in *Science and Technology of Rubber*, Eirich, F. R. (Ed.), Academic Press, New York, 1978, p. 182.
9. Janmey, P. A., Amis, E. J. and Ferry, J. D., Rheology of fibrin clots. VI. Stress relaxation, creep, and differential dynamic modulus of fine clots in large shearing deformations, *J. Rheol.*, 1983, **27**, 135–53.
10. Doi, M. and Kuzuu, N. Y., Nonlinear elasticity of rodlike macromolecules in condensed state, *J. Polymer Sci., Polymer Phys. Ed.*, 1980, **18**, 409–19.

11. Nelb, G. W., Kamykowski, G. W. and Ferry, J. D., Rheology of fibrin clots. V. Shear modulus, creep, and creep recovery of fine unligated clots, *Biophys. Chem.*, 1981, **13**, 15–23.
12. Tobolsky, A. V., *Properties and Structure of Polymers*, Wiley, New York, 1960, p. 234.
13. Lodge, A. S., *Elastic Liquids*, Academic Press, New York, 1964.
14. Laudano, A. P. and Doolittle, R. F., Synthetic peptide derivatives that bond to fibrinogen and prevent the polymerization of fibrin monomers, *Proc. Natl. Acad. Sci. USA*, 1978, **75**, 3085–9.
15. Schindlauer, G., Bale, M. D. and Ferry, J. D., Interaction of fibrinogen-binding tetrapeptides with fibrin oligomers and fine fibrin clots, *Biopolymers*, 1986, **25**, 1315–36.
16. Bale, M. D., Müller, M. F. and Ferry, J. D., Effects of fibrinogen-binding tetrapeptides on mechanical properties of fine fibrin clots, *Proc. Natl. Acad. Sci. USA*, 1985, **82**, 1410–13.
17. Shimizu, A., Schindlauer, G. and Ferry, J. D., Interaction of the fibrinogen-binding tetrapeptide gly-pro-arg-pro with fine clots and oligomers of α-fibrin; comparisons with $\alpha\beta$-fibrin, *Biopolymers*, in press.
18. Budzynski, A., 10th International Congress on Thrombosis and Hemostasis, San Diego, California, 1985.
19. Bale, M. D., Müller, M. F. and Ferry, J. D., Rheological studies of creep and creep recovery of unligated fibrin clots: comparison of clots prepared with thrombin and ancrod, *Biopolymers*, 1985, **24**, 461–82.

3

NON-GAUSSIAN ELASTIC PROPERTIES IN BIOPOLYMER NETWORKS

JOHN M. GOSLINE, ROBERT E. SHADWICK,* M. EDWIN DEMONT†
and MARK W. DENNY‡

*Department of Zoology, University of British Columbia,
Vancouver, BC, Canada V6T 2A9*

ABSTRACT

Highly extensible biomaterials containing rubber-like, crosslinked protein networks usually exhibit non-linear stress–strain curves that are matched to functional requirements, and it is commonly accepted that non-linear elastic properties arise from a composite of rubber-like protein reinforced by stiff fibres. It is possible, however, for protein rubbers to function as non-Gaussian networks and directly provide the non-linear elastic properties. This paper summarizes work on the analysis of non-Gaussian elastic properties in two biomaterials.

The elastic fibres in cephalopod arteries contain network chains that are relatively inflexible when compared with those in elastin, the rubber-like protein of the vertebrates. The equivalent random link in the cephalopod protein is about 12 amino acid residues long, whereas it is only about 8 amino acid residues for elastin. Although the crosslink density of these two protein networks is essentially identical, the difference in backbone flexibility creates significant differences in mechanical properties. Cephalopod elastic fibres exhibit non-Gaussian elastic properties at small strains that match the range of extensions seen in the living tissue, and thus contribute to the functionally important non-linear elastic properties. Elastin, in contrast, does not enter its

* Present address: Department of Biology, University of Calgary, Calgary, Alberta, Canada.
† Present address: Department of Zoology, University of Leeds, Leeds, UK.
‡ Present address: Hopkins Marine Station, Stanford University, Pacific Grove, California, USA.

non-Gaussian region until it is stretched by more than 100%, well beyond the biological range of extension, and thus it behaves as a linear elastic system.

Spider's frame silk is a high strength protein fibre that contains a short-chain polymer network reinforced by extremely small crystalline inclusions. When hydrated in water, this silk exhibits rubber-like elastic properties, and analysis of its non-Gaussian properties provides the basis for a molecular model of the material. Network chains are about 15 amino acid residues long and contain two equivalent random links. The crystalline regions in silk occupy about 25% of the volume and appear to be elongated structures with a length–width ratio of about 5.

INTRODUCTION

The term biopolymer is frequently used to describe a macromolecular system derived from a biological source, but otherwise modified to produce a set of properties that are useful for some industrial application. Biopolymers, however, are designed through the process of evolution to serve specific functions in the organisms that synthesize them, and in this biological context it is important to understand how biological design, through the process of natural selection, can match molecular structure to functional mechanical properties. In addition, it is possible that the analysis of structure–function relationships in natural biopolymers may lead to the development of new synthetic polymeric materials.

The structure–function relationships of biomaterials are reasonably well established in many cases (see, for example, Refs 1 and 2). Relatively little, however, is known about the molecular design of crosslinked protein networks. This paper summarizes our recent work with two different types of protein network, the rubber-like proteins that function in elastic arteries and the tough polymeric fibres in spiders' webs, in which the analysis of non-Gaussian properties provides important insights into their molecular design.

ARTERIAL ELASTIC TISSUES

Although the skeletal mechanics of animals are dominated by rigid structures, made of bone or other stiff biomaterials, all animals require soft, flexible materials (i.e. connective tissues) that can hold together the various soft organs of the body and can provide a deformable skin to isolate and protect these soft tissues from the external environment. This example illustrates aspects of the design of biomaterials that function by being deformed to large strains. Extensible connective tissues are usually complex composites that contain rubber-like elastic fibres and high stiffness collagen fibres, all embedded in a gel of highly hydrated proteoglycan molecules. These composites virtually always exhibit non-linear elastic properties.[1] Figure 1(A) shows a typical stress–strain diagram for such a material, in this case an elastic artery from a dog. The important feature of this diagram is the 'J-shape' to the stress–strain curve. The initial low modulus allows easy extension of the tissue, usually to an extension of about 50–80%, but this extension is limited by a rapidly increasing stiffness at high strains that prevents excessive deformation and rupture.[3] Elastic arteries are designed so that at normal physiological pressures the arterial wall material is loaded to that portion of the curve where the stiffness rises very rapidly. Specifically, the arterial wall material is designed so that its incremental stiffness increases 5- to 10-fold during the roughly 10% expansion that occurs with each beat of the heart, and this rapid increase in wall stiffness allows elastic arteries to expand uniformly along their length without forming aneurysms.[4] Thus, the major arteries provide a large, distributed elastic reservoir that smooths the pressure pulse and reduces the work of the heart, without creating large local deformations that would lead to rapid fatigue and the rupture of arteries.

   Non-linear elastic properties in connective tissues are commonly attributed to the large difference in elastic modulus and extensibility of the elastic fibres (elastin, in the case of vertebrate tissues) and collagen fibres (Fig. 1(B)), but in reality the non-linear behaviour is due to a complex interaction of tissue architecture and material properties, involving the sequential recruitment of different classes of elastic fibres, as well as the extension and reorientation of crimped or folded collagen fibres.[6,7] It is possible, however, for homogeneous polymer networks to exhibit non-linear elastic properties similar to those described above if the networks are constructed so that they function where non-Gaussian mechanical properties are dominant. Figure 2 shows how the non-Gaussian elastic properties of polymer networks are affected by the length of the network chains, where length is indicated as the number of equivalent random links, N, in the random chains between crosslinks. By inspection, it is obvious that networks having mechanical properties similar to those required for arteries will of necessity be very short-chain networks, so that stiffness rises continuously with extension (e.g. the network with $N = 5$ in Fig. 2). In uniaxial tension, long-chain networks have an initial region of decreasing

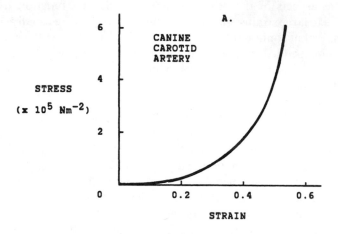

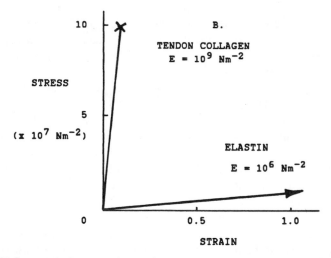

Fig. 1. (A) Stress–strain curve for canine carotid artery, showing the 'J-shaped' curve characteristic of all arterial elastic tissues and most other soft, connective tissues (after Ref. 5). (B) Typical stress–strain curves for collagen and elastin, the major structural components of vertebrate connective tissues. The 'J-shaped' stress–strain curve is usually attributed to the transfer of stress from low stiffness elastin fibres to high stiffness collagen fibres.

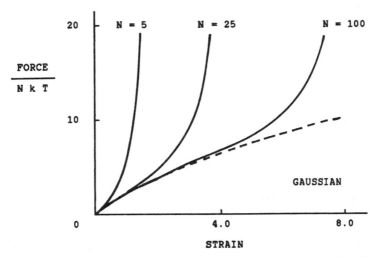

Fig. 2.   Typical non-Gaussian force extension for polymer networks with different chain lengths, where chain length is expressed as the number of equivalent random links, N, per chain.

stiffness, followed by a variable region of essentially constant stiffness. As a consequence, rubber tubes or cylindrical rubber balloons made from lightly crosslinked rubbers always form aneurysms. The question to consider, therefore, is whether evolutionary design has opted for complex tissue architecture or has employed molecular design of the protein chains and crosslinking mechanisms to achieve the desired non-linear elastic properties.

This question can be answered quite simply by measuring the uniaxial force–extension behaviour of natural protein rubbers to determine if the properties of the molecular network are sufficient to account for the properties of the intact tissue. The process, unfortunately, is not that simple. Although elastic structures in animals can be quite large, all of these structures are built from a lattice of fine elastic fibres that range in diameter from about 0·1 μm to a maximum of about 8 μm. In order to determine the mechanical properties of the molecular network rather than the properties of the fibre lattice, it is necessary to test individual elastic fibres.

To deal with these small fibres we have developed simple microscope-based test methods that allow us to determine the mechanical, optical and thermodynamic properties of single elastic fibres. The method is based on the use of slender glass rods (typically 50–150 μm in diameter and 1–2 cm long) that function as force transducers. These glass rods are mounted on microscope slides, and the force is calculated from standard bending equations for the deflection of the cylindrical beams. The deflection of the glass rod is either measured manually with a filar-micrometer eyepiece or with a Video Dimension Analyzer (VDA, Instruments for Physiology & Medicine, La Jolla, California, USA) attached to the microscope (Fig. 3). The VDA can easily follow movements of the glass rod as small as 1 μm. We have used this system to measure forces of the order of 10^{-9} N in thermoelastic experiments with single elastic fibres, but in principle with longer, thinner glass rods it should be possible to increase the sensitivity well beyond this. The only real difficulty is the isolation of single elastic fibres and their attachment to the test frame. This can be achieved by attaching small bundles of fibres to the glass rod at one end and to a movable support at the other end. Then the bundle can be carefully dissected to leave a single fibre.

Elastin is the protein rubber found in the elastic tissues of the vertebrate animals, from fish to man.[8] The elastic and optical properties of single

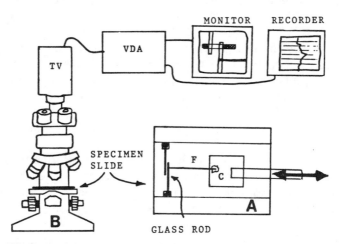

Fig. 3. Diagram of the apparatus used to determine the mechanical, thermodynamic and optical properties of single elastic fibres. (A) Detail of specimen slide—the test fibre (F) is attached at one end to a thin glass rod and at the other end to a movable cover-slide (C). The force is determined by measuring the deformation of the glass rod. (B) Microscope movement detection system—the specimen slide is mounted on a microscope, and a television camera (TV) and monitor are used to follow the deflection of the glass rod. A Video Dimension Analyzer (VDA) generates an electronic window that can follow the movement of the contrast boundary, created by the surface of the glass rod, relative to some fixed reference marker. It provides an analogue voltage that is proportional to the movement.

elastin fibres from the elastic neck ligament of cows were measured by Aaron and Gosline,[9] using the techniques described above. In these experiments the extensions were limited to a maximum of about 150%, but both the force–extension and the stress–birefringence data showed significant deviations from Gaussian behaviour by about 100% extension. Attempts to fit these data to non-Gaussian curves indicate that the elastin network is constructed of protein chains that, on average, behave like random walks of about 10 steps or an ideal random coil made from 10 equivalent random links. The elastic modulus, G, for these fibres is about $4 \cdot 1 \times 10^5 \, \text{Nm}^{-2}$ when swollen in water at room temperature to a volume fraction, v_2, of $0 \cdot 65$, making the average molecular weight of random chain between crosslinks, M_c, about 7100 daltons. The molecular weight of the uncrosslinked precursor is known to be $7 \cdot 2 \times 10^4$ daltons,[10] and correcting for randomly distributed free ends gives an alternative estimate of $M_c =$ 6000 daltons.

The amino acid composition of elastin is known, and from these data it is possible to estimate the number of amino acid residues that together provide sufficient backbone flexibility to act as an ideal link in the random chain. The average weight of the amino acids in elastin is about 85 daltons. Thus, on average, each random chain contains about 70–85 amino acids. This means that each of the equivalent random links requires seven to eight amino acids to achieve the flexibility of a random chain. These aspects of the molecular structure of the elastin network are summarized in Table 1.

Returning to the question of matching the molecular properties of polymer networks to properties in intact elastic tissues, comparison of the shape of the stress–strain curve for the dog's artery with that for isolated elastin fibres (Fig. 4(A)) will indicate if the non-Gaussian properties of

TABLE 1

Summary of the Molecular Properties of Elastin and Octopus Arterial Elastomer
(Elastin Data from Ref. 9; OAE Data from Ref. 11)

Property	Elastin	Octopus arterial elastomer
Elastic modulus, G (Nm^{-2})	$4 \cdot 1 \times 10^5$	$4 \cdot 6 \times 10^5$
M_c (Daltons)	6 000–7 000	6 500
Average residue weight (Daltons)	85	110
Amino acid residues per random chain	70–85	60
Random links per chain	$c.$ 10	$c.$ 5
Amino acid residues per equivalent random link	7–8	12

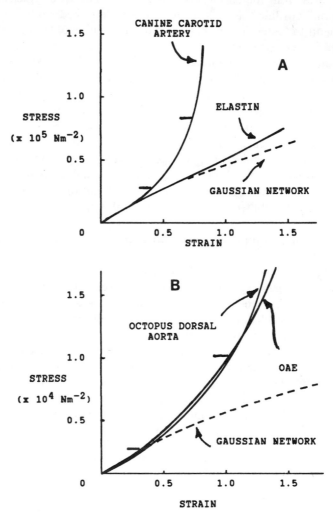

Fig. 4. (A) Comparison of the elastic properties of vertebrate elastic tissue and isolated elastin fibres, showing that the non-Gaussian properties of elastin cannot explain the non-linear properties of the intact tissue. (B) Comparison of the octopus aorta and isolated OAE fibres from the aorta. The non-Gaussian properties of the rubber network make a dominant contribution to the non-linear behaviour of the intact tissue. Horizontal bars next to arterial curves in both (A) and (B) indicate approximate range of *in vivo* stress.

elastin contribute to the 'J-shaped' stress–strain curve seen for the artery. In this figure the stress–strain curves for isolated, purified elastin and for intact arterial elastic tissue are plotted together, with the stress axes adjusted so that the initial slopes are equal. The stress axes differ because the artery wall contains only about 30% elastin by volume, with smooth muscle and collagen contributing the majority of the remaining tissue. It is clear from this comparison that the molecular properties of elastin make little, if any, contribution to the non-linear elastic properties of the intact artery. Thus, the elastic arteries of vertebrates require complex tissue architecture as well as reinforcement by high stiffness collagen fibres in order to generate their functional mechanical properties.

This conclusion, however, does not hold for all organisms. We have studied the mechanics of elastic arteries in other animals, and have discovered that the cephalopod molluscs (i.e. octopus, squid, Nautilus) have elastic arteries that are mechanically very similar to those in the vertebrates.[12,13] That is, they exhibit 'J-shaped' stress–strain curves (Fig. 4(B)), and the properties are adjusted so that the stiffness climbs dramatically in the physiological range of expansion.[14,15] Cephalopod arteries also contain a lattice of fine (2–5 μm diameter) elastic fubres, but in this case the protein rubber is not elastin.

Mechanical tests on the elastic fibres isolated from the artery of the giant pacific octopus, *Octopus dofleini*, reveal that this material exhibits non-Gaussian elastic properties at relatively low extensions, and it is likely that the evolution of this octopus arterial elastomer (OAE) has selected a protein in which the non-Gaussian properties are matched to the mechanical requirements of the intact artery. This possibility is illustrated in Fig. 4(B), where the uniaxial force–extension data for single OAE fibres are plotted together with stress–strain data for whole octopus aortic elastic tissue. Again, the stress axes have been adjusted so that the initial slopes of the two curves are identical, and in this case it is obvious that to a large degree the non-linear tissue elasticity can be explained by the non-Gaussian properties of the molecular network. This should not be interpreted to mean that the octopus aorta derives its mechanical properties entirely from the elastic fibres, but it strongly suggests that the molecular structure of the elastic fibres was a major factor in the evolutionary design of this system. Inspection of this figure, however, indicates that the artery stiffness increases more rapidly than that of the elastic fibres at extensions beyond the physiological level, and it is likely that aspects of tissue architecture as well as the transfer of stress to collagen fibres also contribute to the mechanical properties.

A more detailed analysis of the non-Gaussian elastic properties provides an indication of the molecular basis of this apparent 'design'.[11,14] The non-linear elastic properties of the OAE fibres are best fit by non-Gaussian curves for networks in which the random chains contain on average five equivalent random links. The elastic modulus, G, for the fibres is about $4.6 \times 10^5 \, Nm^{-2}$, a value that is virtually identical to that for elastin, and on the basis of the amino acid composition and the swelling of the hydrated network, the average molecular weight between crosslinks, M_c, can be calculated at about 6500 daltons. The average residue weight of the OAE protein is about 110 daltons, so that each random chain, on average, contains about 60 amino acid residues. Finally, because each random chain contains about five equivalent random links, this means that each equivalent random link is about 12 amino acid residues long. These figures are summarized and compared with similar values for elastin in Table 1.

In retrospect, we see that these two protein networks, elastin and the octopus arterial elastomer, have virtually identical elastic moduli, and yet one seems to exhibit non-Gaussian properties more strongly than the other. The reason for this difference undoubtedly lies in differences in the flexibility of the protein backbone brought about by specific differences in amino acid sequences in these proteins. Unfortunately, we know very little about the sequence design of polypeptides that exhibit rubber-like elasticity. In fact, we cannot even explain why some proteins, like elastin and OAE, exhibit kinetically free, random-coil conformations in dilute aqueous media, when virtually all other proteins have unique, stable conformations in aqueous media and become kinetically free only when treated in strong denaturing agents. The limited sequence information available for elastin (see Ref. 10) does not really provide any explanations, and we are left with simple inferences drawn from the amino acid composition data for the small number of rubber-like proteins that have been identified and studied to date.

Table 2 provides some summary information on the amino acid composition of elastin, OAE, resilin (a protein rubber from insects) and abductin (a protein rubber from the hinge of scallops). Perhaps the most striking feature of the amino acid compositions of elastin, resilin and abductin is that small amino acids (glycine + alanine + serine) make up from 55% to 70% of the total composition. The high glycine content (from 31% to 58%) is probably significant because glycine has the smallest side chain and is known to impose the least restriction to rotational freedom in a peptide backbone.[16] Thus, the presence of glycine and the other small amino acids probably contributes to the overall flexibility of the peptide

TABLE 2

Summary of Amino Acid Composition for Protein Rubbers (Data Expressed as
Residues per Thousand)

	Elastin[a]	Abductin[a]	Resilin[b]	OAE[c]
Glycine	313	627	422	85
Small amino acids	569	738	620	229
Non-polar	590	250	208	393
Polar	94	123	362	519
Average residue weight (Daltons)	85	79	89	110

[a] Porcine arterial elastin and *Aquepectin irridians* abductin data from Ref. 8.
[b] Dragonfly tendon resilin data from Ref. 17.
[c] *Octopus dofleini* OAE data from Ref. 11.

chains in rubber-like proteins. Beyond this simple correlation there is very little else that one can infer from these data. Elastin is rich in non-polar amino acids: abductin and resilin have a mix of polar and non-polar residues. Comparison of individual amino acids is even less revealing. In fact, it seems that there is no simple way to explain why the peptide backbone remains flexible and mobile.

It is clear, however, that the peptide backbone of elastin is quite inflexible, particularly when compared with other elastomers. Our analysis indicates that the equivalent random link for elastin is seven to eight amino acids long. Since each amino acid contains two single bonds with rotational freedom, it takes about 15 points of partial freedom to create an equivalent random link. The situation for the octopus elastomer is even more extreme. The equivalent random link for OAE contains about 12 amino acid residues, or about 24 points of partial freedom. This increased stiffness of OAE correlates well with its amino acid composition. The glycine content is low (8·5%), and small amino acids account for only 23% of the total. As a consequence, the protein is dominated by amino acids with large side chains that will more strongly restrict the rotational freedom of the peptide backbone. It is tempting to speculate that evolutionary design has selected a peptide sequence with high backbone stiffness to obtain the desired non-Gaussian properties, while still somehow maintaining the kinetic freedom necessary for rubber-like elasticity. The large portion of polar amino acids may contribute to this process, but additional information is needed before any firm conclusions can be drawn. It is clear, however, that the design of peptide sequences that give rise to rubber-like materials is an important area for future research, and it is possible that molecular techniques for determining peptide sequences from cloned genes may provide important advances in this area.

SPIDER SILK

The second example deals with a protein network that functions as a high stiffness fibre rather than as a rubber–elastic one. The material is the silk that orb-web weaving spiders use for their draglines (i.e. the threads upon which they lower themselves) and for the guys, frame and radial strands of their orb-web (Fig. 5(A)). This silk has a very high stiffness ($4 \times 10^9 \, \text{Nm}^{-2}$) and strength (about $1.4 \times 10^9 \, \text{Nm}^{-2}$; Refs 18 and 19). This level of stiffness and strength makes spider frame silk amongst the stiffest and strongest of natural biomaterials, comparable to cellulose fibres and much stiffer and stronger than the collagen fibres in human tendons.[1] Frame silk, unfortunately, does not match the stiffness and strength of the super-strength polymeric fibres such as the Kevlars, but perhaps this is not surprising because the main function of frame silk is not just to be stiff and strong. Of equal or perhaps even greater importance is the ability of frame silk to absorb energy before it breaks. That is, silk must be tough, and mechanical studies indicate that silk is exceptionally tough. Denny[18] measured the breaking energy for frame silk at about $1.5 \times 10^8 \, \text{Jm}^{-3}$, a value that is an order of magnitude or more greater than the breaking energy of most other high strength polymeric materials. This comparison is illustrated in Fig. 6, where stress–strain curves for cellulose fibres, frame silk and collagen tendon are plotted together. The energy to break, or toughness, is indicated by the area under the stress–strain curve, and the graph illustrates the dramatic difference in toughness very clearly. Thus silk is not only reasonably stiff and respectably strong, but also it is capable of undergoing quite large deformations and absorbing large quantities of energy before it breaks.

The functional importance of extensibility in frame silk goes beyond the absorption of energy in simple tension. This is because the frame and radial members of the orb-web typically must absorb the largest amounts of energy from insects that collide with the structure by flying at right angles to the axis of individual fibres (Fig. 5(B)). The ability of a fibre to resist such transverse loading will be strongly dependent on the extensibility of the fibre, because at a small deflection angle, θ, the tension, T, in the fibre will be much larger than the applied force, F. There is an optimum extensibility for

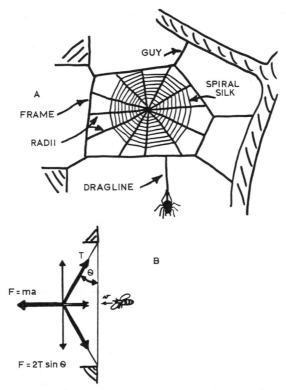

Fig. 5. (A) A typical orb-web, showing the guy, radii, frame and dragline fibres that are produced by the major ampullate gland. The catching spiral silk is made from a different silk that is produced by a different gland, and this spiral silk is coated with a sticky glue that is produced by a third gland. The major ampullate gland silk is also used by the spider to form its dragline. (B) Flying insects typically collide with frame members in an orb-web as shown, with the force of impact, F, acting at right angles to the long axis of any fibre. In this case the tension, T, developed in the fibre is determined by the deflection angle, θ. The optimum extensibility for supporting the maximum force in this type of situation will be about 40%. The geometry shown here is for an approximately 30% extension of the fibre, a value that is typical of frame silk.

fibres deformed in this manner at which a fibre with a given strength will support the maximum transverse force, and for fibres that deform with constant volume this optimum is at about 40% extension.[18] It is undoubtedly no accident that the extensibility of frame silk, which is about 30%, allows this material to function to within a few per cent of this optimum. Fibres with strengths much greater than that of frame silk, but

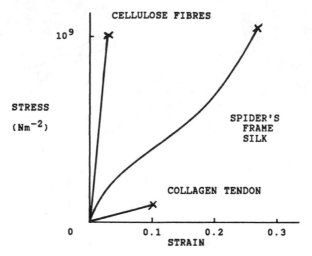

Fig. 6. Typical stress–strain curves for several high stiffness biopolymeric fibres. The cellulose curve is based on data for plant fibres, such as ramie, flax or hemp.[1] The frame silk data are from Denny,[18] and the tendon data are from Woo et al.[20]

with small breaking extensions, will resist much smaller transverse loads than frame silk. Thus, extensibility is an extremely important functional property of this material. In addition, Denny[18] showed that frame silk exhibits a high, first-cycle mechanical hysteresis, such that almost 70% of the energy absorbed in an extension is dissipated viscoelastically and is not available to catapult a trapped insect back out of the web.

The biological question, then, is what aspects of molecular design allow for this balance between stiffness, strength, extensibility and hysteresis? Insect and spider silks have long been known to contain crystalline regions, in which the polypeptide chains fold back and forth on each other to form anti-parallel β-structure crystals.[1] As with other crystalline polymers, however, only part of the polymeric chains exist in these crystalline regions, and a considerable portion of each chain exists in amorphous regions. A reasonable amount is known about the crystal unit cell in silks from a variety of different insects and spiders from X-ray diffraction studies,[21] but virtually nothing is known about the size, shape and volume fraction of the crystals or about the organization of the protein chains in the amorphous regions. Considering the importance of extensibility and hysteresis to the functional properties, an understanding of the amorphous chains and their interaction with the crystals seems essential. However, in its normal dry state it is very difficult to distinguish the contribution of the crystal and amorphous phases to the material properties.

Work [22] has shown, however, that when frame silk is immersed in water it absorbs water, swells to slightly more than twice its initial volume and contracts to about 60% of its initial length. This hydration-induced contraction is associated with a dramatic decrease in initial stiffness from about 10^{10} Nm^{-2} to about 10^7 Nm^{-2} (Fig. 7), and we have used thermoelastic experiments to demonstrate that contracted frame silk exhibits entropy-driven, or rubber-like, elasticity.[23] X-ray diffraction studies of contracted frame silk [24] suggest that the β-structure crystals do not melt in this contraction. Thus, contracted frame silk contains a crosslinked rubber network, where the crosslinks are presumably provided by the crystals; it should be possible to use the kinetic theory of rubber elasticity to characterize the network structure and to reveal basic information about the molecular organization of the protein chains in the amorphous regions of the silk, as well as their interaction with the crystals.

Contracted frame silk exhibits a 'J-shaped' stress–strain curve (Fig. 7) in which the modulus increases by about two orders of magnitude at extensions of only about 80%. This suggests non-Gaussian behaviour in a

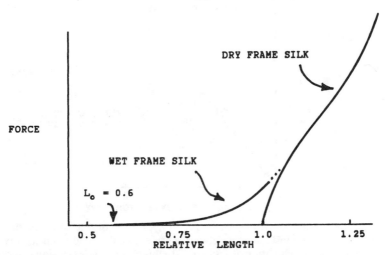

Fig. 7.   The hydration-induced contraction of frame silk. When major ampullate gland silk is immersed in water it absorbs water, swells and shrinks to about 60% of its initial length. The hydrated, contracted silk behaves like a rubber network, but one that enters the non-Gaussian region at very small strains, as shown. The initial modulus of the wet silk is about 10^7 Nm^{-2}, but when extended by 80% relative to the wet, resting state the stiffness reaches about 8×10^8 Nm^{-2}.

very short-chain network. We are currently attempting to analyse the mechanical test data for contracted silk with non-Gaussian network models, and the data presented here give an indication of our initial results.

We find that the mechanical properties of contracted silk can be matched very well by non-Gaussian network curves, using Treloar's[25] series expansion approximations to the inverse Langevan function (Fig. 8). But the values for the number of equivalent random links, N, and elastic modulus, G, derived from these comparisons seem unreasonable. First, G values of the order of $10^4 \, Nm^{-2}$ are obtained, which correspond to an average molecular weight between network crosslinks, M_c, of about 2×10^5 daltons. Thus, the amorphous chains between crosslinks are on average about 2600 amino acid residues long. Next, the N values are less than 1 (see Table 3), which means that each equivalent random link is about 8000 amino acids long. This seems ridiculous, considering that the equivalent

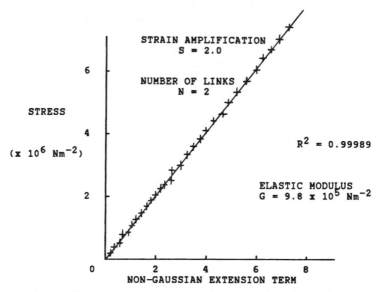

Fig. 8. The non-linear stress–strain curve of hydrated, contracted frame silk can be accurately fit to non-Gaussian network curves, as calculated from the inverse Langevin series approximation of Treloar.[25] In this process the extension values are substituted into the series expansion, yielding a non-Gaussian extension term. Linear regression analysis is then used to determine the number of equivalent random links, N, that gives the best linear fit between the non-Gaussian extension term and the stress. The best fit values of N typically occur with R^2 values greater than 0·9997. The slope of the regression with the highest R^2 gives the elastic modulus, G.

TABLE 3

Summary of Calculated Values from the Non-Gaussian Analysis of Contracted Frame Silk. For each Value of Strain Amplification Factor, S, Best Estimates of the Number of Equivalent Random Links, N, the Apparent Elastic Modulus, G, the Average Molecular Weight between Crosslinks, M_c, and the Number of Amino Acid Residues per Random Link, R, are Derived from the Fitting of Mechanical Data to Non-Gaussian Network Curves

S	N	G (Nm^{-2})	M_c (daltons)	R
1·0	0·3	6×10^3	2×10^5	8 500
1·5	1·3	5×10^5	2 200	22
2·0	2·0	10×10^5	1 350	9
2·5	2·7	11×10^5	1 180	5

random link in the two rubber-like proteins described above contained between seven and 12 amino acids. The analysis outlined above ignores the crystals, however, and treats contracted silk as a homogeneous, amorphous network. Crystals are present, and they will act as reinforcing fillers and have a major effect on the mechanical properties.

The effect of fillers in rubber networks is frequently attributed to a strain amplification of the amorphous network. That is, due to the interaction of the random chains and the rigid filler particles, the strain of individual network chains is larger than the macroscopic strain of the sample by some amplification factor.[26] The analysis of filled networks is very complex,[27] but we attempt to deal with this problem by simply applying an amplification factor to the extension scale of our mechanical data. This 'amplified' data is then fitted to non-Gaussian curves to obtain estimates of the elastic modulus, G, and the number of equivalent random links, N. This process seems to work reasonably well, in that the estimates of G and N are somewhat more reasonable (Table 3). But the strain amplification factor is an adjustable parameter, and we need some way to evaluate what level of strain amplification is appropriate. At present the only way we have of making this decision is to assess the predicted number of amino acid residues per equivalent random link (i.e. column 5 in Table 3). We know from our analysis of elastin and OAE, which have no rigid filler particles, that the number should be between about 7 and 12. Thus, as a first approximation, the strain amplification factor is probably about 2·0, making the modulus of the network, G, about $10^6\ Nm^{-2}$, the average molecular weight of random chains between crystals, M_c, about 1350 daltons, and the number of equivalent random links per chain about two. This, then, is our current model for the molecular network in frame silk.

One problem with this analysis is that a random network made from chains with only two random links is probably somewhat below the limit of short-chain networks that can be described with the available theory. So we must expect that the predictions are at best only approximations to the correct values, and we must attempt to find more accurate models for short, rigid chains. We can, however, take these approximations and make further predictions that may allow us to test the validity of the non-Gaussian analysis described here. We are doing this in two ways. One involves measuring the strain-induced birefringence of swollen silk to see if the pattern of birefringence change is consistent with an amorphous network of chains made from two random links, and the other involves an analysis of crystalline fillers that account for the predicted strain amplification.

Contracted frame silk exhibits a residual positive birefringence of about 9×10^{-3} and, assuming that the amorphous network chains are optically isotropic, this means that the crystals retain some residual preferred orientation in the axial direction of the fibre. The birefringence increases linearly with stretching to a value of about 2.4×10^{-2} at 70% extension. Clearly, this increase in birefringence is due both to the extension and alignment of the random chains in the amorphous network as well as to the axial reorientation of the crystals. If predicted birefringence changes for a random network[25] are subtracted from the total birefringence change, then the residual birefringence change should account for the crystal reorientation. If the random network is assumed to contain chains with two random links and is deformed at a strain amplification of 2, then the change in residual birefringence suggests that crystal reorientation is essentially complete when the fibre is stretched back to the length that it had before it was hydrated in the first place. This predicted matching of the crystal reorientation to the starting length of the dry silk fibre is not conclusive evidence for our non-Gaussian network model, but it is, at least, consistent with the model.

Calculation of strain amplification factors created by filler particles can provide us with additional information about the silk network. Specifically, it is possible to evaluate structural models for the shape and volume fraction of the filler particles that must account for the predicted strain amplification factor of about 2.0. It is known from electron microscope studies of silkworm silk (see Ref. 21) that the β-structure crystals are elongated structures, and the best evidence available suggests a length:

width ratio of about 5:1. Hydrodynamic calculations of the strain amplification around rod-like particles[27] suggest that the strain amplification factor is given by

$$S = 1 + 0.67f\phi + 1.62f^2\phi^2$$

where f is the length:width ratio and ϕ is the volume fraction of filler particles in the network. On this basis of this relationship, we calculate that for $f = 5$, the volume fraction, ϕ, of crystals in the silk fibres will be about 0.12 for the hydrated, contracted silk. Obviously, the combinations of axial ratio and volume fraction are unlimited, but the important thing is that these are measurable parameters that can be determined directly by electron microscopy. We are currently doing the appropriate analyses and hope to have some direct morphological information in the near future that will allow us to 'test' the molecular model we have proposed.

Finally, our objective in this analysis was to use information about the non-Gaussian network properties of contracted silk to provide insights into the functional organization of the dry silk as it is used by the spider in its web. The first thing, therefore, is to remove the water, shrinking the fibres to slightly less than half their hydrated volume and increasing the crystal volume fraction to about 0.25. Drying will also dramatically alter the mechanical properties of the amorphous network. Protein networks, like elastin, enter their glass transition when they are dried,[4] presumably because strong peptide–peptide hydrogen bonds form in the absence of water. Therefore, we expect that hydrated, contracted frame silk will increase in stiffness by at least two orders of magnitude when it is dried, but this is not the only change needed to bring the swollen, contracted silk back to the state in which it exists in the spider's web. Recall that the silk contracts in length to about 60% of its initial length when it is hydrated. Thus, to regain this initial length, the contracted silk must be extended by about 75% before it is dried. Indeed, this process of stretching followed by drying undoubtedly describes the final stages of the process by which the spider spins these fibres. Thus, the crosslinked silk network, which presumably forms as the silk secretion is sheared in its transit through the spinneret, is extended well into the non-Gaussian region before it dries and is locked into this extended conformation. We can now see that spider's frame silk is probably best thought of as a fibre-reinforced composite, and its properties are due to three major components. (1) The high stiffness crystals are aligned essentially parallel to the fibre axis to give a high degree of axial reinforcement. This probably accounts for the high stiffness of frame silk. (2) The amorphous network is highly extended, so that stress is carried almost directly by the covalent bonds in the extended polymer chains. This probably explains the great strength of silk. (3) The amorphous network is shifted into its glass transition region so that any further change in crystal orientation or amorphous chain extension is strongly resisted by frictional forces. This probably contributes to the controlled extensibility, the extremely high breaking energy and the high hysteresis that are so important to the function of the silk in the web. It is unlikely, however, that these most important aspects of silk's properties are due entirely to glassy amorphous chains. The pulling apart of the folded-chain crystals should occur at high stresses, and this will consume a great deal of energy. Regardless of which component is dominant, it is clear that the amorphous protein network plays an extremely important role in the formation of silk fibres and in determining the functional mechanical properties of this exceptional biomaterial.

REFERENCES

1. Wainwright, S. A., Biggs, W. D., Currey, J. D. and Gosline, J. M., *Mechanical Design in Organisms*, Princeton Univ. Press, Princeton, New Jersey, 1982.
2. Vincent, J. F. V. *Structural Biomaterials*, John Wiley & Sons, New York, 1982.
3. Gordon, J. E., Biomechanics, the last stronghold of vitalism, in *Mechanical Properties of Biological Materials*, Vincent, J. F. V. and Currey, J. D. (Eds), Society for Exptl. Biology, Cambridge, 1980, pp. 1–12.
4. Gosline, J. M., Elastic properties of rubber-like proteins and highly extensible tissues, in *Mechanical Properties of Biological Materials*, Vincent, J. F. V. and Currey, J. D. (Eds), Society for Exptl. Biology, Cambridge, 1980, pp. 331–57.
5. Bergel, D. H., Static elastic properties of the arterial wall, *J. Physiol.*, 1961, **156**, 445–68.
6. Wolinsky, H. and Glagov, S., Structural basis of the static mechanical properties of the aortic media, *Circ. Res.*, 1964, **14**, 400–15.
7. Lanir, Y., Rheological behaviour of the skin—experimental results and a structural model, *Bio-rheology*, 1979, **16**, 191–202.
8. Sage, E. H. and Gray, W. R., Evolution of elastin structure, in *Elastin and Elastic Tissue*, Sandberg, L. B., Gray, W. R. and Franzblau, C. (Eds), Plenum Press, New York, 1977, pp. 291–312.
9. Aaron, B. B. and Gosline, J. M., Elastin as a random-network elastomer: Mechanical and optical analysis of single elastin fibres, *Biopolymers*, 1981, **20**, 1247–60.
10. Gosline, J. M. and Rosenbloom, J., Elastin, in *Extracellular Matrix Biochemistry*, Piez, K. A. and Reddi, A. H. (Eds), Elsevier, New York, 1984, pp. 191–228.
11. Shadwick, R. E. and Gosline, J. M., Physical and chemical properties of rubber-like elastic fibres from the octopus aorta, *J. Exp. Biol.*, 1985, **114**, 239–57.

12. Shadwick, R. E. and Gosline, J. M., Elastic arteries in invertebrates: mechanics of the octopus aorta, *Science*, 1981, **213**, 759–61.
13. Gosline, J. M. and Shadwick, R. E., Biomechanics of the arteries of Nautilus, Nototodarus and Sepia., *Pacific Science*, **36**, 283–96.
14. Shadwick, R. E. and Gosline, J. M., Molecular biomechanics of protein rubbers in Mollusks, in *The* Mollusca, *Vol. I, Metabolic Biochemistry and Molecular Biomechanics*, Hochachka, P. W. and Wilbur, K. W. (Eds), Academic Press, New York, 1983, pp. 399–430.
15. Shadwick, R. E. and Gosline, J. M., Mechanical properties of the octopus aorta, *J. Exp. Biol.*, 1985, **114**, 259–84.
16. Flory, P. J., *Statistical Mechanics of Chain Molecules*, John Wiley & Sons, New York, 1969.
17. Andersen, S. O., Resilin, in: *Comprehensive Biochemistry*, Vol. 26 C, Florkin, M. and Stotz, E. H. (Eds), Elsevier, Amsterdam, 1971, pp. 633–57.
18. Denny, M. W., The physical properties of spider's silk and their role in the design of orb-webs, *J. Exp. Biol.*, 1976, **65**, 483–506.
19. Denny, M. W., Silks—Their properties and functions, in *Mechanical Properties of Biological Materials*, Vincent, J. F. V. and Currey, J. D. (Eds), Society for Exptl. Biology, Cambridge, 1980, pp. 247–71.
20. Woo, S., Gomez, M. A., Amiel, D., Ritter, M. A., Gelberman, R. H. and Akeson, W. H., Effects of exercise on the biomechanical and biochemical properties of swine digital flexor tendons, *J. Biomechanical Engineering*, 1981, **103**, 51–6.
21. Lucas, F. and Rudall, M. K., Silks, in *Comprehensive Biochemistry*, Vol. 26B, Florkin, M. and Stotz, E. H. (Eds), Elsevier, Amsterdam, 1968, pp. 475–588.
22. Work, R. W., Dimensions, birefringences, and force–elongation behaviour of major and minor ampullate silk fibres from orb-web-spinning spiders—The effects of wetting on these properties, *Text. Res. J.*, 1977, **47**, 650–62.
23. Gosline, J. M., Denny, M. W. and DeMont, M. E., Spider silk as rubber, *Nature, Lond.*, 1984, **309**, 551–2.
24. Work, R. W. and Morosoff, N., A physico-chemical study of the supercontraction of spider major ampullate silk fibres, *Text. Res. J.*, 1982, **52**, 349–56.
25. Treloar, L. R. G., *Physics of Rubber Elasticity*, Clarendon Press, Oxford, 1975.
26. Mullins, L., Theories of rubber-like elasticity and the behaviour of filled rubber, in *Mechanical Properties of Biological Materials*, Vincent, J. F. V. and Currey, J. D. (Eds), Society for Exptl. Biology, Cambridge, 1980, pp. 273–88.
27. Fedors, R. F., Uniaxial rupture of elastomers, in *Stereo Rubbers*, Saltman, W. M. (Ed.), John Wiley & Sons, New York, 1977, pp. 679–803.

4

FIBRINOGEN AND FIBRIN STUDIED BY SMALL-ANGLE NEUTRON SCATTERING

KELL MORTENSEN

*Department of Physics, Risø National Laboratory,
DK-4000 Roskilde, Denmark*

ROGERT BAUER

*Department of Physics, Veterinary and Agricultural University,
Copenhagen, Denmark*

and

ULF LARSSON

*Department of Medical Biophysics, Karolinska Institute,
Stockholm, Sweden*

$60\,^{9}$

ABSTRACT

The small-angle neutron scattering (SANS) technique has been used to study the structure of fibrinogen and fibrin. Fibrinogen has an elongated 50-nm-long open structure. Upon polymerization, long thin fibrin fibers are formed. The SANS data suggest that the fibrin fiber is like a 300-nm-diameter cable composed of 40-nm-diameter fibrin sub-fibers. .34

1 INTRODUCTION

The key event in the complex cascade of protein interactions which leads to blood clotting is the formation of a network of fibrin from the plasma protein, fibrinogen. Studies of fibrin and fibrinogen have accordingly attracted a lot of interest since their discovery in the 1940s.[1] The structure of fibrinogen has within the last few decades been highly controversial. At

79

present, most experiments point to a trinodular form, which was first proposed by Hall and Slayter.[2] Upon polymerization, fibrinogen molecules line up in parallel, forming long thin fibrin fibers.

The most conclusive studies on the structure of fibrinogen and fibrin clots are at present based on electron microscopy, which in the last few years has made very high resolution possible.

Electron microscopy requires, however, relatively drastic sample treatment, which may affect the results. It is therefore of great importance to verify the structures obtained by electron microscopy by methods which are usable for the untreated proteins. Small-angle scattering (SAS) techniques are ideal for such a purpose, because they can be used on the raw protein in solution. (Neutron scattering may, though, require deuterated or partly deuterated water.) SAS has previously been used for studies on fibrinogen. X-ray studies (SAXS) suggested a straight rod-like structure,[3] whereas the interpretation of neutron scattering (SANS) results was that of a bend form.[4]

In this paper we present small-angle neutron scattering data on fibrinogen and on thrombin-induced fibrin.[5] The data are in agreement with the trinodulus structure of fibrinogen proposed by other techniques. The results on fibrin show, moreover, that this 'open structure', as opposed to the more massive rod- and sphere-like structures of fibrinogen, remains in the fibrin fiber. The SANS data on thrombin-induced polymerization suggest, moreover, that the linear fibrin polymer is like a cable composed of fibrin sub-fibers.[6]

2 EXPERIMENTAL TECHNIQUE

In order to get good contrast and low background in the scattering experiment, all measurements were done on proteins dissolved in 95% deuterated water. Human fibrinogen was used in the concentration 10 mg/ml, and for the polymerization studies low concentrations ($\sim 10^{-4}$ NIH units/ml) of thrombin were added. Light scattering studies indicated that the structure of fibrinogen is not affected by the high concentration of D_2O.[7] There are, however, indications that the polymerization process depends on the D_2O content.

The neutron scattering was performed at the SANS facility at Risø National Laboratory, Denmark. The measurements were carried out with neutrons of wavelength 12 Å and 24 Å. The two-dimensional scattering data were corrected for inhomogeneities in the detector efficiency using the incoherent scattering from water, and scattering from the buffer and quartz containers was subtracted. All data shown are radial averaged.

3 RESULTS AND DISCUSSION

In Fig. 1 is shown a $\log I$ vs. q^2 plot of the scattering data of fibrinogen. In the range $q < 0.14 \, \text{nm}^{-1}$, $I(q)$ follows the Guinier formula

$$I(q) = n\Delta\rho^2 V^2 \exp(-R_g^2 q^2/3) \qquad (1)$$

where $\Delta\rho$ is the excess neutron scattering density, V is the volume and n is the concentration of molecules. From the slope, we get the molecular radius of gyration, $R_g = 13.5 \, \text{nm}$.

The deviation from Guinier behaviour at larger q-values (Fig. 1) gives additional details on the fibrin structure. A plot of $\log(Iq)$ vs. q^2, as shown in Fig. 2, shows a linear relationship. This indicates a very elongated shape of fibrinogen. The scattering function of rod-like structures is given by

$$q \times I(q) = n\Delta\rho^2 \pi \times L \times V \times \exp(-q^2 R_c^2/2) \qquad (2)$$

in a region $qR_c < 1$, R_c being the cross-sectional radius of gyration. L is the length of the rod. The ratio between the $q = 0$ intersect of the Guinier behavior (eqn (1)) and of the rod-like behavior (eqn (2)) leads directly to the length of the molecule, as

$$(I)_{\text{Guinier}}/(Iq)_{\text{rod}}|_{q=0} = L/\pi \qquad (3)$$

For fibrinogen the data represented in Figs 1 and 2 give the length of fibrinogen, $L_{\text{fbg}} \simeq 50 \, \text{nm}$, which is in perfect agreement with the value obtained by other methods.[1] The subscript 'fbg' refers to fibrinogen. For a rod-like structure, the radius of gyration is related to the length and diameter by $R_g^2 = (L^2 + 2R_c^2)/12$. For fibrinogen, the right-hand side is markedly dominated by the length, and the statistics of our present data do not allow us to deduce R_c. Using the known volume of $V_{\text{fbg}} \sim 1800 \, \text{nm}^3$,[1] and assuming cylindrical shape, leads to $R_c = 3.5 \, \text{nm}$.

Upon polymerization, the scattering intensity increases dramatically at the smallest angles, whereas no changes occur at large q-values (Fig. 3). The scattering function can be divided into three regions, where each region contributes to the details of the fibrin structure.

The absence of changes in the scattering function for $q > 0.15 \, \text{nm}^{-1}$ between fibrinogen and the fibrin polymer means that on the corresponding length scale there are no intermolecular correlations between individual

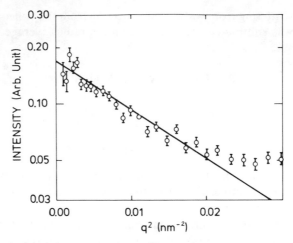

Fig. 1. Guinier plot ($\log I$ vs. q^2) of SANS data of fibrinogen.

fibrin monomers. This agrees with the 'open' structure of the trinodulus model, while it is in direct disagreement with any compact models. The slope of -1 in the $\log I$ vs. $\log q$ plot (Fig. 3) reflects a very elongated nature, i.e. basically a one-dimensional structure,[8] which probably originates from the interconnecting α-helices. The slopes in a $\log(Iq)$ vs. q^2 plot give, according to eqn (2), a cross-sectional radius of gyration $R_c \sim 2$ nm.

In the intermediate q-range, 0.06–0.15 nm^{-1}, the full polymerized fibrin

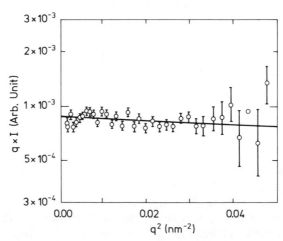

Fig. 2. Cross-sectional plot ($\log Iq$ vs. q^2) of SANS data of fibrinogen.

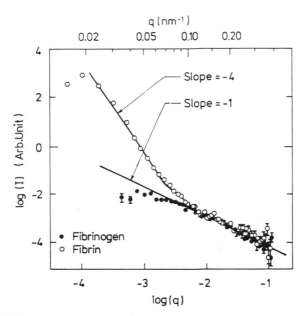

Fig. 3. SANS data of thrombin-induced fibrin polymers on a log I vs. log q scale.

shows $I \propto q^{-2 \cdot 4}$ as seen in Fig. 3. Assuming that the structure factor of fibrin and fibrinogen does not interfere, the structure factors are simply multiplicative. As fibrinogen basically shows $I \propto q^{-1}$, it thus follows that, within the q-range in discussion, fibrin itself has an $I \propto q^{-1 \cdot 4}$ relationship, which closely represents a rod-like polymer. Analyzing the spectrum for fibrin, where the form factor of fibrinogen has been factorized out, leads in a cross-sectional fit (eqn (2)) to an internal diameter of 40 nm.

At the smallest q-values, the fibrin polymer shows an $I \propto q^{-4}$ relationship, which is the Porod behavior typical for objects so large that the scattering profile reflects the properties of the surface only. The Porod scattering function is given by

$$I(q) = n_{fb} \times \Delta\rho^2 \times S_{fb} \times 2\pi \times q^{-4}$$
$$= n_{fb} \times V_{fb} \times \Delta\rho^2 2\pi(S_{fb}/V_{fb}) \times q^{-4} \quad (4)$$

where S_{fb} and V_{fb} are the surface and volume of fibrin. The subscript 'fb' refers to fibrin. As shown previously, the total protein volume is unaffected by the polymerization

$$n_{fb}V_{fb} = n_{fbg}V_{fbg} \quad (5)$$

By using the Guinier approach (eqn (1)) for fibrinogen, eqn (4) leads to

$$V_{fb}/S_{fb} = (2\pi I^0_{fbg})/(V_{fbg} \times I(q) \times q^4) \qquad (6)$$

for the volume to surface ratio. I^0_{fbg} is the Guinier asymptotic $q = 0$ value of fibrinogen. Using the previously obtained parameters for fibrinogen, the results of Figs 1 and 3 lead to $V_{fb}/S_{fb} \simeq 75$ nm. Assuming a cylinder-like fibrin polymer, this ratio gives a fibrin diameter of $d_{fb} = 4V_{fb}/S_{fb} \simeq 300$ nm.

CONCLUSION

In conclusion we have shown that the SANS technique can be used to study details of the structure of fibrinogen and fibrin in solution. Figure 4 shows the results schematically. Fibrinogen is an elongated 50-nm-long molecule. The structure is 'open', as proposed by the trinodulus model.

Fibrin polymers have the form of a 300-nm-wide cable, consisting of 40-nm-wide fibers, which are built up by the 7-nm-wide activated fibrinogen units. The present SANS study confirms that many of the conclusions based on heavily treated fibrinogen and fibrin are also correct for the untreated proteins in solution.

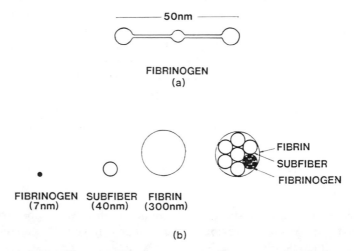

Fig. 4. Schematic representation of the results obtained by the SANS technique: (a) the open (trinodulus) structure of fibrinogen; and (b) the fibrin fiber structure.

ACKNOWLEDGEMENT

This work was motivated by and improved by discussions with R. Rigler and B. Blombäck. The SANS facility at Risø has been established with support from the Danish and Swedish Natural Science Research Councils.

REFERENCES

1. For a recent review, see for example *Molecular Biology of Fibrinogen and Fibrin,* Mosesson, M. W. and Doolittle, R. F. (Eds), Ann. New York Acad. Science **408** (1983).
2. Hall, C. E. and Slayter, H. S., *J. Biophys. Biochem. Cytol.,* 1959, **5**, 11.
3. Lederer, K. and Hammel, R., *Die Macromol. Chem.,* 1975, **176**, 2619.
4. Marguerie, G. and Stuhrmann, M. B., *J. Mol. Biol.,* 1976, **102**, 143.
5. Larsson, U., Rigler, R., Blombäck, B., Mortensen, K. and Bauer, R., in *Structure, Dynamics and Function of Biomolecules,* Springer Series in Biophysics, Vol. 1, Ehrenberg, A., Rigler, R., Gräslund, A. and Nilsson, L. (Eds), Springer-Verlag, New York, 1987.
6. Müller, M. F., Ris, H. and Ferry, J., *J. Mol. Biol.,* 1984, **174**, 369.
7. Larsson, U., Rigler, R. and Blombäck, B., unpublished results.
8. Freltoft, T., Kjems, J. K. and Sinha, S. K., *Phys. Rev.,* 1986, **B33**, 269.

5

THE EFFECT OF GELATION ON WATER–PROTEIN INTERACTION

H. Tenhu, O. Rimpinen

Department of Wood and Polymer Chemistry, University of Helsinki, Meritullinkatu 1 A, 00170 Helsinki, Finland

and

F. Sundholm

Department of Chemistry, University of Helsinki, E. Hesperiankatu 4, 00100 Helsinki, Finland

ABSTRACT

Aqueous gelatin was studied by differential scanning calorimetry (DSC). Samples with varying degrees of crosslinking were prepared using different annealing times below the melting point of the gel. The amount of non-freezable water associated with the protein network was estimated by freezing and thawing the samples. Progressive gelation was shown to decrease the amount of water not capable of crystallising. Chemical crosslinking with glutaric aldehyde produced samples considerably different from the physically crosslinked gel.

INTRODUCTION

Collagen molecules form triple-helical structures which associate to fibrils with considerable strength. The fibrils are further strengthened by crosslinks between the molecular chains. The helix structure seems to be the key to understanding the properties of collagen, as well as the properties of gelatin gels. Gelatin is a product of collagen denaturation, normally carried out with aqueous acid or base. Network formation in dilute gelatin solution is considered as partial renaturation: protein chains tend to associate and to rebuild the original structure.[1]

Water molecules bind strongly to collagen by hydrogen bonds and it is often assumed that water molecules take part in the helix stabilisation ('structural water'). Water may thus participate also in network formation in a gelatin solution. Regardless of whether this is the case or not, water is known to associate strongly to the gelatin network and the mobility of the water molecules decreases when gelation proceeds.[2,3] In previous studies[4-6] we have come to the conclusion that not only the chemical structure but also the physical state of a polymer, expressed for example in segmental mobility, affect its interaction with water. Aqueous samples of hydrophilic polymers (or polymers having hydrophilic groups) contain a considerable amount of water which does not freeze even at very low temperatures, this phenomenon being due to hydrogen bond formation or other intermolecular forces as well as kinetic factors. The amount of non-freezable water can be estimated spectroscopically or with differential scanning calorimetry (DSC) by freezing and melting the sample.

Studies on flexible polymers with random conformation (synthetic polymers or denatured proteins extracted from biological samples) have shown that polymer crosslinking reduces the amount of non-freezable water associated with the macromolecule.[4-6] This can be explained by assuming that the amount of non-freezable water is a measure of the amount of hydration water around the macromolecule. The assumption is reasonable, although simplified—undoubtedly there are several factors affecting the enthalpy of melting of ice detected by DSC.[7,8] One motive for our calorimetric work is to evaluate the applicability of this method to conformational and morphological studies of polymers.

Collagen differs from the above-mentioned flexible polymers because of its highly ordered structure. Chemical crosslinking with well-known crosslinking agents[9,10] such as glutaric aldehyde or benzoquinone does not affect the amount of non-freezable water in aqueous collagen samples. This indicates that although crosslinking has a remarkable effect on the collagen structure on a supermolecular level, the changes on a microscale are less profound and need other methods to be detected. The amount of non-freezable water in collagen can be altered, however, by a more violent treatment combining hydrolysis and crosslinking. In partially hydrolysed collagen the non-freezable water fraction is decreased by chemical crosslinking.[6] For this reason we wanted to extend the investigation to include gelatin. We shall discuss the effect of gelation and that of chemical crosslinking on water–polymer interaction in aqueous gelatin. In order to compare gelatin with collagen we have also prepared gelatin samples oriented in a magnetic field.

EXPERIMENTAL

Materials
Bovine skin gelatin (approximately 225 bloom) from Sigma Chemical Company, St. Louis, USA, was used. Collagen from bovine tendon was also from Sigma; its treatment is described elsewhere.[6]

Gelatin Crosslinking
Crosslinking with glutaric aldehyde was carried out by two methods.

(1) Four millilitres of 12·5% aqueous glutaric aldehyde were added to 8 ml of buffered (pH 7) 10 wt% gelatin solution at 50°C with stirring. After 1 h the sample was allowed to cool to room temperature. The sample was dialysed several times against distilled water.

(2) Using the same amounts of reagents as above, gelatin powder was added to a buffered glutaric aldehyde solution at room temperature. The mixture was stirred for 1 h. The sample was purified as above.

Oriented Gelatin
A 10 wt% aqueous gelatin solution was placed in a 4·7 T magnetic field (of an NMR spectrometer) in a 10-mm-o.d. quartz tube. The solution was kept at 60°C for 1 h and then allowed to gelate at room temperature for 12 h. Slices were cut from the gel obtained and used for the DSC measurements as such.

DSC Measurements
The amount of non-freezable water in different protein samples was estimated with a Perkin Elmer DSC 2. Known amounts of gelatin and water were sealed in aluminium pans and kept at 60°C for 1 h. The samples were stored at 4°C for 2 days before the measurements. Gel samples were cooled in the sample chamber of the calorimeter to −50°C with a cooling rate of 2·5°C/min. After 10 min thermograms were registered heating the samples to room temperature with a heating rate of 10°C/min (melting of *water*). This cycle was repeated three times. After these measurements the samples were heated at 90°C for 5 min (melting of the *gel*). They were then quickly cooled back to room temperature and the freeze-and-thaw cycle was repeated twice. By comparing the area of the melting peak of water to

that of pure water, the amount of freezable water could be calculated. The difference between the total amount of water (which was re-checked by evaporating the water at 120°C and by weighing) is denoted as non-freezable water. The measurements on collagen are used here as a reference, and they are explained elsewhere.[6]

RESULTS AND DISCUSSION

Network formation in solutions of gelatin is a two-step process. The primary process is fast and happens instantaneously after cooling the liquid below the melting point of the gel. The secondary process occurs slowly, in hours or days. Borchard et al.[11] have shown that prolonged annealing below the melting point of the gel gives rise to two endothermic peaks in the melting thermogram of the gel, instead of one peak obtained from a sample with short annealing time.

In our case the calorimetric measurement revealed two distinct melting peaks around the melting point of the gel, indicating that the network structure is not destroyed during repeated freezing and thawing. It should be noted, however, that small changes were observed at low temperatures, seen as a variation in the shape and area of the water melting peak in successive measurements. The variation is complex and occurs in different ways in different concentration regimes. Therefore we have calculated the amount of the non-freezable water from the first melting thermogram and plotted it against protein concentration in Fig. 1(A). Also included in this figure are values obtained when the gel was homogenised at 90°C and frozen immediately after. The corresponding values obtained using insoluble collagen are included as a reference in Fig. 1(B). It can be seen that after homogenisation a slightly larger portion of water remains unfrozen. After homogenisation only the primary gelation process has taken place before the sample is frozen. This is evidenced in melting the gel: only one single melting peak is observed. The result is in accordance with our earlier findings: while gelation proceeds the amount of non-freezable water decreases. Note that in the concentration region between 10 and 20 wt% there are some unexpectedly large values. We try to discuss this matter below.

The shape of the water melting peak in gelatin is very sensitive to the sample treatment prior to the measurement, as well as to the experimental conditions. This has been discussed in detail by Reutner et al.[12] The variation of the melting peak causes severe difficulties in calculating the

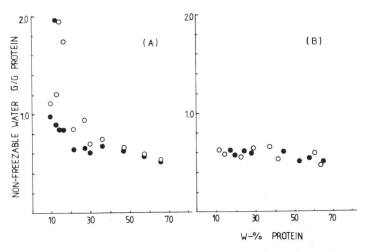

Fig. 1. The amount of non-freezable water per gram of protein as a function of protein content of the sample. (A) Gelatin: ●, first measurement; ○, measured after homogenisation at 90°C. (B) Collagen: ○, untreated collagen; ●, crosslinked with glutaric aldehyde.

enthalpy of melting, which may cause errors in the estimated amount of the non-freezable water. When the gelatin content of the sample exceeds 30 wt%, an abrupt exothermic change is observed during the water melting process. The exotherm is larger, the faster the sample has been cooled. This may be interpreted as crystallisation of supercooled (glassy) water. We suppose, however, that the rapid crystallisation process is accompanied by changes in the network structure. At present we are working to further clarify this point. Thus there are some points in Fig. 1(A) (around 2 g non-freezable water per 1 g of protein) that may be overestimated because of the exothermic process during the water melting.

When comparing gelatin with collagen, it is interesting to study the shape of the water melting peak. In collagen, water seems to be distributed in different surroundings, this giving rise to two melting peaks of different shape. This distinction is not seen in the case of gelatin, especially when the sample is frozen and melted immediately after homogenisation at high temperature. We conclude that in collagen there is an equilibrium between the two fractions of water: one inside the fibre and another outside, in the pores between the fibres. The equilibrium changes during the freeze-and-thaw cycle of the calorimetric measurement although the amount of the non-freezable water is constant. When the gelatin is oriented in a magnetic

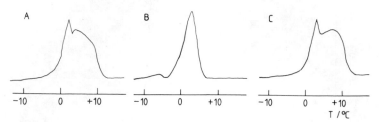

Fig. 2. Melting endotherms of water in different samples (protein content in parentheses): A, collagen (30 wt%); B, gelatin (30 wt%); C, oriented gelatin (10 wt%).

field,[13] the melting peak of water resembles that in insoluble fibrous collagen, as shown in Fig. 2. The relaxation of orientation can be followed by measuring slices of the gel after prolonged annealing at 4°C: the samples gradually begin to behave like the original gelatin over a period of several days.

Surprisingly, the gelatin samples crosslinked with glutaric aldehyde at 50°C contain more non-freezable water than the original untreated samples (Fig. 3). Anyhow, the same trend is to be seen as in pure gelatin: homogenisation at high temperature increases the amount of water that does not crystallise during freezing.

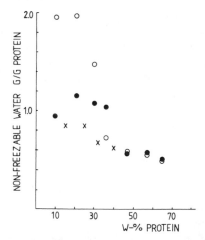

Fig. 3. The amount of non-freezable water per gram of protein in crosslinked gelatin: ●, samples crosslinked at 50°C, first measurement; ○, same samples after homogenisation at 90°C; ×, samples crosslinked at room temperature, first measurement.

It seems evident that crosslinking in a solution at elevated temperature stabilises a random conformation and prevents the helix formation. The samples which were crosslinked at room temperature by adding crystalline gelatin powder to a glutaric aldehyde solution behave in a quite opposite manner. The amount of non-freezable water is of the same order of magnitude as in the original gelatin samples, but decreases after heating to 90°C. In this case only the values obtained in the first measurement are included in Fig. 3. The decrease after heating is 0·1–0·2 g/g.

To conclude, we have shown that the interaction between gelatin and water changes with gelation. Chemical crosslinking in solution has an opposite effect to that of physical crosslinking. This may be due to the bulkiness of the crosslinks produced by glutaric aldehyde[9] which disturbs the helix formation.

REFERENCES

1. Veis, A., *The Macromolecular Chemistry of Gelatin*, Academic Press, London, 1964.
2. Derbyshire, W., in *Water, a Comprehensive Treatise*, Vol. 7, Franks, F. (Ed.), Plenum Press, New York, 1982, Chapter 4, pp. 339–430.
3. Maquet, J., Theveneau, H., Djabourov, M., Leblond, J., Papon, P., *Polymer*, 1986, **27**, 1103.
4. Tenhu, H., Sundholm, F. and Bjorksten, J., *Makromol. Chem.*, 1984, **185**, 2011.
5. Tenhu, H. and Sundholm, F., *Thermochim. Acta*, 1986, **102**, 15.
6. Tenhu, H. and Sundholm, F., *Eur. Polym. J.*, 1986, **22**, 629.
7. Pouchly, J., Biros, J. and Benes, S., *Makromol. Chem.*, 1979, **180**, 745.
8. de Vringer, T., Joosten, J. G. H. and Junginger, H. E., *Coll. Polym. Sci.*, 1986, **264**, 623.
9. Friedman, M. (Ed.), *Protein Crosslinking, Advances in Experimental Medicine and Biology*, Vol. 86A, Plenum Press, New York, 1977.
10. Flory, P. J. and Spurr, O. K., *J. Am. Chem. Soc.*, 1961, **83**, 1308.
11. Borchard, W., Bremer, W. and Keese, A., *Coll. Polym. Sci.*, 1980, **258**, 516.
12. Reutner, P., Luft, B. and Borchard, W., *Coll. Polym. Sci.*, 1985, **263**, 519.
13. Murthy, N. S., *Biopolymers*, 1984, **23**, 1261.

6

STRUCTURE DETERMINATION OF DIFFERENT CASEIN COMPONENTS

ANGELIKA THURN and WALTHER BURCHARD

*Institute of Macromolecular Chemistry, University of Freiburg,
Stefan-Meier Str. 31, 7800 Freiburg, FRG*

ABSTRACT

Casein is the main protein component of milk. It is not a homogeneous polymer but consists of several types of casein, and forms large spherical micelles of remarkable stability. The reason for this stability is not yet fully understood. In co-operation with other laboratories, we have been able to obtain well-separated individual fractions which were studied with regard to their capability for micelle formation and the micelle structure. Studies have been carried out with the α-, β- and κ-caseins. Small-angle neutron scattering and combined static and dynamic light scattering have been applied, which gave information on the particle weight M_w, the radius of gyration R_g and the hydrodynamic radius R_H, and on the particle scattering factor $P_z(q)$, where $q = (4\pi/\lambda) \sin \theta/2$.

The three caseins show strikingly different behaviour. For the β-casein, a star-like structure was detected which corresponds to the aggregation pattern of common soap micelles. The aggregation number is 38. The aggregates of κ-casein appear to be composed of star-like sub-micelles; each sub-micelle contains nine κ-casein molecules and the total aggregation number is 140. α-Casein forms long fibrils which resemble worm-like filaments. The contour length of the cylinders was found to be $L \simeq 1600$ nm, and the filaments appear to be composed of about 12 Kuhn segments of $l_K = 133 \pm 12$ nm in length.

INTRODUCTION

Milk consists of spherical particles of colloidal dimensions. The chemical composition of these particles is about 94% protein and 6% inorganic material, such as calcium, phosphate and citrate ions.[1] The components of milk are: water, 87·0%; protein, 3·3%; milk sugar, 4·8%; fat, 3·8%; ash, 0·7%. Casein (2·7%) is the main protein component of milk and the colloidal stability of the casein micelles is remarkable. For instance, it is well known that milk can be heated up to 100°C without destroying the micellar structure.

Under the influence of milk-clotting enzymes, casein shows the striking behaviour of coagulation. Other groups have already studied this clotting process, neglecting the fact that casein is not a homogeneous protein.[2] We thought it necessary, before this process can be sufficiently understood, to study the individual components separately. The main casein components are: α_s-casein, 45–50%; β-casein, 25–35%; γ-casein, 3–7%; κ-casein, 8–15%. The amino acid sequences of these components are known.[3] Amongst these proteins κ-casein is of particular interest; it is the only component insensitive towards Ca^{2+} ions up to concentrations of 400 mM, and it stabilizes the Ca^{2+}-sensitive fractions against flocculation. It is widely believed that κ-casein acts as a protective colloid similar to detergents in emulsions. The key position that κ-casein holds with respect to the micelle is demonstrated by the fact that the curdling of milk is initiated by specific proteolysis of this protein with milk-clotting enzymes.[4]

The final purpose of this study was to unveil the native milk micelle structure and the distribution of the different components in the micelle. In order to obtain insight into the interaction amongst the different components, it was imperative to study the behaviour of the individual components separately.

For detailed structural information we carried out combined static and dynamic light scattering measurements, where the former gives the molecular weight and the latter the hydrodynamic radius. Further information was obtained by small-angle neutron scattering measurements. We first report the results of β-casein, then κ-casein which shows a far more complicated scattering behaviour, that proved to be difficult to interpret. Finally, we give the first results on α-casein, which shows completely different scattering behaviour.

In co-operation with other laboratories we have been able to obtain the individual well-separated casein components. The preparation of the samples and the instrumental set-up were described previously.[5,6]

RESULTS

β-Casein

The β-casein polypeptide chain is composed of 203 amino acids and has a total molecular weight of 24 000 daltons.[4] According to the primary structure, the macromolecule can be divided into a polar head and an apolar tail, and thus the behaviour of β-casein is expected to be soap-like. Below 4°C β-casein is monomeric and above this temperature this protein shows an endothermic self-assembling.[3,7]

Static and Dynamic Light Scattering, and Small-angle Neutron Scattering
Combined static and dynamic light scattering (LS) measurements were carried out in a temperature range from 10 to 40°C. In this contribution the results are restricted to $T = 15°C$, which is sufficient for a comparison with the measurements from κ-casein. The results of the *static LS* measurements in a concentration range from 0·5 to 8 mg/ml are given in Fig. 1. In this plot the apparent molecular weight at each concentration was corrected for the second virial coefficient;[5,8] the original curve showed a maximum which can be explained as resulting from the influence of A_2. The measured concentration dependence reaches a plateau value above $c = 4$ mg/ml, and this is an indication of the formation of stable micelles at higher concentrations.

The diffusion coefficient, D_z, was determined by *dynamic LS*; it shows very little concentration dependence[5] and the zero concentration value is

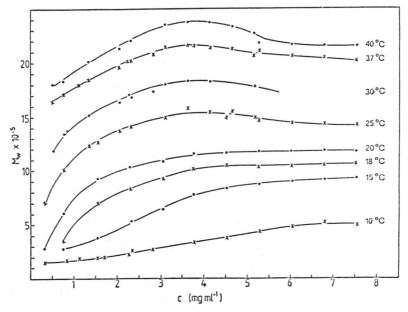

Fig. 1. Concentration dependence of the molecular weight, M_w, corrected for the second virial coefficient.

$D_z = 1.2 \times 10^{-7} \text{cm}^2/\text{s}$. The corresponding hydrodynamic radius, R_H, is defined via the Stokes–Einstein equation as

$$R_H = 1.38066 \times 10^{-16} \, T/(6\pi\eta_0 D_z) \tag{1}$$

and yields $R_H = 154\,\text{Å}$.

SANS measurements were carried out at $T = 15°\text{C}$. Figure 2 shows the Zimm plot for β-casein. The strong upturn of these curves at higher scattering vectors is characteristic of globular structures.[9] The radii of gyration were calculated from the initial slope of the curve in Fig. 3. The concentration dependence of $\langle S^2 \rangle$ is given in Fig. 4. Extrapolating to zero concentration one obtains

$$R_g \equiv \langle S^2 \rangle^{0.5} = 135\,\text{Å}$$

More detailed information on the molecular structure is obtained from a Kratky plot,[9] i.e. $P_z(q)U^2$ vs. U, of the same data (see Fig. 5) with $P_z(q) =$ particle scattering factor, $U = R_g q$, $q = (4\pi/\lambda)\sin(\theta/2)$ and $\lambda =$ wavelength of the neutrons. A pronounced maximum is observed, and both the position and the height of the maximum are very close to those of a star-branched molecule. The ratio of the height of the maximum (1·5) and the minimum (0·1) agrees very satisfactorily with a star molecule with about 36

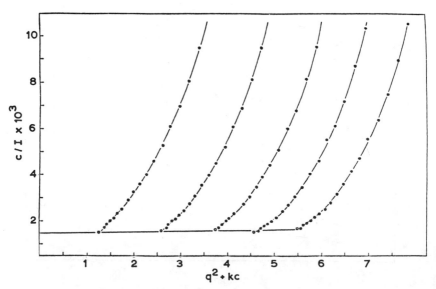

Fig. 2. Zimm plot for β-casein at $T = 15°\text{C}$.

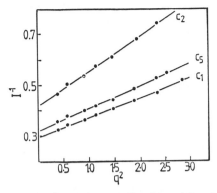

Fig. 3.   Determination of $\langle S^2 \rangle$ from the initial slope of the plot $1/I$ against q^2 from β-casein at three concentrations.

arms, but in the region of $U = 3$–10 there are characteristic deviations from a star molecule with flexible arms[9] joined together at a central point. The remaining small differences can be explained by the finite volume of the amino acid residues, which causes a densely packed star centre and allows only the outer parts of the chains to move freely.

The ratio of the geometric radius, $R_g (\equiv \langle S^2 \rangle^{0.5}$, radius of gyration) to the hydrodynamic radius R_H is defined as the *ρ-value* and is strongly dependent on the structure of a molecule.[9] In Fig. 6(a), R_g as determined by SANS is compared with the hydrodynamic radius R_H obtained from dynamic LS. The corresponding ρ-value (Fig. 6(b)) shows a strong concentration dependence and has to be extrapolated to zero concentration because the measured R_g is only an apparent value.[5] Assuming partial dissociation of the particles at lower concentrations, one obtains the upper limiting value of 1·0. This value is identical to a hollow sphere[9] and is very close to the value expected for a star molecule. In Table 1 the experimental ρ-value for β-casein is compared with ρ-values calculated for various types of macromolecular structures.[9]

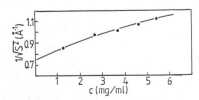

Fig. 4.   Concentration dependence of the radius of gyration at $T = 15°C$.

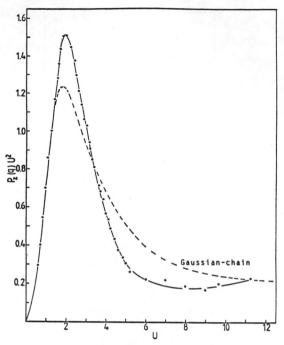

Fig. 5.   Kratky plot of β-casein ($c = 5.88$ mg/ml) at $T = 15°C$ compared with a Gaussian chain star molecule.

TABLE 1

Experimental ρ-Value for the Three Casein Components, in Comparison with ρ-Values Calculated for Various Types of Molecules

Model	ρ-Value[a]
Hard sphere	0·775
Hollow sphere	1·0
Star molecule	1·1–1·7
Statistical aggregate	1·78
Polydisperse coil	1·78
Experiment	
β-casein	(0·75)–1·0
κ-casein	(1·85)–2·05
α-casein	2·78

[a] The values in brackets correspond to the dashed line extrapolations in Figs 6b and 16b respectively.

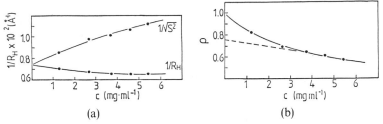

(a)   (b)

Fig. 6. (a) Comparison of the radius of gyration (SANS) and the hydrodynamic radius (QELS) for β-casein at $T = 15°C$. (b) Concentration dependence of the ρ-value for β-casein. Extrapolation to $c = 0$ yields $\rho = 0.75$–1.0.

κ-Casein

As already mentioned, κ-casein is of special interest since it is the only component insensitive towards Ca^{2+} ions up to concentrations of 400 mM, and it stabilizes the Ca^{2+}-sensitive fractions against flocculation.[10] κ-Casein has a molecular weight of 19 023 daltons and is composed of 169 amino acids. From the primary structure alone, κ-casein cannot be clearly divided into a pronounced hydrophilic or hydrophobic part. One section of the molecule, i.e. the p-κ-casein, consists of 105, mainly hydrophobic, amino acids. The other part, the macropeptide, does not contain aromatic amino acids but has a remarkably high content of serin, threonin, asparagin and glutamin. This N-terminal section of the molecule is more hydrophilic than the C-terminal p-κ-casein and may be expected to form the surface of the micelle.

Static and Dynamic LS, and SANS

Figure 7 shows the result of the *static LS* measurements at $T = 15°C$ for a κ-casein sample in simulated milk ultrafiltrate[8] at pH 6.7. Measurements were carried out at this temperature only, since at higher temperatures a

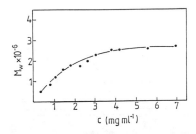

Fig. 7. Concentration dependence of the molecular weight corrected for the second virial coefficient.

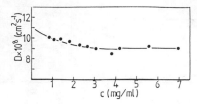

Fig. 8. Concentration dependence of the apparent diffusion coefficient at $T = 15°C$.

time-dependent aggregation process is observed.[11] The data were corrected for the second virial coefficient, as was done for the β-casein.[8] The curve apparently reaches a plateau value of 2.5×10^{-6} daltons at concentrations above 0.4%. This would indicate the formation of stable micelles composed of about 130 monomers at higher concentrations. A similar concentration dependence of the apparent diffusion coefficient (Fig. 8) is observed.

For comparison *SANS* was carried out at the same temperature and in the same range of concentration as for the LS measurements. The Zimm plot (Fig. 9) obtained from SANS shows an upturn at higher scattering vectors, which is a strong indication of globular structures. The radius of gyration, $\langle S^2 \rangle$, was evaluated from the initial part in the plot $(1/I)$ vs. q^2

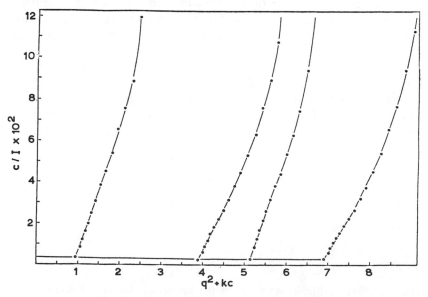

Fig. 9. Zimm plot for κ-casein at $T = 15°C$.

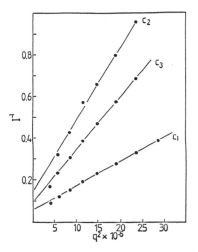

Fig. 10. Calculation of $\langle S^2 \rangle$ from the slope of the plot I^{-1} vs. q^2.

(see Fig. 10). The concentration dependence of $\langle S^2 \rangle$ is given in Fig. 11. Extrapolating to zero concentration one obtains

$$R_g \equiv \langle S^2 \rangle^{0.5} = 358 \,\text{Å}$$

As already mentioned, more detailed information on the shape and the branching structure of a molecule is obtained if the data are plotted according to Kratky[9] (note: $P_z(q) = I/I_0$). In Fig. 12 a Kratky curve of κ-casein for $c = 5 \cdot 18 \,\text{mg/ml}$ is compared with the theoretical curves for a sphere and a statistical aggregate, respectively. Instead of the pronounced maximum ($P_z(q)U^2 = 1 \cdot 1$ at $U = 1 \cdot 9$) expected for spheres, a broad maximum at about $U = 4$ is found. The height of $2 \cdot 9$ is close to a value of $3 \cdot 0$,

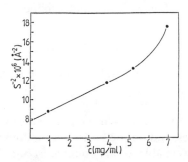

Fig. 11. Concentration dependence of $\langle S^2 \rangle^{0.5}$ for κ-casein.

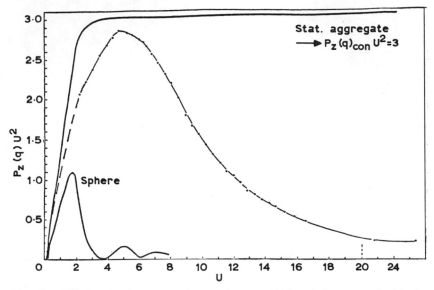

Fig. 12.   Kratky plot for a κ-casein sample at $c = 5 \cdot 18$ mg/ml compared with the theoretical curves of a sphere and a statistical aggregate.

which is predicted for random aggregation processes.[9] This may be taken as an indication that the particle scattering factor, $P_z(q)$, is composed of two different types. The one contribution comes from a sub-micelle, $P_z(q)_{sub}$, and the other from the connectivity between the particles, $P_z(q)_{con}$. Thus, according to Debye,[12] $P_z(q)$ can be written

$$P_z(q) = P_z(q)_{sub}P_z(q)_{con}$$

Applying the Guinier approximation for $P_z(q)$ it is possible to determine the *radius of gyration of a sub-micelle* (for mathematical details see Refs 5 and 9), and one obtains

$$R_{g,sub} \equiv \langle S^2 \rangle_{sub}^{1/2} = 70 \text{ Å}$$

Thus, with the value R_g of the total micelle, one finds

$$R_g/R_{g,sub} = 5 \cdot 1$$

Under the assumption that the κ-casein aggregates via the ends of the f-functional sub-micelles, and on the basis of a random aggregation process of these particles, we were able to carry out *model calculations* for the radius of gyration and the particle scattering factor.[5,8] Figure 13 shows the

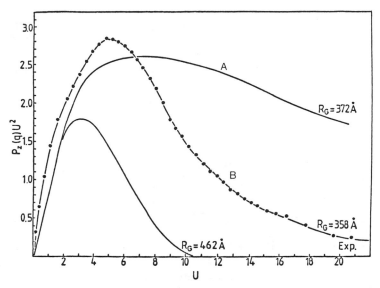

Fig. 13. Comparison of the experimentally measured particle scattering factor with that calculated for a star molecule (A) and a hard sphere (B). R_G = radius of gyration, calculated for $r = 0$.

calculated $P_z(q)$ of the total molecule in a Kratky plot. For the star-like sub-micelle we found a broad maximum at $U = 6$–8, while the maximum for the sphere-like sub-micelle at $U = 3$ is not high enough compared with the experiment. The experimental curve lies in between the calculated curves with a maximum at $U = 5$, and this leads to the conclusion that the sub-micelle is not equivalent to an ideal Gaussian star (flexible arms). If we take into account the finite volume of the individual protein repeating units, we come to the model suggested in Fig. 14. The centre of the sub-micelle is

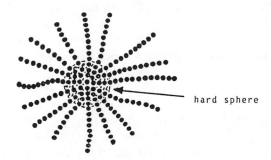

Fig. 14.   Suggested model for a κ-casein sub-micelle.

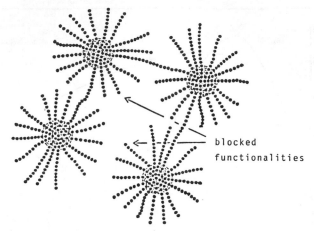

Fig. 15. Schematic model suggestion for the aggregation of κ-casein sub-micelles.

equivalent to a densely packed sphere: only the outer parts of the sub-micelle behave like a star molecule. This model lies in between the two extreme models which were the basis for our calculations. Figure 15 shows schematically the aggregated sub-micelles.

As already mentioned, the *ρ-value* is characteristically dependent on the structure of a molecule. Figure 16(a) shows the comparison of $\langle S^2 \rangle^{0.5}$ ($\equiv R_g$) (SANS) with R_H (dynamic LS), and Fig. 16(b) the concentration dependence of the corresponding ρ-value. The strong concentration dependence of R_g, which is the main reason for the concentration dependence of ρ, is striking. The measured R_g is only an apparent quantity, and the observed concentration dependence is caused by excluded volume effects. Thus ρ has to be extrapolated to $c = 0$. Assuming partial dissociation of the particles, one finds the upper limiting value of $\rho = 2 \cdot 05$. Table 1 gives the ρ-value of κ-casein in comparison with the ρ-values calculated for different types of molecules. At lower concentrations R_g is apparently much higher than R_H. This large difference can be explained by assuming loosely packed aggregates which allow the solvent to penetrate into the molecule to a certain extent.

α_s-Casein

The genetic variant B of α_{s1}-casein consists of a polypeptide chain with 199 amino acids and a molecular weight of 23 600 daltons. The molecule contains eight phosphoserin residues, of which seven are located in positions 43–80, and contains an additional 12 carboxyl groups. Thus, this

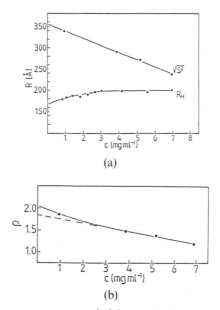

Fig. 16. (a) Comparison between $\langle S^2 \rangle^{0.5}$ and R_H. (b) Concentration dependence of ρ. Extrapolation to $c = 0$ yields $\rho = 1.85$–2.05.

section of the protein is extremely polar. Prolin is distributed over the whole molecule and hinders the formation of regular structures. In any case, a secondary structure could not be found.[3] The positions 100–199 are strongly apolar and induce a pronounced tendency to associate. In the presence of Ca^{2+} ions, α_{s1}-casein forms insoluble Ca^{2+} salts.

Static and Dynamic LS
α-Casein shows completely different aggregation behaviour from the two other casein components. Figures 17 and 18 represent the results of the LS measurements at $T = 35°C$ for an α-casein sample in 0.2M phosphate buffer, pH 6.7. Measurements were carried out only at this temperature since no temperature influence on the molecular weight was found. The molecular weight (Fig. 17(a)) and the radius of gyration (Fig. 17(b)) show no detectable concentration dependence and were determined to $M_w = 3.4 \times 10^6$ and $\langle S^2 \rangle^{0.5} = 190$ Å. Corresponding results for the hydrodynamic radius, R_H, obtained from dynamic LS are shown in Fig. 18. As expected, the ρ-value does not show any concentration dependence (Fig. 19) and was found to be $\rho = 2.78$. Such a high value is not found for spherical or star-branched micelles, but corresponds to a rigid rod[13] of about 123 spherical units in

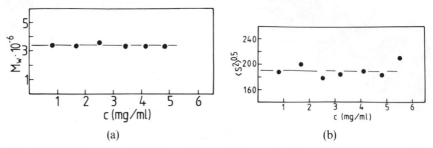

Fig. 17. (a) Concentration dependence of the molecular weight for a κ-casein
sample. (b) Concentration dependence of the radius of gyration.

length. An indication for rigid-rod behaviour is also given by the Kratky
plot (see Fig. 20). The curve reaches a plateau value of $U^2 P_z(q) = 2 \cdot 2$ at $U = 3$ and shows a linear increase at $U = 6$. The kink point at $U = 6$ corresponds
to the characteristic length below which rigid-rod behaviour is effective.

A more quantitative analysis is obtained from a Holtzer plot which is
shown in Fig. 21. The magnitude of the ratio of the maximum height to that
of the asymptotic plateau was shown[14] to be a function of the Kuhn
segments N_K per chain. Assuming monodispersity one obtains, from the
calibration curve given previously[14] with the ratio of $2 \cdot 15$, about 11 Kuhn
segments per worm-like chain. If a polydispersity of $M_w/M_n = 2$ is assumed,
the number of Kuhn segments is about $N_K = 13$. The maximum is found at
$U_{max} = 1 \cdot 52$; theory predicts $U_{max} = 1 \cdot 4$ for monodispersity and $U_{max} = 1 \cdot 73$
if $M_w/M_n = 2$, and this indicates a rather low polydispersity.

Furthermore, the magnitude of the asymptote gives the linear mass
density (M/L) according to the equation[15]

$$q(R_\theta/Kc) \xrightarrow{Lq \gg 1} \pi(M/L)$$

With the molecular weight M_w determined at $Lq \ll 1$, one obtains the

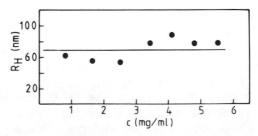

Fig. 18. Concentration dependence of the hydrodynamic radius.

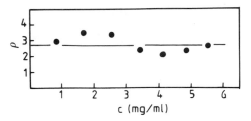

Fig. 19. Concentration dependence of the ρ-value.

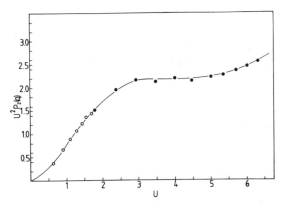

Fig. 20. Kratky plot for an α-casein sample. ●, measurements at angles larger than 20°; ○, interpolated data, obtained from the linear part in the Zimm plot (not shown) in the angular region from 0° to 20°.

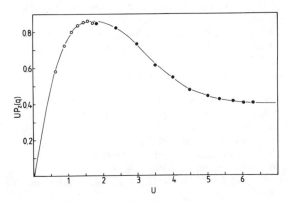

Fig. 21. Holtzer plot for an α-casein sample. Symbols as for Fig. 20.

TABLE 2
Results Calculated for α-Casein[15]

Number of Kuhn segments, N_K	11	13
Mass/length, M/L	2 239 daltons/nm	2 239 daltons/nm
Contour length, L	1 594 nm	1 594 nm
Kuhn length, l_K	145 nm	123 nm

contour length L. Finally, the length of the Kuhn segment follows from

$$l_K = L/N_K$$

The data for the linear mass density M/L, contour length L, the number of Kuhn segments N_K and the length of such a segment are listed in Table 2.

We close this discussion with a speculative remark. In common linear aggregation the degree of aggregation increases with increasing concentration. Experimentally, however, a constant molecular weight over the whole region was found, and this indicates a 'closed' aggregation mechanism. Mostly, such closed aggregation occurs on soap micelle formation, which is definitely not present in the α_s-casein. The only other type of closed aggregation we can think of would consist of a ring formation. The experimental data are indeed in accordance with this interpretation.

REFERENCES

1. Bartholomé, E. et al. (Eds), Ullmanns Enzyclopädie der technischen Chemie, Vol. 16, Verlag Chemie, Weinheim, 1978, p. 689.
2. Linderstrom-Lang, K., Compt. Rend Trav. Carlsberg, 1929, 17(9), 1.
3. Belitz, H. D. and Grosch, W., Lehrbuch Lebensmittelchemie, Springer-Verlag, Heidelberg, 1982.
4. Swaisgood, H. E., CRC Critical Reviews in Food Technology, June 1973, 375–414.
5. Thurn, A. and Burchard, W., Colloid & Polym. Sci., 1987, 265, 653.
6. Bantle, S., Schmidt, M. and Burchard, W., Macromolecules, 1982, 15, 1604.
7. Payens, T. A. J. and Vreeman, H. J., Solution Behaviour of Surfactants, Vol. 1, Plenum, New York, 1982, pp. 549–71.
8. Thurn, A., PhD Dissertation, Freiburg, 1986.
9. Burchard, W., Adv. Polym. Sci., 1983, 48, 1.
10. Jakubke, H.-D. and Jeschkeit, H. (Eds), Lexikon Biochemie, Verlag Chemie, Weinheim, 1981.
11. Vreeman, H., personal communication.

12. Debye, P., *Ann.-Phys.*, 1915, **46**, 809.
13. Semlyen, J. A., *Cyclic Polymers*, Elsevier Applied Science Publishers, London, 1986.
14. Schmidt, M., Paradossi, G. and Burchard, W., *Macromol. Chem., Rapid Commun.*, 1985, **6**, 767.
15. Thurn, A. and Burchard, W., *Colloid & Polym. Sci.*, 1987, **265**, 897.

7

GALACTOMANNAN–BORATE SYSTEMS: A COMPLEXATION STUDY

E. Pezron, A. Ricard, F. Lafuma and R. Audebert

Laboratoire de Physico-Chimie Macromoléculaire de l'Université Pierre et Marie Curie, UA 278, ESPCI, 10 rue Vauquelin, 75231 Paris Cedex 05, France

ABSTRACT

We study the formation of weak reversible gels of galactomannans induced by borate complexation. Phase diagrams observed in water and in 1M NaCl are presented. To obtain information about the crosslinking mechanism involved in the gel formation, interactions of galactomannan and glycoside models with borate are studied by ^{11}B-NMR. Evidence of the existence of five-membered and six-membered ring mono- and di-complexes is given. The corresponding complexation constants and peak assignments are reported. The mono-complexation constant, K_1, has been determined for guaran in 1M NaCl from dialysis experiments: $K_1 = 11\cdot4$ litres mol^{-1}. This value is close to those reported for glycoside models.

1 INTRODUCTION

Under suitable conditions (concentration, temperature, pH, ionic strength), galactomannans such as guar gum or its derivatives are gelled by boric acid or borax. This kind of gel is widely used as stimulation fluid in the petroleum industry. To optimize borate-crosslinking systems and to understand the rheological properties of such gels, it seems necessary to study the crosslinking mechanism at a molecular level. So after the characterization of purified galactomannan samples and a presentation of the phase diagrams observed, the complexation equilibria involved in these polysaccharide–borate systems are studied by ^{11}B-NMR and dialysis.

113

2 EXPERIMENTAL

2.1 Materials and Procedure

Hydroxypropylguar and guar were kindly provided by Etudes et Fabrications Dowell Schlumberger, France. These galactomannans were purified by dissolution in water at room temperature, centrifugation and precipitation in ethanol 96°. Absolute concentrations of polymers were determined by total organic carbon analysis (DC 80, Dohrmann/Xertex). Borax ($Na_2B_4O_7 . 10H_2O$), sodium chloride, methyl α-D-galactopyranoside and methyl α-D-mannopyranoside were commercial materials of analytical grade. Galactomannan–borax samples (25 ml) were prepared at room temperature by mixing appropriate amounts of the aqueous solutions of borax and galactomannans. The system was then vigorously stirred and allowed to rest for 48 h.

2.2 Characterization of Galactomannan Samples

Weight-average molar weights were determined by the Zimm method from light scattering data (11 angles from 30° to 150° and five concentrations of polymers lower than 0·1%), using a Fica photometer with a 6328 Å Helium–Neon laser as a light source. The increment index dn/dc values were measured at 6328 Å with a Brice Phoenix differential refractometer and were $0·155\,\mathrm{ml\,g^{-1}}$ for guaran and hydroxypropylguaran in water and $0·149\,\mathrm{ml\,g^{-1}}$ in 1M NaCl. The mannose to galactose ratios were determined by acid hydrolysis, followed by thin-layer chromatography on crystalline cellulose and spectrophotometric titration at 490 nm.[1] The ratios also estimated from ^{13}C-NMR spectra of the galactomannan samples were in good agreement with those obtained by the analytical method. A Contraves 30 rheometer was used for viscosity measurements at 25°C.

2.3 ^{11}B-NMR

Spectra were recorded at 23°C using a Bruker WP 250 spectrometer at 80·25 MHz with Et_2OBF_3 as an external reference. Samples were prepared by dissolving the appropriate amounts of borax and polyols in deionized water (milli-Q system of Millipore) and D_2O. The pH was adjusted with NaOH. Resonance peak assignment was made thanks to a preliminary study in very good agreement with results of Van Duin et al.[2] The association constants were calculated from a measurement of the relative area of each signal.

2.4 Dialysis Experiments

Bags were prepared from carefully washed cellulose sausage casing and were filled with a measured amount of a borax solution. The bag was then immersed in a glass-stoppered tube which contained the same volume of a predialysed guaran solution. It takes about 8 days to reach the equilibrium at room temperature $(23 \pm 1°C)$. The control run with only pure water outside the bag was carried out to take into account the amount of borax bound to the bag. At equilibrium, free borate concentrations were determined by hydrochloric acid titration. The complex concentration was calculated from the total amount of borate and the measured free borate concentration. The experiments were performed at two different concentrations of guaran and five concentrations of borax. As the polymer concentrations were low (0·1% and 0·2%), we considered that only monodiol-type complexes were formed so that the monodiol–borate complex formation constant K_1 was directly calculated:

$$K_1 = [\text{complexed borate}]/([\text{free borate}] \times [\text{galactomannan}])$$

3 GALACTOMANNANS: MAIN CHARACTERISTICS AND SOLUTION PROPERTIES

The galactomannans are a family of neutral polysaccharides found in legume seed and composed of a $(1 \rightarrow 4)$-linked β-D-mannan chain substituted with $(1 \rightarrow 6)$-linked α-D-galactopyranosyl side-groups.[3] Guaran, the galactomannan obtained from guar, has a mannose to galactose ratio rather less than 2 (see Fig. 1). Hydroxypropylguar is a derivative of guaran where some of the hydroxyl groups have been substituted by hydroxypropyl groups upon ether formation: the molar substitution of our sample is 0·48. In fact, the D-galactose distribution on the D-mannan backbone is not regular,[4] and the model generally considered for the galactomannan structure[5] points out the existence of highly branched or comb-like zones and no-branched or smooth zones (see Fig. 2). The tendency to aggregation increases with decreasing substitution, i.e. higher mannose to galactose ratios, while solubility decreases.[6]

The preparation of our galactomannan samples, involving steps of dissolution in cold water and centrifugation, therefore leads to low mannose to galactose ratios,[7] 1·2 for hydroxypropylguaran and 1·3 for guaran. Under these conditions, light scattering allows reproducible

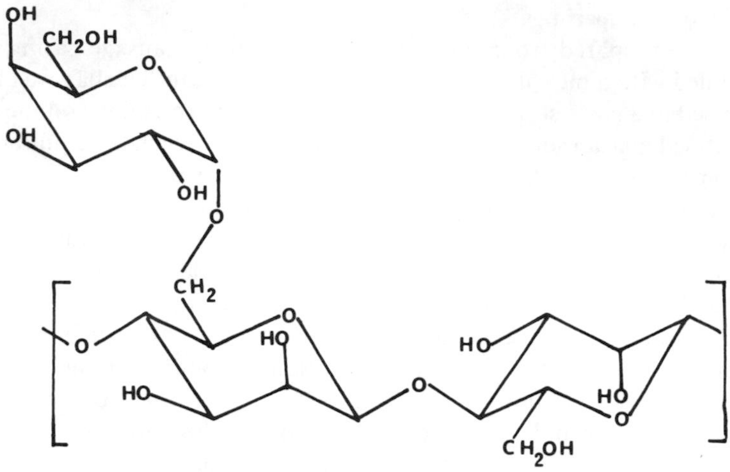

Fig. 1. Proposed average unit structure of guaran.[3]

estimations of weight-average molar weight, 2.8×10^6 g for hydropropyl-guaran and 4.5×10^6 g for guaran. In addition, galactomannans are well known for their viscous properties. They are, for example, widely used as thickeners in the food industry. Many studies have been carried out on their solution properties and flow behaviour.[1,8-10] It is possible to estimate the concentration at which the molecules in solution start to overlap.[11] The determination of this critical concentration C^* is experimentally accomplished by plotting the low-shear specific viscosity against the reduced concentration $C[\eta]$, where C is the polymer concentration and $[\eta]$ the intrinsic viscosity. A transition in the solution behaviour of the polymer takes place at the overlap concentration C^*, beyond which polymer–polymer entanglements start to become significant. Figure 3 shows the master plot obtained for guaran and hydroxypropylguaran in water or 1M NaCl. It is found that $C^* \simeq 1.4/[\eta]$. This overlap concentration is of great importance for gelation problems.

$$M—\begin{bmatrix} M \\ | \\ G \end{bmatrix}_a—[M]_b—M—M—[M]_b—$$
$$\qquad\qquad\qquad\qquad\qquad | \\ \qquad\qquad\qquad\qquad\qquad G$$

Fig. 2. Galactomannan structure.[4] (a) comb-like zone; (b) smooth zone.

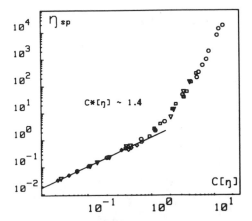

Fig. 3. Variation of the low-shear specific viscosity with the overlap parameter $C[\eta]$. For galactomannan samples: ○, guaran in pure water, $[\eta] = 1.6$ litres g^{-1}; □, guaran in 1M NaCl, $[\eta] = 1.25$ litres g^{-1}; ▽, hydroxypropylguaran in pure water, $[\eta] = 1.3$ litres g^{-1}; ◇, hydroxypropylguaran in 1M NaCl, $[\eta] = 1.1$ litres g^{-1}.

4 SOL–GEL AND PHASE DIAGRAMS

It is well known that in pure water some polyhydroxy compounds thicken or gel in the presence of boric acid or borax.[12] Most studies are concerned with the rheological properties of polyvinyl alcohol–borax solutions, while controversy still exists about the crosslinking mechanism of this system.[13–18] However, very little is published about thickening phenomena, gelation and interactions of galactomannan–borax systems.

The sol–gel diagrams obtained for guaran or hydroxypropylguaran with borax in pure water are very similar (see Fig. 4). In both cases no gelation is observed for concentrations of galactomannans lower than the overlap concentration C^*. Besides, two reasons may explain the shift between the two sol–gel transition curves drawn. First, the molar weight of hydroxypropylguaran is lower than that of guaran. Secondly, hydroxypropylation reduces the number of diol sites available for crosslinking, as will be developed further, so that more borax is needed to gel an aqueous solution of hydroxypropylguaran.

The phase diagram observed in a high ionic strength medium is quite different (see Fig. 5). Again, no gelation is observed for polymer concentrations below C^*, but the collapse of the gel is observed at quite low concentrations of borax. The borax concentration necessary to promote

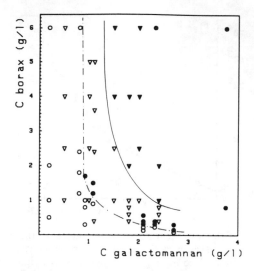

Fig. 4. Sol–gel transition curves of aqueous galactomannan–borax systems: hydroxypropylguaran—sol (▽), gel (▼); guaran—sol (○), gel (●).

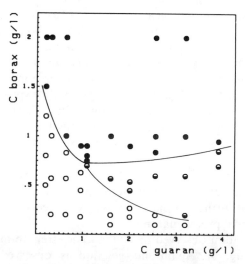

Fig. 5. Phase diagram of aqueous guar–borax 1M NaCl system: sol, ○; gel, ◐; collapse, ●.

the collapse slightly increases with the galactomannan concentration. In pure water no gel collapse was observed; this may be due to an electrostatic effect since the repulsion between the negatively charged borate complexes on the polymer chain prevents the formation of many cross-links—studies on polyvinyl alcohol–borate systems in dilute solutions[19] or in gels[14] by different techniques show similar effects. Therefore, to interpret such collapse and sol–gel transition curves, and to be able to determine the number and types of complexes formed, it seems crucial to study the glactomannan–borax complexation from the stoichiometric and thermo-dynamic points of view.

5 GALACTOMANNAN–BORATE COMPLEXES[20]

5.1 Diol–Borate Chemistry

At low borate concentration, borax $(Na_2B_4O_7 \cdot 10H_2O)$ is completely dissociated into two monomeric species, borate ions and boric acid:

$$B_4O_7^{2-} + 7H_2O \rightleftharpoons 2B(OH)_3 + 2B(OH)_4^-$$

Boric acid and borate especially are well known for their ability to form monodiol and didiol complexes. Several studies involving pH-metry or ^{11}B-NMR have been carried out on the interactions of simple linear diols and sugars with borate.[21–25] The dissociation of boric acid and the two successive reactions involved in the diol–borate complexation are represented by the following equilibria:

$$B(OH)_3 + H_2O \overset{K_a}{\rightleftharpoons} \underset{(B^-)}{B(OH)_4^-} + H^+ \qquad K_a = [B^-][H^+]/[HB] = 10^{-9.2}$$

$$K_1 = [BP^-]/[B^-][P]$$

$$K_2 = [BP_2^-]/[BP^-][P]$$

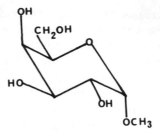

Fig. 6. Methyl α-D-galactopyranoside.

5.2 ^{11}B-NMR Study of Glycoside–Borate Interactions

5.2.1 Borax Solutions

Spectra of dilute borax solutions show a pH-dependent average signal due to the rapid exchange of boric acid (resonance peak observed at $\delta = 18.4$ ppm) and borate ($\delta = 0.6$ ppm). For the concentrations we used (total boron concentration of 2.5×10^{-2}M), monomeric species largely predominate[26,27] and no polyborate was observed.

In aqueous solution, borax is dissociated to equal amounts of borate and boric acid so that pH = pK_a = 9.2. Adding a diol to the borax solution lowers the pH since borate forms a complex, while no boric acid diol complex has been detected.

5.2.2. Methyl α-D-Galactopyranoside and Methyl α-D-Mannopyranoside– Borate Interactions

These glycosides are interesting to consider for complex formation because they are models of galactomannan sugar units and they allow a study of the diol sites available for borate complexation in guaran and hydroxypropyl-guaran (see Figs 6, 7 and 1). Ring sizes of methyl α-D-galactopyranoside and

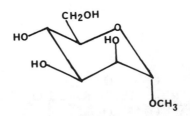

Fig. 7. Methyl α-D-mannopyranoside.

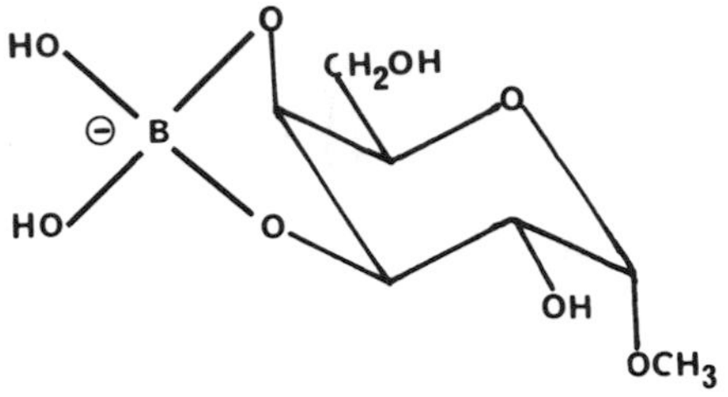

Fig. 8.   Methyl α-D-galactopyranoside–borate complex (1,2 diol–borate complex: $BP_{1,2}^-$).

methyl α-D-mannopyranoside are fixed so that for each of these glycosides only two kinds of monodiol–borate complexes were observed: one five-membered ring complex corresponding to the 1,2-diol site and one six-membered ring complex corresponding to the 1,3-diol site (see Figs 8 and 9). For methyl α-D-galactopyranoside, six- and five-membered ring didiol complexes are also observed, but five-membered ring 1,2-didiol borate complexes largely predominate. Figure 10 sums up these facts: when increasing amounts of diol are added to a borax solution, more and more complexes are observed. We may distinguish four resonance peaks relative to monodiol and didiol–borate complexes of the 1,3- and 1,2-type. At the same time pH decreases so that the peak relative to the boric acid/borate exchange is shifted downfield. The chemical shift variation with pH was

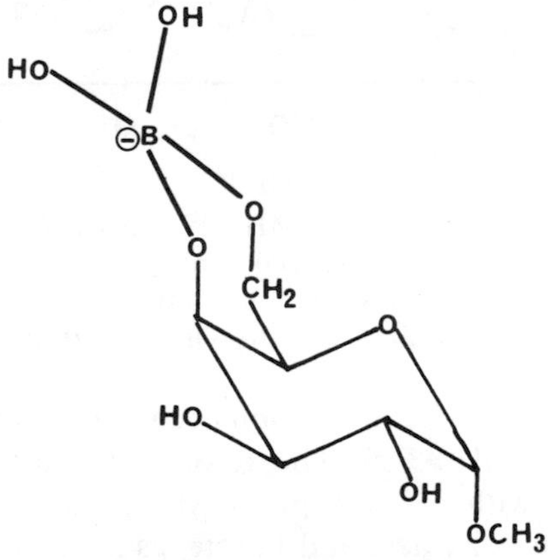

Fig. 9.   Methyl α-D-galactopyranoside–borate complex (1,3 diol–borate complex: $BP_{1,3}^-$).

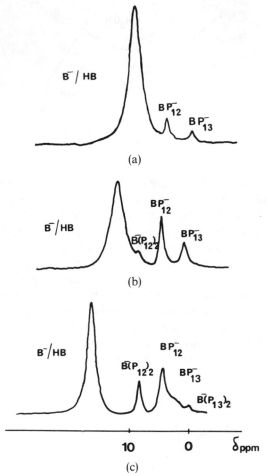

Fig. 10. ^{11}B-NMR spectra at 80·25 MHz of borax ($5·6 \times 10^{-3}$M)—methyl α-D-galactopyranoside solutions at different concentrations. They show the evolution of pH and complex formation with sugar concentration. (a) Sugar 10^{-2}M, pH = 9·15; (b) sugar 0·1M, pH = 8·90; (c) sugar 0·25M, pH = 8·30.

also studied (see Fig. 11). Complexation occurs only when borate ions are present in solution (pH > 7·5). The resonance peaks observed do not significantly move with pH, except for the peak relative to the inter-conversion between boric acid and borate, as already noticed, and those relative to the 1,3-diol–borate complexes, which means that they are less stable than the 1,2-diol–borate complexes.

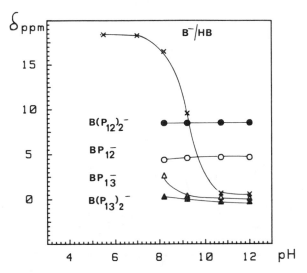

Fig. 11. ^{11}B-NMR data at 80·25 MHz of boric acid ($2·6 \times 10^{-3}$M)—methyl α-D-galactopyranoside (0·2M) solutions at different pH values.

5.3 Hydroxypropylguaran and Guaran–Borate Interactions

The quality of the spectra obtained with model borax solution allowed determination of complex formation constants (see Table 1). In the case of guaran or hydroxypropylguaran, to observe chemical shifts corresponding to complexes we had to prepare samples with galactomannan concentrations of at least 10 g litre^{-1}, i.e. $6·2 \times 10^{-2}$ mol litre^{-1} of sugar units,

TABLE 1

Comparison of Formation Constants Obtained from ^{11}B-NMR Data for Models and from Dialysis Experiments for Guaran[a]

Compounds	Diol sites involved	K_1 (litres mol^{-1})	K_2 (litres mol^{-1})	Conditions
Guaran	1, 2 + 1, 3	11·4	—	Room temperature, 1M NaCl
Methyl α-D-mannopyranoside	1, 2	10·9	2·2	23°C, 1M NaCl
	1, 2	18·2	2·3	23°C, water
Methyl α-D-galactopyranoside	1, 2	12·1	1·7	23°C, 1M NaCl
	1, 2	10·6	2	23°C, water
	1, 3	5·9	1·5	23°C, 1M NaCl
	1, 3	6	1·2	23°C, water

[a] Constants for guaran–borate complexes are calculated on the basis of moles of sugar units per litre.

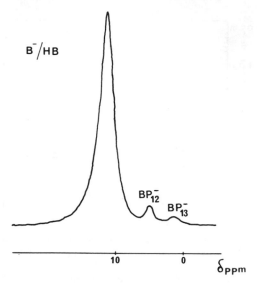

Fig. 12. ^{11}B-NMR spectra at 80·25 MHz of hydroxypropylguaran (18·7 g litre^{-1}) and borax (2·15 g litre^{-1}) in 1M NaCl.

so that these samples were gels. Under these conditions no peak relative to didiol complexes was observable. But the ^{11}B-NMR spectra of these systems show unambiguously two resonance peaks relative to the two kinds of monodiol–borate complexes (see Fig. 12). Nevertheless, no accurate complex formation constant determination can be expected from these spectra showing low complex concentrations. However, it was possible to get the monodiol–borate formation constant value, K_1, in 1M NaCl using dialysis experiments.[20] In Table 1, this result is compared with data of glycoside models calculated from ^{11}B-NMR spectra. The monodiol–borate complex formation constant for guaran is in the range of the values obtained for the two models. In the case of the galactomannan, K_2 was not determined but this constant is around 2 litres mol^{-1} for both glycoside models.

6 CONCLUSION

From a complexation point of view (number and type of complexes), galactomannans behave as glycoside models towards borate ions. This result reveals itself very useful for the interpretation of phase diagrams[28] and may already qualitatively explain some experimental features such as the thickening of galactomannan–boric acid solutions observed only for a pH higher than 7·5–8, the importance of the polyelectrolyte effect[29] and the reversibility of galactomannan–borax gels with dilution, pH or temperature.

ACKNOWLEDGEMENTS

We are grateful to C. Allain for fruitful discussions and to F. Rondelez for his interest in this study. This work is supported by Etudes et Fabrications Dowell Schlumberger (St Etienne, France).

REFERENCES

1. Sabater, A., Thesis, University of Paris XI, 1979.
2. Van Duin, M., Peters, J. A., Kieboom, A. P. E. and Van Bekkum, H., *Tetrahedron*, 1985, **41**, 3411.
3. Whistler, R. L., Li, T. K. and Dvonch, W., *J. Amer. Chem. Soc.*, 1948, **70**, 3144.
4. McCleary, B. V., Clark, A. H., Dea, J. C. M. and Rees, D. A., *Carbohydr. Res.*, 1985, **139**, 237.
5. Courtois, J. E. and Le Dizet, P., *Bull. Soc. Chem. Biol.*, 1969, **52**, 15.
6. Dea, I. C. M., Morris, E. R., Rees, D. A. and Welsh, E. J., *Carbohydr. Res.*, 1977, **57**, 249.
7. Hui, P. A. and Neukom, H., *Tappi*, 1964, **47**(1), 39.
8. Doublier, J. L., Thesis, University of Paris VI, 1975.
9. Robinson, G., Ross-Murphy, S. and Morris, E., *Carbohydr. Res.*, 1982, **101**, 17.
10. Morris, E., Cuttler, A., Ross-Murphy, S. and Rees, D., *Carbohydr. Polym.*, 1981, **1**(1), 5.
11. Graessley, W., *Adv. Polym. Sci.*, 1974, **16**, 1.
12. Deuel, H. and Neukom, A., *Makromol. Chem.*, 1949, **3**, 13.
13. Nickerson, R. I., *J. Appl. Polym. Sci.*, 1971, **19**, 111.
14. Shultz, R. K. and Myers, R. R., *Macromolecules*, 1969, **2**, 281.
15. Bolewsky, K. and Rychly, B., *Kolloid Z. Z. Polym.*, 1968, **48**, 288.
16. Murakami, I., Fujino, Y., Ochiai, H. and Tadokoro, J., *J. Polym. Sci. Polym. Phys. Ed.*, 1980, **18**, 2149.
17. Ochiai, H., Fujino, Y., Tadokoro, Y. and Murakami, I., *Polymer*, 1980, **21**, 486.
18. Maerker, J. M. and Sinton, S. W., Proceedings IX International Congress on Rheology, Mexico, 1984.
19. Ochiai, H., Fujino, Y., Tadokoro, Y. and Murakami, I., *Polym. J.*, 1982, **14**, 423.
20. Pezron, E., Lafuma, F., Ricard, A. and Audebert, R., *Macromolecules*, in press.
21. Roy, G. L., Laferriere, A. L. and Edwards, J. O., *J. Inorg. Nucl. Chem.*, 1957, **4**, 106.

22. Conner, J. M. and Bulgrin, U. C., *J. Inorg. Nucl. Chem.*, 1967, **29**, 1953.
23. Henderson, W. G., How, M. J., Kennedy, G. R. and Mooney, E. F., *Carbohydr. Res.*, 1973, **28**, 1.
24. Kennedy, G. R. and How, M. J., *Carbohydr. Res.*, 1973, **28**, 13.
25. Makkee, M., Kieboom, A. P. G. and Van Bekkum, H., *Rec. Trav. Chim. Pays-Bas*, 1985, **104**, 230.
26. Mommii, R. K. and Nachtrieb, N. H., *Inorg. Chem.*, 1967, **6**, 1189.
27. Anderson, J. L., Eyring, E. M. and Whittaker, M., *J. Phys. Chem.*, 1964, **68**, 1128.
28. Pezron, E., Leibler, L., Ricard, A. and Audebert, R., *Macromolecules*, in press.
29. Liebler, L., Pezron, E. and Pincus, P. A., *Polymer*, in press.

8

FORMATION OF THERMALLY REVERSIBLE
NETWORKS FROM STARCH POLYSACCHARIDES

W. VORWERG, F. R. SCHIERBAUM, F. REUTHER
and B. KETTLITZ

*Central Institute of Nutrition of the Academy of Sciences of the GDR,
Arthur-Scheunert-Allee 114/116, 1505 Bergholz-Rehbrücke, GDR*

ABSTRACT

The main components of homopolymer mixtures in starches are different in chemical structures. The capacity to aggregate is determined by distribution of molecular weight of amylose and amylopectin and certainly by the length of exterior chains of the branched amylopectin molecules. Methods applied for measurement of the physical properties are time dependence of the dynamic viscosity, shear modulus and relaxation. The thermal behaviour has been investigated by the method of Ferry and Eldridge. In the presence of amylopectin the aggregation process of amylose is prevented and thermally reversible networks are formed. Small amounts of added soluble amylose increase the intensity of polymer interaction. Amylose molecules with a degree of polymerization < 500 are favourable for a high aggregation velocity. Amylopectin contributes to the rigidity as well as to the viscoelastic properties of the gel, largely depending on its molecular magnitude. The results lead us to assume that cooperative interactions between amylose and amylopectin diminish the extent of junction zones in comparison to amylose aggregates.

1 INTRODUCTION

Most native starch polysaccharides aggregate irreversibly in aqueous solutions or dispersions. These aggregates are thermostable in the temperature range up to 100°C. In the 1970s a new type of starch polysaccharide aggregate was discovered, namely aqueous gels of hydrolysed potato starch which is produced by the action of the bacterial enzyme α-amylase. These gels are susceptible to melting in a hot water bath.[1] The starch hydrolysis product is known as maltodextrin. Here the question of the reason for this change in thermal behaviour arises. It appears conceivable at first sight that the smaller molecular sizes of amylose and amylopectin, which are obtained by hydrolysis, influence the physical properties. However, smaller amylose molecules aggregate more intensely into thermo-irreversible structures.[2] On the other hand, amylopectin, which is the branched component, has a very weak tendency to form aggregated structures,[3] and a lengthening of the amylopectin exterior chains as a result of the α-amylase action is inconceivable and has to be excluded. Oligosaccharides are the third class of substance in the degraded starch product. Sugars such as glucose and maltose disturb polysaccharide aggregation but have no effect on thermo-irreversibility. The extension of this finding to oligosaccharides has been verified. The aim of this research is a description of the *integrated* system. Indeed, structural conditions and external factors were found to act mutually and cause the level of structure formation.

A mutuality is observed, i.e. limited structural interactions between similar and different molecules exist, and these determine the conditions for their reactions. The problems to be solved include the analysis of maltodextrins with respect to their composition, separation into molecular weight fractions, physico-chemical characterization and investigations of the effects of well-defined fractions in complex mixtures.

For characterization of the physical properties, methods were applied which allow measurement of the time dependence of the dynamic viscosity, the shear modulus and relaxation. The thermal behaviour was investigated by the method of Eldridge and Ferry.

2 MATERIALS AND METHODS

2.1 Measuring Thermal Properties of Maltodextrin Gels

On the assumption that a binary association is to be considered as the basis for gel formation, a correlation between polymer concentration and melting temperature was established for gelatine gels.[4] The validity of this correlation was confirmed for maltodextrins by Braudo *et al.*[5] There exists a linear correlation between the logarithm of concentration and the reciprocal melting temperature, T_m, from which the melting enthalpy of the aggregates can be estimated:

$$\log c = \frac{-\Delta H}{R \times T \times 2 \cdot 303} + \text{constant}$$

2.2 Rheological Characterization of Solutions and Gels

2.2.1 Investigation of the Dynamics of Aggregation of Concentrated Solutions

The effects of aggregation were studied by the time dependence of the dynamic viscosity. Freshly prepared polysaccharide solutions at concentrations of 10–35% were filled into a cooled cylinder or cone-plate device of the Rheotest II, VEB Prüfgerätewerke Medingen, rotational viscosimeter. Changes in the dynamic viscosity were followed at the constant shear rates of $1 \cdot 55 \, \text{s}^{-1}$ (cylinder) or $5 \cdot 56 \, \text{s}^{-1}$ (cone-plate) at a constant temperature of 4°C. The aggregation is characterized by a continuous rise in viscosity, finally retaining a stationary viscosity. This method was extended by heating the aggregated solution in the device with a constant temperature gradient of 3°C min^{-1}, during which procedure a continuous decrease of the dynamic viscosity occurs.

2.2.2 Rheological Techniques for Gel Measuring

Gels were also measured in a rotational viscosimeter which was equipped with a torsion bar. The deformation was carried out at very low shear rates of $9 \times 10^{-3} \, \text{s}^{-1}$ to $10^{-2} \, \text{s}^{-1}$ with recording of the relaxing stress afterwards. The maltodextrin solution was, immediately after preparation, filled into the cooled cylinder or cone-plate device. During the measurement on the gel which was obtained in the gap in the shortest possible time, definite shear rate was observed for some seconds. The dependence of shear stress on time was measured by a torsion gauge and a recording instrument. In the calculation, Hooke's Law was used.[6]

2.3 Investigated Substances

2.3.1 Non-fractionated Maltodextrins

The relevant molecular data of non-fractionated maltodextrins are given in Table 1. The DE, $\bar{P}_w$ and $\bar{P}_n$ values show the differences in the degree of degradation. MD 13 is more strongly degraded than the other samples. The $\bar{P}_w/\bar{P}_n$ ratio indicates a large molecular heterogeneity. This result is based on a broad distribution of molecular sizes. The more the substance is degraded the smaller is the heterogeneity. The aqueous maltodextrin

TABLE 1
Analytical Data of Investigated Maltodextrins

Substance	DE^a (%)	$\bar{P}_w{}^b$	$\bar{P}_n{}^c$	$\bar{P}_w/\bar{P}_n$	$\lambda_{max}{}^d$ (nm)
MD 2	5·4	2 400	32	75	556
MD 26	6·2	2 250	27	83	548
MD 13	6·8	1 000	23	43	529

[a] Dextrose equivalent calculated from the reducing value according to Ref. 7.
[b] Estimated from light scattering measurements in aqueous solutions.
[c] Estimated from the reducing ability.
[d] Maximum of absorption of the polysaccharide–iodine complex (iodine concentration 10^{-4}N).

solutions were analytically separated by gel permeation chromatography, using Sephadex G 200.

The gel chromatogram of MD 26 (Fig. 1) confirms the broad distribution of molecular sizes. The absorption maxima of the degraded maltodextrin–iodine complex are shifted towards larger wavelengths, which means that a fairly large amount of amylose is present in MD 26. Comparison of the molecular composition of potato starch and the maltodextrin substances further clarifies the effects of degradation (Table 2). Molecular size ($\bar{P}_n$) and the mass ratio of branched and linear components are changed. Specifically, the amount of amylose is decreased

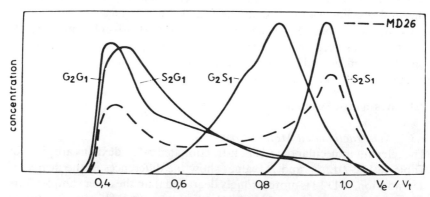

Fig. 1. Gel chromatograms of maltodextrins. The dashed line corresponds to the original maltodextrin MD 26; the curves labelled G_2G_1, S_2G_1, G_2S_1 and S_2S_1 refer to fractions prepared from MD 26.

TABLE 2
Molecular Composition of Maltodextrin MD 2 in Comparison with Potato Starch

Sample	Branched components		Linear components	
	$\bar{P}_n$	Amount (%)	$\bar{P}_n$	Amount (%)
Native potato starch				
($\bar{P}_n$ 1 800)	2 000	75–80	1 000–1 300	20–25
MD 2[a] ($\bar{P}_n$ 32)	80–600	36	100–150	3·3
	25–80	23	50	5·5

[a] MD 2 includes oligosaccharides G_1–G_{20} with linear and branched structures of about 20–22%. About 10% of the substance of MD 2 was not caught by using the fractionation method.

by about 50%. Moreover, it was estimated that a relatively high amount of oligosaccharides of about 20% was present.

2.3.2 Fractions of Maltodextrins

The next step in the investigation to find correlations between structure and physical properties consisted of a preparative fractionation of malto-dextrin MD 26 in aqueous solutions with methanol as precipitant. Four fractions were produced in equal mass portions. The fractions differ in their distributions of molecular sizes, as shown in the gel chromatograms of Fig. 1.

Fraction G_2G_1 probably contains an enrichment of relatively high molecular weight, branched maltodextrins since the gel chromatogram exhibits a broad distribution which may be attributed to the large differences in molecular sizes between amylopectin and amylose. According to the molecular data of Table 3, fractions $S_2 G_1$ and $G_2 S_1$ appear to be

TABLE 3
Molecular Indices of Maltodextrin Fractions

Fraction	Percentage of the fraction in the maltodextrin	$\bar{P}_n$[a]	λ_{max} (nm)[b]
G_2G_1	20–30	250	549
S_2G_1	20–30	90	552
G_2S_1	20–25	90	544
S_2S_1	25–30	10	503

[a] Estimated from the reducing ability.
[b] Maximum of absorption of polysaccharide-Iod-complexes.

quite similar in their composition. However, the gel chromatograms disprove this assumption. All three G-fractions are capable of aggregating, as is obvious from the λ_{max} values. The fraction $S_2 S_1$ mainly includes oligosaccharides.

2.3.3 Amylose Fractions

To study the influence of amylose fractions on aggregation, high molecular weight amylose (Serva Feinbiochemica, Heidelberg, FRG) was hydrolysed and fractionated. Amylose fractions with P_n 21, 35, 140, 320 and P_n 500 were included as additives in the investigation of rheological properties.

2.3.4 Acetylated Maltodextrins

The interactions of maltodextrins can be hampered by the introduction of acetyl groups. Depending on the content of acetyl groups, aqueous maltodextrin solutions can be brought to remain stable.[8] This observation aroused the question of whether small additives of amylose are able to induce super-molecular structure formation in solutions of acetylated maltodextrins.

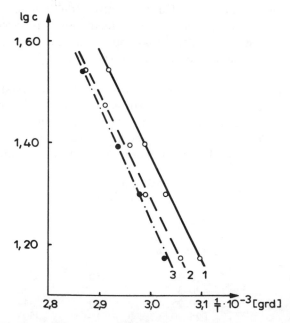

Fig. 2.   Melting enthalpies of maltodextrin gels: (1) MD 2, $\Delta H = 38.3$ kJ mol^{-1}; (2) MD 26, $\Delta H = 39.9$ kJ mol^{-1}; (3) MD 13, $\Delta H = 42.8$ kJ mol^{-1}.

3 RESULTS AND DISCUSSION

3.1 Melting Enthalpies of Maltodextrin Gels

Maltodextrin gels melt in the temperature region of 40–90°C. The values of the melting enthalpies are, as expected, in the scale of several hydrogen bonds. For the maltodextrins MD 2, MD 26 and MD 13, the melting enthalpies are about 40 kJ mol^{-1} (see Fig. 2). There is no dependence on the degree of degradation. The lower limit of maltodextrin concentration, which produces gels, was observed to be 10–12%.

In the present study the possible influence of substances like sugars, oligosaccharides, selected electrolytes and organic substances on melting enthalpy was explored. Gels containing oligosaccharides have higher melting enthalpies (see Table 4). This reveals that oligosaccharides are actively involved in structure formation. This effect is especially noticeable with sample MD 13, which is more strongly degraded than the other maltodextrins which contain the largest amount of oligosaccharides. The other added substances were investigated at concentrations where gel

TABLE 4
Melting Enthalpies of Maltodextrin Gels

Sample	$-\Delta H(kJ\,mol^{-1})$	Without oligosaccharides $-\Delta H(kJ\,mol^{-1})$
MD 2	38·3	34·7
MD 26	39·9	39·7
MD 13	42·8	30·5

formation just occurs. Electrolytes influence gelation according to the Hofmeister lyotropic ion series. Organic substances such as alcoholic precipitants for maltodextrins and hydrogen bond breaking and complexing agents influence the gel properties. A publication of detailed results is under preparation. In this context it is important to mention that all investigated additives change the melting enthalpies only gradually. The strength of bonds apparently cannot alter fundamentals in behaviour. The strength of the interactions which cause the formation of stable aggregates is similar in all solvent systems, provided that the systems permit gel formation. These remarks only concern melting enthalpies. Other physical properties such as dynamics of aggregation and mechanical rigidity are more dependent on solvent effects.

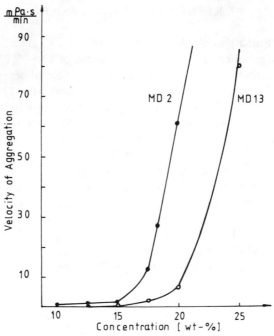

Fig. 3. Dependence of velocity of aggregation on concentration of various maltodextrin solutions (cylinder device).

3.2 Rheological Properties

3.2.1 Influence of the Degree of Degradation

With increasing degradation of starch polysaccharides, the velocity of aggregation and the rigidity of gels are reduced, as shown in Figs 3 and 4. The shear moduli of MD 2 and MD 26 are only slightly different. MD 2, the less degraded sample, has a more solid consistency than MD 26. The

TABLE 5
Rheological Investigations of Maltodextrin Solutions (Cone-plate Device, 4°C)

Sample	Concentration (%)	Velocity of aggregate (mPa s min^{-1})	Aggregation time (min)	Dynamic viscosity (10^3 mPa s)
MD 26	20	15	25	45·6
G_2G_1	21	2	514	126·0
S_2G_1	22	4	172	95·5
G_2S_1	25	Non-gelling under these conditions		
S_2S_1	30	Non-gelling	–	–

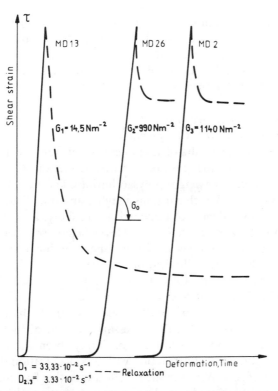

Fig. 4. Shear modulus (———) and relaxation (–––) measurements of aqueous maltodextrin gels (cylinder device). G_1, G_2 and G_3 denote the shear moduli of MD 13, MD 26 and MD 2, respectively. The applied corresponding shear rates were $D_1 = 33\cdot33 \times 10^{-2}\,\text{s}^{-1}$ and $D_2 = D_3 = 3\cdot33 \times 10^{-2}\,\text{s}^{-1}$.

TABLE 6
Gel Properties of Maltodextrins (Cone-plate Device, 4°C)

Sample	Concentration (%)	Shear modulus (Nm^{-2})	Relaxation (%)
MD 26	20	21	29
G_2G_1	21·2	20	27
S_2G_1	20·8	5	39
G_2S_1	26·5	2	72
S_2S_1	30·0	Non-gelling	—

compliance of these samples is considerably lower than that of MD 13 which has undergone more degradation. When the shearing is stopped, a shear stress relaxation is observed which is much more pronounced for MD 13 than in gels of MD 2 and MD 26. These results allow the conclusion that the degree of degradation determines properties of elasticity and rigidity.

3.2.2 Dependence of Aggregation Process, Shear Modulus and Reversible Deformation on Structural Composition

In Tables 5 and 6 the results of investigations of fractions in comparison with source substance MD 26 are presented. Fractions $G_2 G_1$ and $S_2 G_1$ include higher molecular weight polysaccharides. Their gel chromatograms are similar, but $S_2 G_1$, which contains a higher amount of substances in the middle region of the degree of polymerization, shows a stronger tendency of aggregation than fraction $G_2 G_1$. Nevertheless, solutions of fraction $G_2 G_1$ produce the highest viscosities and shear moduli. For the formation of consistency, the ability of hydration and the mobility of macromolecules are, besides structure and molecular size, decisive criteria. The process of structure formation could possibly be obstructed by a small quantity of aggregating segments of molecules and a relatively high mobility of molecules. In fraction $G_2 S_1$, the necessary portion of higher molecular weight structure-forming substances is probably absent. Under the applied conditions, a solution of this fraction with a concentration of 25% shows some viscosity effects. Without shear conditions, this fraction forms a gel with a very low shear modulus. The relaxation seems to be related to the density of aggregate packs. At weaker structure formation, which means lower values of shear modulus and viscosity, relaxation is increased.

3.2.3 Influence of Added Amylose

Aqueous solutions of maltodextrins requires a relatively long time for structure formation if the concentration is lower than 20%. By using supplemental amylose in a concentration of about 1% in solution, the process of aggregation begins immediately, and the gel-forming time is clearly shortened. A solution of MD 13, for instance, which has no ability to form gels at the concentration of 10%, produces a gel with 1% amylose. The degree of polymerization is important in the region $\bar{P}_n < 500$. The comparative effects with regard to aggregation time can be seen in Fig. 5. Amylose with a polymerization degree of 140 has the strongest influence, but the dynamic viscosity is only increased gradually. Thus, the concentration of maltodextrin substance determines the viscosity of the solution. Added amylose, on the other hand, especially influences the dynamics of aggregation.

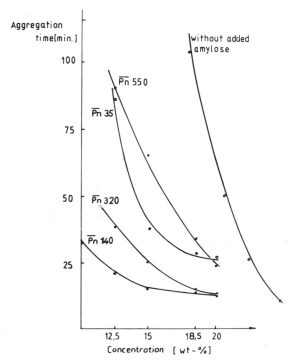

Fig. 5.   Effects of added amylose fractions (1%) on MD 13 solutions in various concentrations.

3.2.4 The Combination of Acetylated Maltodextrins with Amylose

Maltodextrin with an acetyl amount of 1·1% produces a weak gel. Higher acetyl amounts of 3·2% and 5·2% cause a firm solution stability such that even after several days no clouding was observed. Amylose of $P_n = 140$ accelerates the formation of aggregates of the slowly gelling, 1·1% acetyl-containing maltodextrin. Surprisingly, in solutions of maltodextrins with 3·2% and 5·2% acetyl a separation of aggregating amylose does not appear. In acetylated maltodextrins, the added amylose induces an aggregation which leads to gel formation (see Fig. 6). The interesting aspect of this result is that amylose $P_n = 140$ reacts with relatively short unsubstituted chains of acetylated maltodextrins. Amylose molecules evidently form bridges between the chains which do not interact with themselves. There are reasons for making the assumption that amylose not only operates as a crystallization nucleus for the formation of amylose aggregates, but also interacts with segments of branched or substituted molecules. This result is supported by a further experiment. A suspension of freshly aggregated

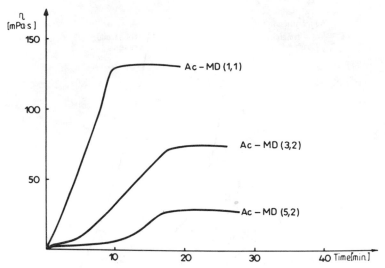

Fig. 6.   Aggregation course in solutions of acetylated maltodextrins with 2% added amylose ($P_n = 140$).

amylose $P_n = 140$ was added to a solution of maltodextrin with 1·1% acetyl amount. The solution became cloudy but a gel was not produced.

4 CONCLUSIONS

The complex combination of the three classes of substances in maltodextrins is the reason for the special properties of maltodextrin gels. The degraded amylose and amylopectin molecules produce a new type of gel, which cannot be made either with amylose or with amylopectin, independent of molecular weight. Neither of the two components can be marked as a major gelling component. It is true that amylose is the most intensive and active structure-forming component, but it is hindered in its interaction with molecules of the same kind by segments of amylopectin which are able to form aggregate structures with amylose segments. Structure formation cannot proceed in the manner of retrogradation, but will lead with a smaller breadth of interaction to aggregates having a lower stability. They are therefore thermo-reversible.

The third class of substance, the oligosaccharides, favours structure formation. The oligosaccharides immobilize the solvent water and increase the stability of the aggregates. Perhaps oligosaccharides fill gaps in zones of bonds when segments of various lengths react. The stability of the aggregates is only gradually changeable under the influence of solvent or temperature. If the conditions for the gel formation are such that the influence of solvents and temperature partially prevents the gelation process, then the rheologically established rate of aggregation, aggregation time and consistency of the gels are changed because of the reduced quantity of segments which are able to associate. The energy content of the aggregates remains constant. The type of physical bonds is not influenced. Under the various conditions, changes in properties can be related to the quantity of segments able to aggregate or to form aggregates.

REFERENCES

1. Richter, M., Schierbaum, F., Augustat, S. and Knoch, K.-D., US Patent 3 962 465, 1976.
2. Pfannemüller, B., Mayerhöfer, H. and Schulz, R. C., Conformation of amylose in aqueous solution: Optical rotatory dispersion and circular dichroism of amylose–iodine complexes and dependence on chain length of retrogradation of amylose, *Biopolymers*, 1971, **10**, 243–61.
3. Ring, S. G., Miles, M. J., Morris, V. J. and Orford, P. D., Recent observations on starch retrogradation, in *New Approaches to Research on Cereal Carbohydrates*, Hill, R. D. and Munck, L. (Eds), Elsevier, Amsterdam, 1985, p. 109.
4. Eldridge, J. E. and Ferry, J. D., Studies of the crosslinking process in gelatin gels. III. Dependence of melting point on concentration and molecular weight, *J. Phys. Chem.*, 1954, **58**, 992–5.
5. Braudo, E. E., Belavtseva, E. M., Titova, E. F., Plashchina, J. G., Krylov, V. L., Tolstoguzov, V. B., Schierbaum, F. and Berth, G., Struktur und Eigenschaften von Maltodextrin-Hydrogelen, *Stärke*, 1979, **31**, 188–94.
6. Wulf, K. and Philipp, B., Zum Auftreten mechanischer Kippschwingungen bei Scherbeanspruchung von Zellulosedispersionen, *Z. phys. Chemie, Leipzig*, 1975, **256**, 478–86.
7. Willstätter, R. and Schudel, G., *Ber. Dtsch. Chem. Ges.*, 1918, **51**, 780.
8. Kettlitz, Bernd, Herstellung und funktionelle Eigenschaften acetylierter Maltodextrine, PhD Thesis, GDR, 1984.

9

INSECT CUTICLE AS A COVALENTLY CROSSLINKED PROTEIN NETWORK

S. O. ANDERSEN

August Krogh Institute, University of Copenhagen,
13 Universitetsparken, 2100 Copenhagen Ø, Denmark

ABSTRACT

Insects are surrounded by an exoskeleton, the cuticle, a composite structure consisting of chitin filaments embedded in a protein matrix. The physical properties of cuticles vary according to the insect species and the body parts from which they are derived. This variability is mainly due to differences in the matrix proteins and in their secondary modifications (crosslinks). Recent progress is presented and discussed, showing that insects have been experts in developing structural materials which combine strength, flexibility and lightness, well adapted to their functions in the living animal.

1 INTRODUCTION

The cuticle of insects and of other arthropods is an extracellular layer composed of lipids, chitin and proteins, which covers the whole animal.[1] The thickness as well as the mechanical properties can vary widely, depending upon the part of the body from which the sample is derived. Two common types are soft cuticle covering most of the body of larvae of flies and moths, and hard cuticle from the legs and thorax of beetles and locusts. Other types of stiffness and flexibility can be found, such as the very hard mandibles of plant-eating insects,[2] abdominal intersegmental cuticle which can be stretched to more than 10 times its unstrained length,[3] and cuticles such as resilin having nearly perfect rubberlike elasticity.[4]

Despite all this diversity in properties, all cuticles are made of composite materials consisting of chitin filaments (2·8 nm in diameter and of indefinite length) embedded in a matrix of proteins, and the differences in properties

141

appear mainly to be due to the relative amount and orientation of chitin filaments, the nature of proteins present, and secondary modifications (sclerotization) of the proteins.[1] Below will be given a brief view of the subject, emphasizing the main problems, which are now under active investigation.

2 CHITIN

The chitin filaments will typically account for 20–25% of the dry weight in hard cuticles and for 50–60% in soft cuticles. Values as low as 3–5% have been reported for some highly stretchable cuticles,[5] and some resilin-containing samples are virtually devoid of chitin.[4] As a rule the chitin filaments run parallel to the cuticular plane and accordingly to the surface of the epidermal cells, which have secreted the cuticle. The most common arrangement of the filaments is a helicoidal pattern, where the filaments in one layer run at a slight constant angle to those in the layer above it, giving a gradually rotating direction as one traverses the cuticle.[1] When cuticle is sectioned for electron microscopy, such a helicoidal arrangement will appear as a parabolic pattern as shown in Fig. 1. Sometimes a sort of

Fig. 1. Electron micrograph of section of unhardened femur cuticle from adult *Locusta migratoria*. The sample was obtained while the animal was in the process of moulting. The parabolic pattern is due to oblique sectioning through a helicoidal arrangement of chitin filaments. Magnification: the bar equals 1 μm. (Photo by B. Kristensen.)

plywood structure is found, where the layers are rather thick and the filament angle changes by about 90° from layer to layer. Often there is a combination of patterns, where thick unidirectional layers alternate with helicoidal layers.[1] Presumably both chitin architecture and chitin interactions with the protein matrix will be of importance to the mechanical properties of the cuticle, and we know little about the precise interaction between the two components in the cuticle.

3 PROTEINS

We are gradually obtaining some knowledge of the protein matrix, although most questions still remain unanswered. During their growth, insects periodically shed the old cuticle having produced a new one beneath the old. The new cuticle is expanded to a larger size immediately after the moult, and at this stage a major fraction of the cuticular proteins can readily be extracted for purification and characterization. A fraction of the proteins, in locust leg cuticle amounting to about 10% of the total protein content, cannot be extracted without degradation, and it has been suggested that this fraction is covalently bound to the chitin scaffold.[6] In the locust leg cuticle the protein extractability decreased drastically during the first hours after the moult, and after 1 day only 10–20% of the proteins could be extracted. This decrease in extractability is connected to an incorporation of low molecular weight phenolic compounds into the cuticle,[7] where they presumably form stable crosslinks between the protein molecules. The cuticle is thereby made stiffer, depending upon how much phenolic material is incorporated. In cuticles which remain soft and flexible, little, if any, crosslinking takes place.

The only known exception to the above picture is resilin-containing cuticle.[4,8] Resilin is a special, cuticular protein (or a mixture of proteins) which is deposited in small patches of cuticle and possesses nearly perfect rubber-like elasticity. The protein is secreted together with chitin and is crosslinked immediately after deposition. No crosslinking agent is incorporated, but the crosslinks are formed by dimerization of tyrosine residues located in separate protein chains.[9] Apparently the reaction proceeds via a free-radical mechanism. The properties of resilin agree with such a molecular network of randomly coiled protein chains crosslinked at regular intervals, and the elastic force is derived nearly exclusively from the entropy decrease upon stretching.

Proteins can be extracted from soft cuticles and from not yet hardened cuticle in sufficient amounts for further characterization. It is characteristic that many proteins are obtained from all types of cuticle,[10] and that clear-cut differences are found between soft cuticles and cuticles destined to become hardened. Figure 2 shows a two-dimensional separation of proteins extracted from the soft cuticle of a silk moth larva, *Hyalophora cecropia*. Several proteins are present, all with isoelectric points in the acidic range (about pH 5), and the molecular weights are between 10 and 30 kilodaltons. In contrast, a sample of not yet hardened locust leg cuticle (Fig. 3) shows a significantly larger number of proteins, most of which have rather alkaline pI values (about pH 8–10). These proteins have also rather low molecular weights (8–40 kilodaltons).

Cuticle from various regions of the same animal gives different protein

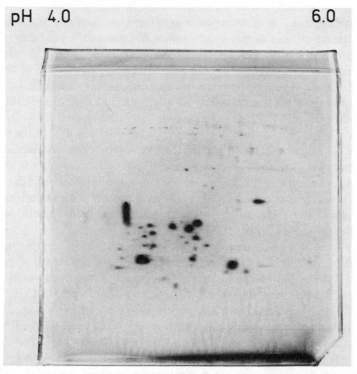

Fig. 2. Two-dimensional separation of proteins extracted from fifth-stage larval cuticle of the American silk moth, *Hyalophora cecropia*. Separation in the horizontal direction by isoelectric focusing in the pH interval 4–6, and in the vertical direction by electrophoresis in Na-dodecylsulphate containing buffer.

patterns, whereas the pattern from a given region is constant as long as it comes from the same species, indicating that the regional differences may be functional.[11]

We have purified a number of locust cuticular proteins and obtained some primary amino acid sequences[12,13] (Figure 4). It is evident that the various proteins are structurally related, that they are independent gene products, and that they cannot be derived from a single precursor protein by various secondary modifications. The individual proteins appear to be divided into segments or regions of varying length. Most of the proteins are dominated by alanine-rich regions where alanine amounts to about 50% of the residues; proline and valine are also abundant in these regions. An often repeated sequence is -Ala-Ala-Pro-Ala-. Other regions are very rich in

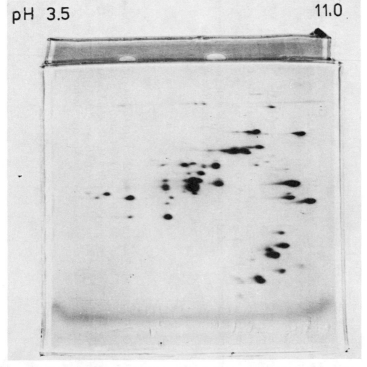

Fig. 3. Two-dimensional separation of proteins extracted from leg cuticle of the migratory locust, *Locusta migratoria*, in the process of moulting. Separation in the horizontal direction by isoelectric focusing in the pH interval 3·5–11, and in the vertical direction by electrophoresis in Na-dodecylsulphate containing buffer.

Prot. 7:	Gly-Val-Ile-Ser-Ala-Gly-Tyr-Ala-Ala-Ala-Pro-Ala-Tyr-Ala-Ala-Ala-Pro-Pro-Ala-Tyr-Ala-Ala-Ala-Ala-
Prot. 8:	Ile-Glu-Asp-Glu-Tyr-Gly-Asn-Val-Leu-His-Ala-Ser-Gly-Gly-Tyr-Ala-Ala-Pro-Ala-Ala-Tyr-Ala-Ala-Pro-Val-Ala-Tyr-Ala-Ala-Pro-Val-Ala-Lys-Ala-Val-Val-Ala-Ala-Pro-Ala-Val-Ala-Tyr-Ala-Ala-Pro-Val-Ala-Lys-Ala-Val-Val-Ala- ? -Pro-Val-
Prot. 37:	Gly-Leu-Leu-Gly-Leu-Gly-Tyr-Gly-Gly-Tyr-Gly-Tyr-Gly-Ala-Ala-Leu-Ala-Ala-Pro-Ala-Ala-Val-Ser-Tyr-Ala-Ala-Pro-Ala-Ile-Ala-Ala-Ala-Pro-Ala-Val-Ser-Tyr-Ala-Ala-Pro-Ala-
Prot. 54:	Gly-Tyr-Leu-Gly-Gly-Tyr-Ala-Ala-Pro-Ala-Ile-Ala-Ala-Ala-Pro-Ala-Ile-Ala-Ala-Leu-Pro-Ala-Ala-Ser-Ser-Ile-Ala-Ala-Asn-Gly-Tyr-
Prot. 64:	Gly-Leu-Leu-Gly-Leu-Gly-Tyr-Gly-Gly-Tyr-Ser-Gly-Tyr-Gly-Gly-Tyr-Gly-Gly-Tyr-Gly-Pro-His-Leu-Ala-Leu-Ala-Tyr-Ala-Pro-Leu-Ala-Leu-Ala-Tyr-Pro-Ala-Val-Val-Ala-Ala-Pro-Ala-Ile-Ala-Gln-Ala-Pro-Ala-Val-Ala-Val-Ala-Pro-Ala-Val-Ala-? -Ala-Pro-

Fig. 4. N-terminal sequences of selected proteins from the cuticle of *Locusta migratoria.*[12] A question mark (?) indicates uncertainty in the identification of residue.

Gly-Tyr-Leu-Gly-Gly-Ile-Ala-Ala-Pro-Val-Gly-Tyr-Ala-Ala-Pro-Ala-Val-Gly-Tyr-Ala-Ala-Pro-Ala-Ile-Ala-Ala-Ala-Pro-Val-Ala-Val-Ala-His-Ala-Val-Ala-Pro-Ala-Ala-Ala-Ser-Val-Ala-Asn-Thr-Tyr-Arg-Ile-Ser-Gln-Thr-Ala-Arg-Val-Leu-Ala-Ala-Pro-Ala-Ala-Tyr-Ala-Ala-Pro-Ala-Val-Ala-Ala-Ala-Pro-Ala-Ile-Gly-Tyr-Ala-Ala-Pro-Ala-Ile-Ala-Ala-Ala-Pro-Ala-Leu-Gly-Tyr-Ala-Arg-Tyr-Ala-Ala-Ala-Ala-Pro-Val-Ala-Val-Ala-His-Ala-Ala-Val-Pro-Ala-Ala-Ala-Ser-Val-Ala-Asn-Thr-Tyr-Arg-Ile-Ser-Gln-Thr-Ala-Arg-Leu-Leu-Ala-Ala-Pro-Ala-Val-Ala-His-Ala-Pro-Val-Ala-Tyr-Ala-Ala-Pro-Ala-Ala-Tyr-Ala-Ala-Pro-Ala-Ile-Gly-Tyr-Gly-Tyr-Gly-Gly-Leu-Ala-Tyr-Gly-Ala-Ala-Pro-Val-Ala-Lys-Val-Tyr.

Fig. 5. Total amino acid sequence of protein No. 38 from the cuticle of *Locusta migratoria.*[13]

glycine and also enriched in leucine and tyrosine; the N-terminal region of the proteins often belongs to this type.

So far we have only obtained the total amino acid sequence for one protein from locust cuticle (Fig. 5).[13] We have used some of the published methods[14-16] for calculating the likely way for this protein to be folded. The results indicate that the glycine-rich regions will preferably be in the extended β-sheet configuration, whereas the alanine-rich regions may be in a rather unstable α-helical configuration. We have so far no experimental determinations of the relative amounts of the various configurational types in this protein.

The number of polar groups is unusually low in these proteins. In the fully sequenced protein there is only a single negative charge (the C-terminal residue), seven positive charges at pH 7, and a few uncharged polar groups, such as glutamine, asparagine, serine and threonine. This indicates that the interaction between the proteins will be dominated to a large extent by hydrophobic interactions. This is supported by the observation that the proteins are more soluble at 4°C than at room temperature.[11]

Amino acid sequences have also been obtained for some of the proteins in fruit fly larval cuticle.[17] These sequences show no resemblance to those from locust cuticle: there is no tendency for residues to occur in regions or in repeats, the alanine content is moderate and there are plenty of polar residues. The differences in sequence between the two types of proteins may be of functional importance as the fruit fly larval cuticle is a typical, soft cuticle and the locust protein is derived from a type of cuticle destined to be hardened.

4 SCLEROTIZATION

As mentioned above it is a common phenomenon that cuticular proteins undergo secondary modifications (hardening) after the insect has moulted. The chemistry of the reactions involved has been studied for many years, but hardened cuticle is an extremely resistant material to investigate, and we have probably only obtained a very incomplete understanding so far.[18] Locust cuticle seems a relatively simple system, as only a single sclerotizing agent has been described, N-acetyldopamine. When hardening is initiated the epidermal cells secrete N-acetyldopamine into the preformed cuticle, where it encounters two enzymatic activities: a desaturase which introduces a double bond into its side-chain, whereby dehydro-N-acetyldopamine is formed, and a diphenoloxidase, which catalyzes the oxidation of both N-acetyldopamine and dehydro-N-acetyldopamine to the corresponding quinones. The diphenoloxidase is much more active towards dehydro-N-acetyldopamine than towards N-acetyldopamine, so when sufficient desaturase is present nearly all available N-acetyldopamine will be processed via the unsaturated intermediate, whereas most of the N-acetyldopamine will be oxidized directly to quinone if the desaturase is present only in low amounts. Both quinones can react with nucleophilic groups in proteins, but the unsaturated quinone is more reactive and appears to be the best candidate for forming crosslinks between peptide chains.[19]

We have not yet purified intact crosslinks, and we can only make assumptions regarding their structure. We have isolated the dimeric structure[20]

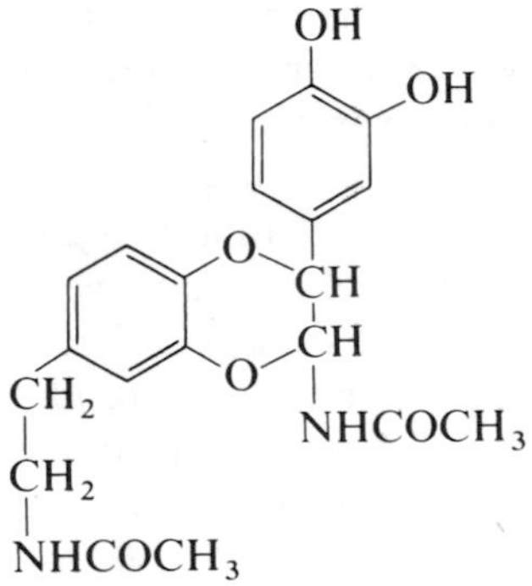

presumably formed by reaction between N-acetyldopamine and the unsaturated quinone:

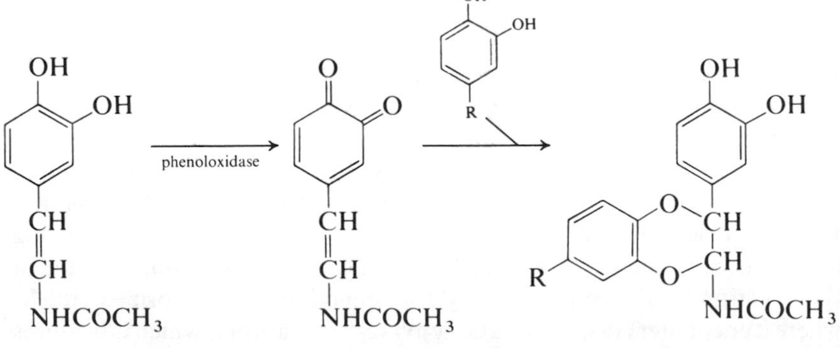

This indicates that the unsaturated quinone can react with two closely spaced phenolic groups.[19] Hypothetically, two phenolic groups from separate tyrosine residues in the proteins could be linked together in a similar manner, if the two tyrosines could come sufficiently close to each other. The proteins appear to be closely packed in the cuticle, presumably due to their hydrophobic nature, and residues in neighboring proteins can be in close contact. The total water content of freshly moulted locust cuticle is about 43%, which is approximately the water content of many protein crystals. During sclerotization the water content drops to about 32%, indicating a closer packing of the proteins. This could be due either to the formation of crosslinks, to increased hydrophobicity after incorporation of aromatic materials, or to filling the interspaces between the proteins with polymerized phenols, but it is presumably a combination of all these factors.

REFERENCES

1. Neville, A. C., *Biology of the Arthropod Cuticle*, Springer, Berlin, 1975.
2. Hillerton, J. E., Robertson, B. and Vincent, J. F. V., The presence of zinc or manganese as the predominant metal in the mandibles of adult, stored-product beetles, *J. stored Prod. Res.*, 1984, **20**, 133–7.
3. Vincent, J. F. V. and Prentice, J. H., Rheological properties of the extensible intersegmental membrane of the adult female locust, *J. Mater. Sci.*, 1973, **8**, 624–30.
4. Andersen, S. O. and Weis-Fogh, T., Resilin. A rubberlike protein in arthropod cuticle, *Adv. Insect Physiol.*, 1964, **2**, 1–65.
5. Hackman, R. H., Expanding abdominal cuticle in the bug *Rhodnius* and the tick *Boophilus*, *J. Insect Physiol.*, 1975, **21**, 1613–23.
6. Andersen, S. O., Arthropod cuticles: their composition, properties and functions, *Symp. Zool. Soc. London*, 1977, **39**, 7–32.
7. Andersen, S. O. and Barrett, F. M., The isolation of ketocatechols from insect cuticle and their possible role in sclerotization, *J. Insect Physiol.*, 1971, **17**, 69–83.
8. Andersen, S. O., Resilin, in *Comprehensive Biochemistry*, Vol. 26C, Florkin, M. and Stotz, E. H. (Eds), Elsevier, Amsterdam, 1971, pp. 633–57.
9. Andersen, S. O., Covalent crosslinks in a structural protein, resilin, *Acta Physiol. Scand.*, 1966, **66** (suppl. 263), 1–81.
10. Willis, J. H., Regier, J. C. and Debrunner, B. A., The metamorphosis of arthropodin, in *Current Topics in Insect Endocrinology and Nutrition*, Bhaskaran, G., Friedman, S. and Rodriguez, J. G. (Eds), Plenum Press, New York, 1981, pp. 27–46.
11. Andersen, S. O., Højrup, P. and Roepstorff, P., Characterization of cuticular proteins from the migratory locust, *Locusta migratoria*, *Insect Biochem.*, 1986, **16**, 441–7.
12. Højrup, P., Andersen, S. O. and Roepstorff, P., Isolation, characterization, and N-terminal sequence studies of cuticular proteins from the migratory locust, *Locusta migratoria*, *Eur. J. Biochem.*, 1986, **154**, 153–9.

13. Højrup, P., Andersen, S. O. and Roepstorff, P., Primary structure of a structural protein from the cuticle of the migratory locust, *Locusta migratoria*, *Biochem. J.*, 1986, **236**, 713–20.
14. Garnier, J., Osguthorpe, D. J. and Robson, B., Analysis of the accuracy and implications of simple methods for predicting the secondary structure of globular proteins, *J. Mol. Biol.*, 1978, **120**, 97–120.
15. Levitt, M., Conformational preferences of amino acids in globular proteins, *Biochemistry*, 1978, **17**, 4277–85.
16. Chou, P. Y. and Fasman, G. D., Prediction of protein conformation, *Biochemistry*, 1974, **13**, 222–45.
17. Snyder, M., Hunkapiller, M., Yuen, D., Silvert, D., Fristrom, J. and Davidson, N., Cuticle protein genes of *Drosophila*: Structure, organization and evolution of four clustered genes, *Cell*, 1982, **29**, 1027–40.
18. Andersen, S. O., Sclerotization and tanning of the cuticle, in *Comprehensive Insect Physiology, Biochemistry and Pharmacology*, Vol. 3, Kerkut, G. A. and Gilbert, L. I. (Eds), Pergamon Press, Oxford, 1985, pp. 59–74.
19. Andersen, S. O. and Roepstorff, P., Sclerotization of insect cuticle—III. An unsaturated derivative of *N*-acetyldopamine and its role in sclerotization, *Insect Biochem.*, 1982, **12**, 269–76.
20. Andersen, S. O., Jacobsen, J. P. and Roepstorff, P., Studies of the sclerotization of insect cuticle. The structure of a dimeric product formed by incubation of *N*-acetyldopamine with locust cuticle, *Tetrahedron*, 1980, **36**, 3249–52.

SECTION 2

FORMATION OF NETWORKS

10

INTRAMOLECULAR REACTION AND NETWORK FORMATION AND PROPERTIES

R. F. T. STEPTO

Department of Polymer Science and Technology,
University of Manchester Institute of Science and Technology,
Sackville Street, Manchester M60 1QD, UK

ABSTRACT

Results from systematic studies of pre-gel intramolecular reaction and gel points in tri- and tetrafunctional polyurethane polymerisations using hexamethylene diisocyanate(HDI) and polyoxypropylene(POP) triols and tetrols are reviewed. A contrast is made between total fractions of ring structures per molecule in linear and non-linear bulk polymerisations. The variation of gel point with dilution, reactant molar mass and functionality is characterised using the Ahmad–Stepto expression for the gel point.

Previously established correlations between the static shear moduli of the networks formed at complete reaction and gel points are also reviewed. The persistence of intramolecular reaction, even in bulk polymerisations, leads to reductions in shear moduli below those expected for the perfect networks. Trifunctional networks show larger reductions than tetrafunctional networks.

The reductions in moduli are analysed in detail in the present paper in terms of simple loops and it is shown that such loops arise from both pre-gel and post-gel intramolecular reaction. Even in the limit of perfect gelling systems, imperfections will persist simply because reaction between groups in the gel must occur and this must be intramolecular, with some leading to inelastic loops. The phenomenon is essentially a law-of-mass-action effect. Further, by taking the limit of perfect gelling systems having reactants of infinite molar mass, it is possible to predict the finite structures in the gel within which all possibilities of intramolecular reaction must be considered in order to give the numbers of defects deduced experimentally.

INTRODUCTION

An important and well-known relationship from the statistical theory of rubber elasticity in its simplest form is[1]

$$G = nkT = \rho RT / M_c \qquad (1)$$

where G is the static shear modulus, n is the number of elastic chains per unit volume, ρ is the density of the dry network and M_c is the number-average molar mass between elastically effective junction points. Apart from theoretical shortcomings in this relationship, there are difficulties in knowing n or M_c *ab initio* for actual networks. The quantity is of course of key importance in interpreting the physical properties of polymer networks in terms of affine or phantom chain behaviour,[2-5] and whether chain entanglements or interactions contribute additionally to measured moduli.[2,5-7]

With traditional rubbery networks formed from crosslinking preformed polymer chains,[1] there are problems in defining M_c because of the initial distribution of chain lengths and the subsequent non-random reaction of active sites on the chains. For non-linear addition or sequential polymerisations using unsaturated monomers, non-random reaction usually occurs due to the unequal reactivity of reactive sites and excessive intramolecular reaction leading to inactive loops.[8] Recent attention has therefore turned to non-linear random (or end-linking) polymerisations to form model networks.[2-14]

It is often assumed[2-5] that for such networks n is known directly from the stoichiometry, molar masses and functionalities of the reactants. The assumption implies the complete reaction of functional groups, no pre-gel intramolecular reaction, and post-gel intramolecular reaction occurring only when it leads to elastically active chains. Intramolecular reaction is in fact an integral part of random polymerisations simply because reactive groups are spatially connected. The ring structures most likely to form are the smallest and *ipso facto* are inelastic loops.[8,15] The question then arises whether inelastic loop formation can ever be neglected when seeking to evaluate n in eqn (1) or elaborations thereof. The question is much more fundamental than whether or not network imperfections arise from incomplete reaction or side reactions of functional groups in particular systems.[6,14] It is linked with the statistics of the polymerisation used and the resulting molecular species and structures.

There is little hope of analysing networks chemically or spectroscopically for inelastic loops. Their occurrence must eventually be predicted from polymerisation statistics. However, statistical theories of non-linear polymerisation and network formation[8,15-19] are at present inadequate in their prediction of the amount of intramolecular reaction, because drastic simplifications of the infinite structures which may form have to be made. Designed experiments on model reaction systems are needed in which the known parameters affecting intramolecular reaction are varied systematically and any resulting deviations from perfect network behaviour are interpreted in terms of these parameters.[8-13,20] The general importance of inelastic loop formation arising from intramolecular reaction can then be ascertained and hopefully indications given of the significant structures which give rise to inelastic loops. The knowledge of such structures will then indicate the improvements which have to be made in statistical theories of network formation.

The essential thesis is that, because of pre-gel and post-gel intramolecular reaction, model networks formed from non-linear polymerisations can never be assumed to be perfect networks. Their structure and, hence, properties are inexorably linked with the polymerisation statistics of their formation. In this respect, the effects of intramolecular reaction can most usefully be studied using end-linking reactions. Reactants with like groups having equal reactivities can be chosen so that the polymerisation statistics in the absence of intramolecular reaction are clearly defined. The present chapter concentrates on such investigations linking gel-points and network properties and using systems leading to polyurethane networks.[11-13] Background studies cover intramolecular reaction in linear polymerisations,[21] pre-gel intramolecular reaction in non-linear polymerisations[22] and gel-point determinations.[23-27] In all cases, stoichiometric ratios of reactants have been used so that at complete reaction perfect networks would be formed in the absence of inelastic loop formation. The results have been reviewed recently[8,15,20] and only a résumé will be presented here. However, a new interpretation of the data is also given showing the universal occurrence of inelastic loop formation and allowing the significant structures involved in the formation of such loops to be deduced.

PRE-GEL INTRAMOLECULAR REACTION

Figure 1 shows the number of ring structures per molecule (N_r) versus extent of reaction (p) for bulk linear and non-linear polyurethane-forming reactions.[11] The linear polymerisation[21] used hexamethylene diisocyanate (HDI) and a polyoxyethylene (POE) diol and the non-linear polymerisation[22] HDI and a polyoxypropylene (POP) triol. The much larger number of

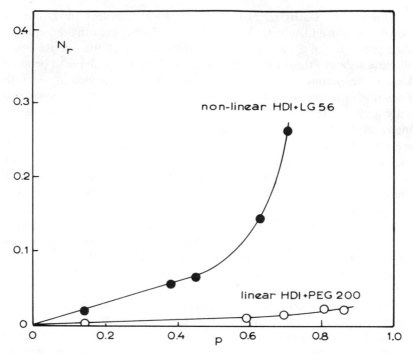

Fig. 1. Number fraction of ring structures per molecule (N_r) as a function of extent of reaction (p) for linear and non-linear polyurethane-forming reactions in bulk with approximately equimolar concentrations of reactive groups.[11] $\bigcirc$ = linear polymerisation[21]—HDI + PEO diol(PEG200) at 70°C, $[NCO]_0 = 5 \cdot 111$ mol kg^{-1}, $[OH]_0 = 5 \cdot 188$ mol kg^{-1}, number-average number of bonds in chain forming the smallest ring structure (v) = 25·2 (see Fig. 3). $\bullet$ = non-linear polymerisation[22]— HDI + POP triol (LG56) at 70°C, $[NCO]_0 = 0 \cdot 9073$ mol kg^{-1}, $[OH]_0 = 0 \cdot 9173$ mol kg^{-1}, $v = 115$.

ring structures per molecule in the non-linear compared with the linear polymerisation is due to the larger number of opportunities per molecule for intramolecular reaction in the former case. The number increases as a non-linear polymerisation proceeds and becomes infinite at the gel point and beyond. The data for the non-linear polymerisation stop just prior to the gel point ($p = 0 \cdot 765$) when about one molecule in every three has a ring structure.

The difference in number of opportunities per molecule for intra-molecular reaction in a linear $RA_2 + RB_2$ and a non-linear $RA_2 + RB_3$ polymerisation is illustrated in Fig. 2. Here, c_{int}, the effective concentration

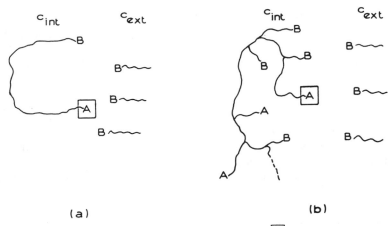

Fig. 2. Concentrations of B groups around reference $\boxed{A}$ groups in (a) an $RA_2 + RB_2$ polymerisation and (b) an $RA_2 + RB_3$ polymerisation. c_{int} is the concentration of B groups from the same molecule and c_{ext} that from other molecules.[15]

of (B) groups around a group ($\boxed{A}$) with which they can react intramolecularly depends on the lengths and structures of chains connecting reacting pairs and the number of such pairs. In the linear case, only one pair ever exists. c_{ext} is the concentration of groups from other molecules which react intermolecularly with $\boxed{A}$.

The probability of intramolecular reaction at any point in a polymerisation depends on the relative values of c_{int} and c_{ext}. c_{ext} decreases as a polymerisation proceeds and its initial value is determined by the amount of solvent present (if any) and the molar masses and functionalities of the reactants. For example, assuming a density of 1, the initial concentration of the B groups of bulk RB_{f_b} monomer or prepolymer is f_b/M_b, where M_b is the molar mass. c_{int} is also characterised by the monomer molar masses and functionalities in terms of the number of bonds (v) in the chain forming the smallest ring structure, as illustrated in Fig. 3. According to Gaussian chain statistics, the mutual concentration of the A and B groups in Fig. 3 (i.e. the concentration of the A end groups around the B end groups or vice versa) can be written as[11,15,25,28,29]

$$Pab = (3/2\pi vb^2)^{3/2}/N \text{ (moles functional groups per unit volume)} \quad (2)$$

where N is the Avogadro constant, b is the effective bond length of the chain of v bonds before reaction occurs and $vb^2 = \langle r^2 \rangle$, the mean-square end-to-end distance of the chain. Pab is essentially c_{int} per pair of reacting groups forming the smallest ring structure. Thus, intramolecular reaction increases as v or molar masses of reactants of given functionalities decrease and as chain stiffness (b) decreases.

In both the linear and non-linear polymerisations, ring structures of approximately v, $2v$, $3v$,..., iv,... bonds may form. In linear polymerisations the numbers of molecules which can form different sizes of ring may be evaluated more accurately as only linear sequences of repeating units occur. In non-linear polymerisations, as may be deduced from Fig. 2, the number of isomeric structures of a given degree of polymerisation (DP) increases with DP, and opportunities for forming ring structures of different sizes vary with the isomer. Accordingly, theoretical interpretations

(a)   (b)

Fig. 3. Repeating units of the chain structures in (a) an $RA_2 + RB_2$ polymerisation and (b) an $RA_2 + RB_3$ polymerisation. The repeating units have v bonds separating groups $\boxed{A}$ and $\boxed{B}$ which may react intramolecularly.[15]

of N_r vs. p data from linear polymerisations at various initial dilutions in solvent[15,21,29-31] are much more satisfactory than those of data from non-linear polymerisations.[15,32-36] The difficulty is the definition of the significant structures for intramolecular reaction within molecular species, which enables reasonable approximations to be made for the numbers of opportunities for forming ring structures of various sizes.

The domination of numbers of opportunities in non-linear polymerisations may be inferred from Fig. 1. Approximate evaluation[11] of the probability of intramolecular reaction per pair of groups which can so react shows that it is 1·7 times larger for the linear polymerisation. The evaluation uses estimates of Pab/c_{ext} for the two systems. Pab is higher for the linear system (v is less) and c_{ext} is also higher (concentration of end groups higher), with the increase in Pab dominating when ratios are taken. Thus, the much larger values of N_r at a given p for the non-linear polymerisation come from factors in the expression for c_{int} which relate to the numbers of opportunities per molecule.

GELATION

The gel point in end-linking polymerisations is much more significant than in the crosslinking of preformed polymer chains. Stockmayer's general relationship for the gel point in the absence of intramolecular reaction of a polymerising mixture of reactants bearing A and B separately is[8,15,37]

$$\alpha_c^\circ(f_{a_w} - 1)(f_{b_w} - 1) = 1 \qquad (3)$$

where α_c° is $(p_a^\circ p_b^\circ)_c$, the critical value of the product of extents of reaction of A and B groups at gelation, and f_{a_w} and f_{b_w} are, respectively, the so-called weight-average functionalities of reactants bearing A and B groups. For $RA_2 + RB_3$ polymerisations $f_{a_w} = 2$ and $f_{b_w} = 3$, hence $\alpha_c^\circ = 0.5$ at gelation.

For the vulcanisation of unsaturated monomers, for example, α_c° is replaced by the critical crosslinking index or fraction of double bonds reacted, $f_{a_w} = 2$ for the vulcanising agent, and f_{b_w} is equal to dp_w, the weight-average DP of the preformed chains.[8] The much higher functionality means that α_c° is often negligibly small and, assuming random reaction, the gel point is passed almost as soon as the polymerisation is started. The higher values of α_c° for end-linking polymerisations mean that the actual α_c may be more accurately determined and its deviation from α_c° used to characterise the total amount of pre-gel intramolecular reaction. For example, the gel point of the non-linear polymerisation shown in Fig. 1 occurred at $p = 0.765$, as compared with $p_c^\circ = (\alpha_c^\circ)^{1/2} = 2^{-1/2} = 0.707$ for no intramolecular reaction. In addition, extents of reaction at gelation are experimentally much more easily determined than values of N_r. The former require only the observation of gelation and the chemical or spectroscopic determination of end groups, whereas the latter require the additional determination of M_n as the polymerisation proceeds.[21,22] As will be seen, the gel point is also useful for characterising network defects in terms of the propensity of a polymerisation to undergo intramolecular reaction.

Gel points in $RA_2 + RB_f$ polymerisations (i.e. values of α_c) may be interpreted in terms of the Ahmad–Stepto relationship[26]

$$(f - 1)\alpha_c = (1 + \lambda'_{ab})^2 \qquad (4)$$

λ'_{ab} is a ring-forming parameter given by the expression

$$\lambda'_{ab} = c_{int}/c_{ext} = (f - 2) \times Pab \times \phi(1, 3/2)/(c'_a + c'_b) \qquad (5)$$

where $\phi(1, 3/2) = \sum_{i=1}^{\infty} 1^i i^{-3/2} = 2.612$ and the factors $(f - 2) \times \phi(1, 3/2)$ represent an approximation to the number of opportunities for intramolecular reaction in the total expression for c_{int}. $(c'_a + c'_b)$ is some

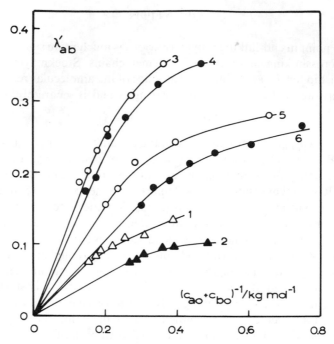

Fig. 4. Analysis of gelation data[11−13,38] according to eqns (2) and (5) with $(c_a' + c_b')$ equated to $(c_{a0} + c_{b0})$. Specifications of the reaction systems corresponding to curves 1–6 are given in Table 1. Curves 1 and 2, HDI + POP triols; curves 3–6, HDI + POP tetrols.

instantaneous value of the concentration of A and B groups in the reaction mixture and represents c_{ext}. Thus, in the theory, constant values of c_{int} and c_{ext} are assumed so that λ_{ab}' is constant. In the interpretation of experimental data, $(c_a' + c_b')$ may be equated to $(c_{a0} + c_{b0})$ and $(c_{ac} + c_{bc})$ as extreme values, where $(c_{a0} + c_{b0})$ is the initial concentration of reactive groups and $(c_{ac} + c_{bc})$ is the concentration at gelation.

Figure 4 shows gelation data for polyurethane-forming polymeris-ations interpreted according to eqns (2) and (5) with $(c_a' + c_b')$ equated to $(c_{a0} + c_{b0})$.[8,11−13,38] The data refer to reactions of HDI with POP triols and tetrols of different molar masses and, hence, values of v, as specified in Table 1. The values of M_n for the triols were about 700 and 1100 g mol^{-1} and for the tetrols they varied from about 660 to 2100 g mol^{-1}. $(c_{a0} + c_{b0})$ was varied by carrying out reactions at different dilutions in solvent, with points at the lowest values of $(c_{a0} + c_{b0})^{-1}$ referring to reactions in bulk. $\lambda_{ab}' = 0$

TABLE 1

Reaction Systems used for the Gelation Results in Fig. 4 and for Network Studies
(Fig. 5)

System	f	$M_c^{\circ\,a}$ $(g\ mol^{-1})$	v^b	v_2/v^c	$b(i)^d$ (nm)	$b(ii)^d$ (nm)
(1) HDI/LHT240	3	635	33	0·303	0·247	0·400
(2) HDI/LHT112	3	1 168	61	0·164	0·222	0·363
(3) HDI/OPPE1	4	500	29	0·345	0·240	0·356
(4) HDI/OPPE2	4	586	33	0·303	0·237	0·347
(5) HDI/OPPE3	4	789	44	0·227	0·234	0·305
(6) HDI/OPPE4	4	1 220	66	0·152	0·215	0·281

[a] M_c° = molar masses of chains between elastic junction points in the perfect networks.
[b] v = Numbers of bonds in chains forming the smallest ring structures.
[c] v_2/v = Fractions of the chains of v bonds coming from the difunctional unit (HDI).
[d] $b(i)$ and $b(ii)$ are the effective bond lengths evaluated using eqns (2) and (5) and initial slopes of plots of λ'_{ab} versus (i) $(c_{a0} + c_{b0})^{-1}$ (Fig. 4) and (ii) $(c_{ac} + c_{bc})^{-1}$.

corresponds to no intramolecular reaction and even for reactions in bulk $\lambda'_{ab} > 0$. The results show that λ'_{ab} is larger for tetrafunctional systems (curves 3–6) than for trifunctional ones (curves 1 and 2), and that λ'_{ab} decreases as v increases in accordance with eqns (2) and (5). However, the predicted direct proportionality between λ'_{ab} and $(c_{a0} + c_{b0})^{-1}$ (or c_{ext}^{-1}) is not observed except for concentrations near bulk. Use of $c_{ac} + c_{bc}$ for c_{ext} gives plots with even more curvature. The curvature results from basic approximations in the theory, as discussed in detail elsewhere.[8,15,26] However, analysis of the slopes of the initial, linear portions of the curves enables values of b to be derived as the parameter representing the influence of chain stiffness on intramolecular reaction. The results are given in the last two columns in Table 1. It can be seen that b decreases as v, or more strictly v_2/v, decreases, where v_2/v is the fraction of bonds in the chain forming the smallest ring structure which comes from the HDI unit. Thus, the HDI chain structure is apparently stiffer than the POP chain structure. Analogous trends in b have been found for polyester-forming systems.[8,15,26] Note that the data cannot be interpreted using a constant value of b. The values of b in the last column (for $c_{ext} = c_{ac} + c_{bc}$) are more in keeping with those expected from solution properties, indicating that most intramolecular reaction occurs near the gel point and the correct average value of c_{ext} (namely, $c_a' + c_b'$) is nearer to $c_{ac} + c_{bc}$ than to $c_{a0} + c_{b0}$.

The behaviour shown in Fig. 4 may be summarised by the following points:

(1) Gel points may be varied by diluting reaction mixtures with solvent. They are always delayed beyond the ideal Flory–Stockmayer value of α_c, even for reactions in bulk.

(2) The amount of pre-gel intramolecular reaction (λ'_{ab}) at a given initial dilution of reactive groups increases primarily with f and secondarily with decreases in v. There is a smaller but significant increase as chain stiffness (b) decreases. (The effect of chain stiffness can in fact be larger for chain structures with aromatic units.[11−13])

(3) The points referring to bulk reactions show that λ'_{ab} for a given functionality is relatively insensitive to molar mass. The decrease in $(vb^2)^{-3/2}$ with increase in molar mass is to a large extent counteracted by the concomitant decrease in $(c_{a0} + c_{b0})$.

CORRELATIONS BETWEEN GEL POINTS AND NETWORK PROPERTIES

Correlations between amount of pre-gel intramolecular reaction, as manifested by the gel point, and the properties of the network formed at complete reaction may be investigated by dividing a reaction mixture into two. One part is used for gel-point determination and one part is allowed to react to completion in moulds at the same temperature to give samples of suitable shapes for the measurement of physical properties. Values of initial moduli and T_g have been investigated in this way.[10−13,38] Present discussion is limited to the variation of initial shear modulus with extent of reaction at gelation.

Shear moduli were measured using cylindrical specimens and a uniaxial-compression apparatus especially designed for measuring small strains.[10] Equilibrium-swollen and dry specimens without sol fraction (usually negligible) were used. The stress–strain data obtained were analysed according to Gaussian rubber-elasticity theory[1−3,10] using the equations

$$\sigma = G(\Lambda - \Lambda^{-2}) \tag{6}$$

and

$$G = ART\rho\phi_2^{1/3}(V_u/V_F)^{2/3}/M_c \tag{7}$$

Here, σ is the stress per unit unstrained area, G the shear modulus, Λ the deformation ratio, ρ the density of the dry network, ϕ_2 the volume fraction of solvent present in a swollen network, V_u the volume of the dry unstrained network and V_F the volume at formation. A has the value $(1-2/f)$ if chains show phantom behaviour and 1 if they show affine behaviour. M_c is the effective molar mass of chains between elastically active junction points. Effects of inelastic loops will be reflected as reductions in G or proportionate increases in M_c. Equation (7) was used to derive values of M_c/A from the values of G evaluated according to eqn (6). In this way, data from the same network in the dry and equilibrium-swollen states could be combined.

The results obtained[11–13,38] for the networks formed from the reaction systems of Fig. 4 and Table 1 are shown in Fig. 5. They are plotted as M_c/AM_c° versus $p_{r,c}$. M_c° is the molar mass between chains of the perfect network and $p_{r,c}$ is the extent of intramolecular reaction at the gel point. M_c° can be calculated directly from the molar masses of the reactants and is essentially the molar mass of a chain of v bonds. The values of M_c° have been given in Table 1. For stoichiometric reaction mixtures, $p_{r,c}$ is defined by the equation

$$p_{r,c} = (\alpha_c)^{1/2} - (\alpha_c^\circ)^{1/2} \tag{8}$$

From eqn (4), $p_{r,c}$ may alternatively be given by

$$p_{r,c} = \lambda'_{ab}/(f-1)^{1/2} \tag{9}$$

The detailed analysis of the Gaussian stress–strain plots and the use of dry and swollen samples to obtain single values of M_c/A for the individual networks have been discussed in detail elsewhere.[10,13] The uncertainties introduced in M_c/AM_c° are much less than the changes with $p_{r,c}$ shown in Fig. 5. The large increases in M_c/AM_c° with dilution also show that the effects are not due to incomplete reaction, which will be less in more dilute systems. Preliminary measurements of NCO groups remaining at 'complete reaction' indicate values of p in excess of 0·99.

For perfect networks with affine chain behaviour, $A = 1$ and $M_c/AM_c^\circ = 1$. For phantom chain behaviour, $M_c/AM_c^\circ = 3$ for $f = 3$ networks and 2 for $f = 4$ networks. Thus, on both these bases the networks show marked increases in M_c (reductions in dry modulus) compared with M_c°. The reductions in dry modulus show that any effects due to entanglements are completely overshadowed. It may also be argued that effects due to entanglements would not be expected from small-strain, static measurements and for networks with such small values of M_c°. The latter reason should be taken with caution as the effective value of M_c is in some cases much larger than M_c°. Assuming affine behaviour, the most likely for small

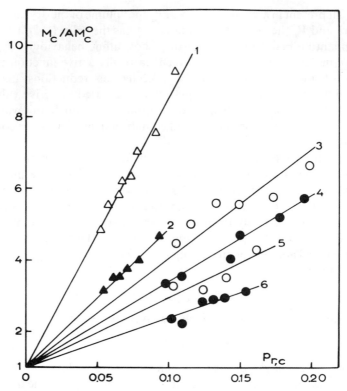

Fig. 5. Ratio of molar mass of chains between elastically active junction points to front factor (M_c/A) relative to molar mass of chains between elastically active junction points for the perfect network (M_c°) versus extent of intramolecular reaction at gelation ($p_{r,c}$) for the polyurethane networks formed from the reaction systems in Table 1 and Fig. 4. Values of M_c/A determined directly from small-strain shear moduli according to eqn (7) using both dry and equilibrium-swollen networks.[11-13,38]

strains[2-5] (and see later), the reductions in modulus are particularly large, even for reactions in bulk. For triol system 1, $p_{r,c}$ is about 5% and about a five-fold reduction in dry modulus occurs. Similarly, for bulk tetrol system 6, $p_{r,c}$ is about 10% and a two-fold reduction results.

The results in Fig. 5 basically show direct correlations between amount of pre-gel intramolecular reaction and network defects existing at complete reaction. The only plausible explanation is that the defects are due to inelastic loops, formed pre-gel and post-gel. The results show the importance of having some knowledge of the polymerisation statistics of the formation of a network in order to be able to understand its structure and interpret its properties. Extents of reaction at gelation constitute the minimal knowledge in this respect. They merely reflect the propensity of a reacting system for pre-gel intramolecular reaction. However, they are useful as they are relatively easily determined compared with other quantities such as N_r, distribution of ring species and modulus and sol fraction as functions of extent of reaction.

The lines in Fig. 5 represent empirical relationships between M_c/AM_c° and $p_{r,c}$. The slopes relate to numbers of inelastic loops formed at complete reaction for given extents of pre-gel intramolecular reaction. For a given value of $p_{r,c}$, more inelastic loops are formed in the triol networks (curves 1 and 2) compared with the tetrol networks (curves 3–6). In addition, as v increases the number of inelastic loops decreases (curves $1 \rightarrow 2$, and curves $3 \rightarrow 6$). Similar trends have been obtained from polyester-forming systems using POP triols and diacid chlorides.[10]

The lines have been extrapolated to $M_c/AM_c^\circ = 1$ at $p_{r,c} = 0$, predicting that perfect networks having affine, small-strain behaviour are formed in the limit of ideal gelling systems. As discussed earlier, this assumption is unlikely to be true as intramolecular reaction must occur post-gel and statistically there is no reason to assume that some smallest ring structures (Fig. 3) do not form. Given equal numbers of opportunities, they are the most readily formed and they must result in elastically ineffective chains. The validity of the limiting prediction has recently been examined in detail[20] and is considered further in the following sections.

EVALUATION OF INELASTIC LOOP FORMATION

Assume for the moment that all the network defects are due to the smallest loops.[8,15,20] The occurrence of such loops in $RA_2 + RB_3$ and $RA_2 + RB_4$ polymerisations excluding sol species is shown in Fig. 6. Structures (a) and (b) show that for trifunctional networks every smallest loop removes two junction points, and structure (c) shows that for tetrafunctional networks only one junction point is lost per loop. Notwithstanding that more complex ring structures will occur, this difference is the basic reason why trifunctional networks are more sensitive to loop defects than tetrafunctional ones (as shown in Fig. 5).

For a perfect network in the dry state, the concentration of elastic chains is given by the equations

$$\rho/M_c^\circ = n^\circ = (f/2)N_B \tag{10}$$

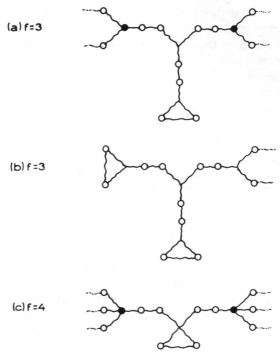

Fig. 6. Smallest loop structures in the gel in, (a) and (b), $RA_2 + RB_3$ polymerisations and, (c), $RA_2 + RB_4$ polymerisations: ●, elastically active junction points; ○, reacted pairs of groups (—AB—).

where N_B is the concentration of RB_f units (junction points) in the $RA_2 +$ RB_f polymerisation. Structures (a) and (b) show that for $f = 3$ the actual concentration will be

$$\rho/M_c = n = (f/2)N_B(1 - 2fp_{r,e}) \tag{11}$$

where $p_{r,e}$ is the extent of intramolecular reaction at the end of the reaction, with $N_B \times f \times p_{r,e}$ the number of intramolecularly reacted B (or A) groups. Hence,

$$M_c/M_c^\circ = 1/(1 - 6p_{r,e}) \tag{12}$$

Equation (12) may be rearranged to allow evaluation of $p_{r,e}$ from measured values of M_c/A with

$$p_{r,e} = \frac{1}{6}\left(1 - \frac{1}{A}(AM_c^\circ/M_c)\right) \tag{13}$$

Similar arguments for $f = 4$ give

$$p_{r,e} = \frac{1}{4}\left(1 - \frac{1}{A}(AM_c^\circ/M_c)\right) \qquad (14)$$

as only one junction point is lost per smallest loop. Although the detailed arguments leading to eqns (13) and (14) are based on smallest loops, the equations themselves are valid provided each inelastic loop causes the loss of two junction points when $f = 3$ and one junction point when $f = 4$, *irrespective* of the size of the loop.

To evaluate $p_{r,e}$, a value of A has to be assumed and used together with the experimental value of M_c/AM_c°. Accordingly, Figs 7 and 8 show $p_{r,e}$ plotted versus $p_{r,c}$ for affine ($A = 1$) and phantom ($A = 1 - 2/f$) chain behaviour using the results in Fig. 5. One condition that should be obeyed in the plots is that $p_{r,e} > p_{r,c}$, as $p_{r,e}$ includes pre-gel and post-gel reaction. The lines $p_{r,e} = p_{r,c}$ are indicated in the figures and it is apparent that this condition is met only for affine behaviour (Fig. 7), as expected for the small-strain measurements used. The condition $p_{r,e} > p_{r,c}$ need not be met if a significant number of complex ring structures are formed pre-gel which then become elastically active when incorporated in the gel. The present interpretation of the plots in Figs 7 and 8 makes the simplifying assumption that all pre-gel ring structures remain elastically inactive in the final network.

In keeping with the discussion on post-gel intramolecular reaction at the end of the previous section, the points in Fig. 7 do not show a tendency for $p_{r,e}$ to tend to zero as $p_{r,c}$ tends to zero. That is, even in the limit of a perfect gelling system, inelastic loops are formed post-gel. Extrapolation to $p_{r,c} = 0$ gives $p_{r,e}^\circ$, the extent of reaction leading to inelastic loops at complete reaction in the perfect gelling system. The values of $p_{r,e}^\circ$ range from about 9% to 18% for the systems studied. As expected from considerations of pre-gel intramolecular reaction, the values of $p_{r,e}^\circ$ are smaller for $f = 3$ compared with $f = 4$, and they increase as v decreases.

The derived values of $p_{r,e}^\circ$ may be reconverted to values of M_c/M_c° for $p_{c,r} = 0$ using eqns (13) and (14). Figure 9 shows M_c/M_c° versus $p_{r,c}$, as in Fig. 5, but with $A = 1$ and the values of M_c/M_c° at $p_{r,c} = 0$ consistent with those of $p_{r,e}^\circ$ from Fig. 7. The curves give just as satisfactory a fit to the data as the straight lines in Fig. 5. Because it is not possible to have concentrations of reactive groups higher than those in bulk, points at lower values of $p_{r,c}$ than those shown cannot be obtained for the particular systems studied. Thus, there are uncertainties in the values of $p_{r,e}^\circ$ from Fig. 7 and in the intercepts shown in Fig. 9. However, merely the existence of non-zero intercepts

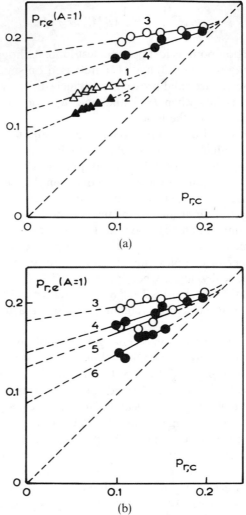

Fig. 7. Extent of intramolecular reaction at complete reaction leading to inelastic loops ($p_{r,e}$) versus $p_{r,c}$: (a) systems 1–4 of Table 1; (b) systems 3–6 of Table 1. Values of $p_{r,e}$ are derived from modulus measurements assuming affine behaviour and each loop leading to two elastically active junction points lost for $f = 3$ networks (systems 1 and 2) and one junction point lost for $f = 4$ networks (systems 3–6). Intercepts at $p_{r,c} = 0$ define values of $p_{r,e}$ for ideal gelling systems, denoted $p_{r,e}^\circ$.

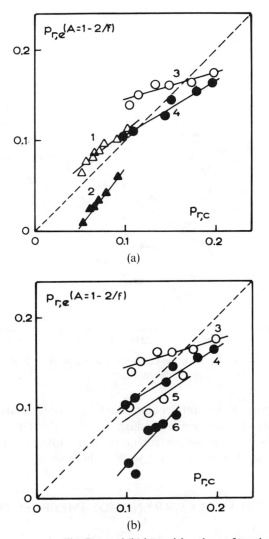

Fig. 8. $p_{r,e}$ versus $p_{r,c}$. As Fig. 7(a) and (b) but with values of $p_{r,e}$ derived assuming phantom chain behaviour.

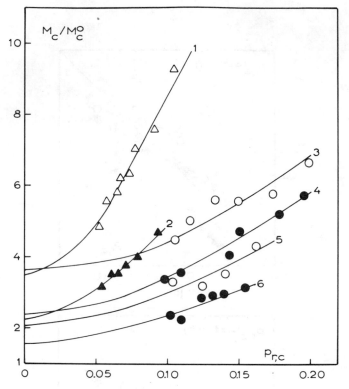

Fig. 9. M_c/M_c° versus $p_{r,c}$ with curves drawn consistent with the values of $p_{r,e}^\circ$ from Fig. 7.

shows that post-gel intramolecular reaction indeed leads to network imperfections. At their present stage of development, theories and calculations of polymerisation including intramolecular reaction[16-19] cannot treat this phenomenon. They all predict that $p_{r,e} = 0$ when $p_{r,c} = 0$.

UNIVERSAL OCCURRENCE OF IMPERFECTIONS

The reaction systems discussed do not have particularly large values of v. Hence, it may be argued that, for triols and tetrols of higher molar mass, networks with $M_c = M_c^\circ$ would have been obtained, at least for bulk reaction systems. The decrease of $p_{r,e}^\circ$ for a given functionality as v increases (Fig. 7) indicates a trend in this direction. However, for $M_c = M_c^\circ$, $p_{r,e}^\circ$ must be equal to zero. That such a value is unlikely to be obtained, irrespective of v, is emphasised by using the present values of $p_{r,e}^{\circ}$ to predict a value for infinite v. Thus, Fig. 10 shows $p_{r,e}^{\circ}$ versus $(f-2)/(vb^2)^{3/2}$. The abscissa is proportional to c_{int} of eqn (5) and relates to the probability of pre-gel intramolecular reaction. The data used in Fig. 10 are given in Table 2, together with the extrapolated values $(p_{r,e}^{0,\infty})$. The values of b are those from Table 1 with $c_{ext} = c_{a0} + c_{b0}$. If the values of b with $c_{ext} = c_{ac} + c_{bc}$ from Table 1 are used, then linear behaviour is again found with approximately the same intercepts at $(f-2)/(vb^2)^{3/2} = 0$. The last two columns in Table 2 give the intercept values of M_c/M_c° in Fig. 9 and the corresponding values of shear moduli of the dry networks relative to those of the perfect networks. The values of G/G° for $v = \infty$ are estimates of the maximum values obtainable for model end-linking networks formed from reactants of high molar mass. It is clear that significant reductions in modulus will remain.

The positive intercepts in Fig. 10 show that post-gel (inelastic) loop formation is influenced by the same factors as pre-gel intramolecular reaction but is not determined solely by them. The important conclusion is that imperfections still occur in the limit of infinite reactant molar masses or very stiff chains $(vb^2 \to \infty)$. They are a demonstration of a law-of-mass-action effect. Because they are intercepts in the limit $vb^2 \to \infty$, spatial correlations between reacting groups are absent and random reaction occurs. Intramolecular reaction occurs post-gel simply because of the unlimited number of groups per molecule in the gel fraction. The present

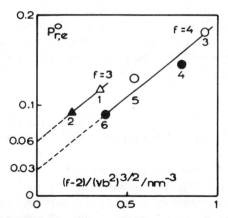

Fig. 10. Dependences of post-gel intramolecular reaction leading to inelastic loops in the limit of ideal gelling systems $(p_{r,e}^{\circ})$ on the parameters affecting pre-gel intramolecular reaction $((f-2)/(vb^2)^{3/2})$.

TABLE 2

Values of parameters Characterising Pre-gel Intramolecular Reaction $(v, b, (f-2)/(vb^2)^{3/2})$ and the Extents of Post-gel Intramolecular Reaction in the Limit of Ideal Gelling Systems which Lead to Inelastic Loop Formation at Complete Reaction $(p_{r.e}^{\circ})^a$

System	v	$b(i)$ (nm)	$(f-2)/(vb^2)^{3/2}$ (nm^3)	$p_{r.e}^{\circ} \rightarrow M_c/M_c^{\circ} \rightarrow G/G^{\circ}$		
$f = 3$						
(1) HDI/LHT240	33	0·247	0·350	0·119	3·50	0·29
(2) HDI/LHT112	61	0·222	0·192	0·092	2·23	0·45
	∞		0	0·060	1·57	0·64
$f = 4$						
(3) HDI/OPPE1	29	0·240	0·926	0·181	3·62	0·28
(4) HDI/OPPE2	33	0·237	0·792	0·145	2·38	0·42
(5) HDI/OPPE3	44	0·234	0·534	0·129	2·07	0·48
(6) HDI/OPPE4	66	0·215	0·376	0·090	1·56	0·64
	∞		0	0·030	1·14	0·88

^{a} The values of $p_{r.e}^{\circ}$ define the indicated values of M_c/M_c° and the reductions in shear moduli of dry networks relative to those of the perfect networks $(G/G^{\circ} = M_c^{\circ}/M_c)$. The values of $p_{r.e}^{\circ}$ in the limit of reactants of infinite molar mass $(v = \infty)$ are noted by $p_{r.e}^{\circ,\infty}$ in the text.

values of $p_{r.e}^{\circ,\infty}$ (0·06 for $f = 3$ and 0·03 for $f = 4$) are derived from modulus measurements, assuming two junction points lost per inelastic loop in $f = 3$ networks and one junction point lost per loop in $f = 4$ networks. The assumption is strictly correct if only the smallest loops are formed. However, as the intercepts relate to random reaction, classical polymerisation theory may be used to investigate whether they can be consistent with defects arising only from simple loops.

SIGNIFICANT STRUCTURES FOR INELASTIC LOOP FORMATION UNDER CONDITIONS OF RANDOM REACTION

The conditions of random reaction mean that intramolecular reaction occurs only post-gel. This corresponds exactly to the condition treated by Flory,[39] whereby intramolecular reaction occurs only between groups on the gel. Accordingly, various types of pairs of reacting groups may be defined, namely sol–sol, sol–gel and gel–gel. Only the last can lead to inelastic loops. Hence, the first task is to find the total amount of gel–gel reaction which occurs. This is evaluated for an RA_f polymerisation. The equations are simpler than those for an $RA_2 + RB_f$ polymerisation and to within the accuracy required here the expressions derived may be used for an $RA_2 + RB_f$ polymerisation after a suitable transformation.

The number of x-mers in the sol (N_{xs}) is given by the equation

$$N_{xs} = N_f \omega'_x ((1-p)^2/p) \beta^x \tag{15}$$

where

$$\beta = p(1-p)^{f-2} = p^*(1-p^*)^{f-2} \tag{16}$$

and

$$\omega'_x = f((f-1)x)!/x!((f-2)x+2)! \tag{17}$$

p is the extent of reaction of A groups and p^* is the smallest root of eqn (16) for a given value of β. For the pre-gel regime, $p \le p_c = 1/(f-1), p = p^*$. The number of unreacted groups on x-mers in the sol (Γ_{xs}) is simply equal to $N_{xs}(fx - 2x + 2)$. Hence, the number of unreacted groups on sol molecules (Γ_s) is given by the equation

$$\Gamma_s = \sum_1^\infty N_{xs}(fx - 2x + 2) \tag{18}$$

From eqn (15),

$$\Gamma_s = N_f((1-p)^2/p) \times ((f-2)\sum x\omega'_x\beta^x + 2\sum \omega'_x\beta^x) \tag{19}$$

and standard evaluation of the summations[39] gives

$$\Gamma_s = N_f((1-p)^2/p) \times (p^*/(1-p^*)^2) \times ((f-2) + 2(1 - p^*f/2)) \tag{20}$$

or

$$\Gamma_s = fN_f(1-p)((1-p)/p) \times (p^*/(1-p^*)) \tag{21}$$

The total number of unreacted groups on sol and gel is simply

$$\Gamma = fN_f(1-p) \tag{22}$$

and the number of unreacted groups on the gel (Γ_g) may be found as the difference $\Gamma - \Gamma_s$. The probability at extent of reaction p that a randomly chosen unreacted group is on the gel is given by

$$\gamma_g = \Gamma_g/\Gamma = 1 - \Gamma_s/\Gamma \tag{23}$$

and from eqns (21) and (22)

$$\gamma_g = 1 - ((1-p)/p)(p^*/(1-p^*)) \tag{24}$$

Hence, given random reaction, the number of gel–gel pairs of groups reacting at p, i.e. the number of ring structures (loops) forming at p, is

$$\Gamma_{gg} = \Gamma \times \gamma_g \times \gamma_g = fN_f(1-p)\gamma_g^2 \tag{25}$$

Equation (25) may be normalised to give the number of loops forming at p per initial number of groups

$$p_r = (1-p)\gamma_g^2 \tag{26}$$

The total number of loops formed by the end reaction is given by

$$\Gamma_{gg,e} = \int_{1/(f-1)}^{1} \Gamma_{gg}\, dp = fN_f \int_{1/(f-1)}^{1} p_r\, dp \tag{27}$$

$$= fN_f \times p_{r,e} \tag{28}$$

where $p_{r,e}$ represents the total extent of intramolecular reaction between gel and complete reaction. Considering the whole polymerisation, the same reasoning as that leading to eqn (27) shows that the total number of pairs of groups reacted is

$$\Gamma_e = \int_0^1 fN_f(1-p)(\gamma_g^2 + 2\gamma_g\gamma_s + \gamma_s^2)\, dp = fN_f/2 \tag{29}$$

$\gamma_s = 1 - \gamma_g$ is the fraction of groups on the sol and the terms in γ_g and γ_s represent the changing binomial distribution of randomly reacting pairs between gel and sol.

Figure 11 shows p_r vs. p for RA_3 and RA_4 polymerisations. The change with functionality arises from the different values of p^* for a given value of p. The integrals under the curves give $p_{r,e} = 0.0604$ for $f = 3$ and 0.1306 for $f = 4$. For a stoichiometric $RA_2 + RB_f$ polymerisation, $p_{r,e}^{\circ,\infty}$ may be equated approximately to $p_{r,e}$ for the RA_f polymerisation. The former is the (number of loops/initial number of A (or B) groups) and the latter is the (number of loops/initial number of groups). The approximation follows[40] from the statistical equivalence between $(p_a p_b)$ in the $RA_2 + RB_f$ case for branched species and p in the RA_f case, the fact that rings in the $RA_2 + RB_f$ case contain $2, 4, 6, \ldots$ A (or B) groups as in the RA_f case, and the assumption that the gel molecules have equal numbers of A and B groups. It is accurate to about 10%.

The calculated values of $p_{r,e}^{\circ,\infty}$ can now be compared with the experimentally deduced values in Fig. 10. For $f = 3$, the values are equal (0.06) and imply that random gel–gel reaction leads to an average of two elastically active junctions or three elastically active chains lost per pair of

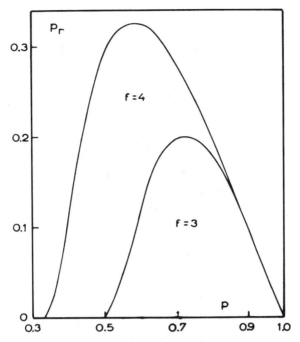

Fig. 11. Number of gel–gel reacting pairs per initial number of groups (p_r) versus extent of reaction (p) in RA_3 and RA_4 polymerisations. The integrals under the curves give $p_{r,e} = 0.0604$ (RA_3) and 0.1306 (RA_4).

gel–gel groups reacted. For $f = 4$, the experimental value of 0.03 and calculated value of 0.13 indicate an average of $0.03/0.13 = 0.23$ elastically active junctions or 0.46 elastically active chains lost per pair of gel–gel groups reacted. The agreement between the experimental and calculated values of $p_{r,e}^{\circ,\infty}$ for $f = 3$ and the relatively large ratio of 0.23 for $f = 4$ indicates that the sizes of loop which need to be considered for inelastic loop formation are indeed small.

Figure 12 shows two-membered (the second smallest) loops in RA_3 and RA_4 polymerisations, together with the numbers of elastic junction points and elastic chains lost. The double chains formed are treated as single elastic chains, which is correct on an entropic but not strictly so on a free-energy basis. The loops in Fig. 12 together with one-membered loops (cf. Fig. 6) are taken as an approximation to the network defects which occur. Larger loops contain elastically active junction points and the numbers of definitely inelastic chains are difficult to assign. The question now arises,

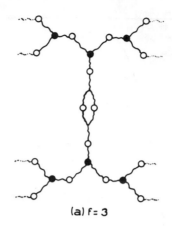

(a) f = 3

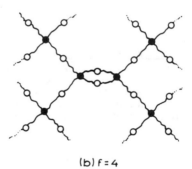

(b) f = 4

Fig. 12. Two-membered loop structures in (a) an RA_3 network and (b) an RA_4 network: ●, elastically active junction points; ○, reacted pairs of groups (—AA—). RA_3—two elastically active junctions or $2 \times (f/2) = 3$ elastically active chains lost per loop. RA_4—$2 \times f = 4$ junctions transform to $2 \times f = 3$ junctions so that one elastically active chain is lost per loop. One-membered loop structures may be deduced from Fig. 6 if the structure —○—○—(—BA—AB—) is reduced to —○—, denoting —AA—.

what are the probabilities of forming such structures given random reaction? Of the many isomeric structures in the gel which have unreacted groups, two extreme types may be delineated regarding the probabilities for forming one-membered loops, namely, linear and symmetric isomers. They are illustrated in Fig. 13 for the RA_3 case. Linear isomers can form the smallest number of one-membered loops and symmetric isomers the largest

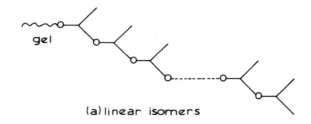

(a) linear isomers

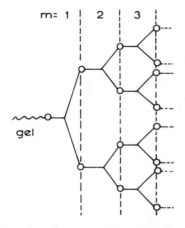

(b) symmetric isomers

Fig. 13. Isomeric structures in RA_3 gels: (a) linear; and (b) symmetric. Each contains n units, divided, in the symmetric case, into m generations.

number. For linear isomers of n units, the total number of pairs for intramolecular reaction is

$$(f-2)((f-1)(n-\tfrac{1}{2})+(f-2)(n-1)(n-2)/2+(f-3)(n-1)/2), \quad n>1 \tag{30}$$

or

$$\binom{f-1}{2}, \quad n=1$$

The number of pairs which can form one-membered loops is

$$\binom{f-1}{2}+(n-1)\binom{f-2}{2} \tag{31}$$

and the number for two-membered loops

$$(f-1)(f-2) + (n-2)(f-2)^2, \quad n > 1 \tag{32}$$

For symmetric isomers of m generations or $((f-1)^m - 1)/(f-2)$ units, the total number of pairs is

$$(f-1)^m((f-1)^m - 1)/2 \tag{33}$$

and the number for one-membered loop formation is

$$\binom{f-1}{2}(f-1)^{m-1} \tag{34}$$

Two-membered loops cannot form.

Expressions (30)–(34) can now be used to estimate the fractions of one- and two-membered loops which form from the two types of isomer for different values of n and m, i.e. assuming, on average, equal populations of the linear and isomeric structures between gel point and complete reaction, the fractions of gel–gel reaction leading to inelastic loops. Comparison with ratios of experimental and calculated values of $p_{r,e}^{\circ,\infty}$ can then be made, remembering that the experimental values $((p_{r,e}^{\circ,\infty})_{\text{exptl}})$ were evaluated on the basis of one loop leading to two junctions lost for $f=3$ and one loop leading to one junction lost for $f=4$, and the calculated values $((p_{r,e}^{\circ,\infty})_{\text{calc}})$ gave the total amounts of intramolecular reaction. From Figs 6 and 12 the experimental basis for $f=3$ corresponds directly to the fractions of one- and two-membered loops. For $f=4$, the number of junctions lost is one for one-membered loops and $\frac{1}{2}$ ($=$ one chain) for two-membered loops. The various fractions of one- and two-membered loops are given in Table 3. The fractions in the final columns give the values to be compared directly with the ratios $(p_{r,e}^{\circ,\infty})_{\text{exptl}}/(p_{r,e}^{\circ,\infty})_{\text{calc}}$, namely 1 for $f=3$ and 0·23 for $f=4$. The values of n which correspond to these values give the sizes of structures in the gel within which all intramolecular reaction has to be considered to give the experimentally deduced concentrations of inelastic junction points or chains. The structures are illustrated in Fig. 14. They are proposed as the minimum ones within which probabilities of intramolecular reaction have to be evaluated by theories if network defects are to be accounted for and absolute values of moduli or concentrations of elastic chains predicted. Intramolecular reaction within these structures is significant even in the limit of reactants of infinite molar mass ($v \to \infty$). For actual systems, finite values of v, the numbers of inelastic loops actually forming will depend on the detailed chain statistics within the structures. The structures should be

TABLE 3

Total Numbers of Loops and Numbers and Fractions of One- and Two-membered Loops Formed by Random Intramolecular Reaction within Linear and Isomeric Gel Structures of Different Numbers of Units (n)

n	Total number	$1-$	$2-$	Fraction $1-$	Fraction $2-$	Fraction $(1- + 2-)$
Linear isomers, $f = 3$						
1	1	1	0	1	0	1[a]
2	3	1	2	0·33	0·67	1[a]
3	6	1	3	0·17	0·50	0·67

m	n	Total number	$1-$	Fraction $1-$
Symmetric isomers, $f = 3$				
1	1	1	1	1[a]
2	3	6	2	0·33

n	Total number	$1-$	$2-$	Fraction $1-$	$\frac{1}{2}$(Fraction $2-$)	(Fraction $1-$)+ $\frac{1}{2}$(fraction $2-$)
Linear isomers, $f = 4$						
1	3	3	0	1	0	1
2	10	4	6	0·40	0·30	0·70
3	21	5	10	0·24	0·24	0·48
4	36	6	14	0·17	0·19	0·36
5	55	7	18	0·13	0·16	0·29
6	78	8	22	0·10	0·14	0·24[a]
7	105	9	26	0·09	0·12	0·21[a]
8	136	10	30	0·07	0·11	0·18

m	n	Total number	$1-$	Fraction $1-$
Symmetric isomers, $f = 4$				
1	1	3	3	1
2	4	36	9	0·25[a]
3	13	351	27	0·08

[a] These fractions agree with the experimentally deduced concentrations of inelastic junction points or chains on the basis of one- and two-membered loops. Symmetrical isomers cannot form two-membered loops.

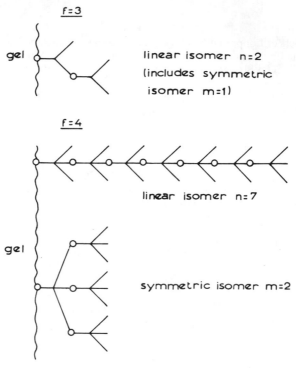

Fig. 14. Proposed significant structures for the formation of inelastic chains in trifunctional and tetrafunctional networks. For $f = 3$, the linear isomer $n = 2$ *includes* the linear isomer $n = 1$ and the symmetric isomer $m = 1$ as its terminal unit. For $f = 4$, the linear isomer $m = 6$ is *included* in the linear isomer $n = 7$ shown.

considered in the sol and the gel fractions and changes in their concentrations with extent of reaction also included.

CONCLUSIONS

Intramolecular reaction always occurs in non-linear polymerisations and because of this end-linking polymerisations never lead to perfect networks. The amount of pre-gel intramolecular reaction may be small for reactants of high molar mass. However, significant post-gel intramolecular reaction always occurs. In the actual reaction systems discussed here, intramolecular reaction from both sources leads to large reductions in shear moduli below the values expected for perfect networks.

With end-linking polymerisations, the monitoring of network formation may easily be achieved through the determination of gel points. Such monitoring is important for the interpretation of networks structure and properties. For example, the systematic variation of the gel point by varying the dilution of given reactants is particularly useful. Reductions in modulus are found to depend on the same factors as pre-gel intra-molecular reaction. They increase similarly with dilution and with decrease in reactant molar masses. However, they are less for networks of higher functionality, whereas intramolecular reaction is higher. This distinction is understandable, simply in terms of numbers of elastic chains lost per one-membered and two-membered loop.

The absolute values of the reductions in moduli, or increases in M_c, can be interpreted in terms of small inelastic loops. The small-strain static moduli measured are consistent with affine chain behaviour, showing, on the basis of one-membered loops, that between about 10 and 20% of groups react to form inelastic loops by the end of a polymerisation. For the more concentrated systems of lower molar mass, a significant proportion of this amount comes from post-gel intramolecular reaction ($p_{r,e} - p_{r,c}$).

Extrapolation of $p_{r,e}$ to the limit of zero pre-gel intramolecular reaction for given reaction systems shows that post-gel intramolecular reaction always results in network defects, with significant increases in M_c above M_c°. Such post-gel intramolecular reaction is characterised as $p_{r,e}^\circ$. The variation of $p_{r,e}^\circ$ with intramolecular-reaction parameters shows that even in the limit of infinite molar mass, i.e. no spatial correlation between reacting groups, inelastic loops will be formed. The formation may be considered as a law-of-mass-action effect, essentially the random reaction of functional groups. Intramolecular reaction under such conditions ($p_{r,e}^{\circ,\infty}$) must be post-gel and may be treated using classical polymerisation theory. The case of RA_f polymerisations has been evaluated and the experiment-ally deduced values of $p_{r,e}^{\circ,\infty}$ interpreted in terms of one- and two-membered loop formation within structures of defined size. The knowledge of such structures is important for theories of non-linear polymerisation and network formation. Theories have to simplify in some way the infinite numbers of structures which actually occur and it is important that those significant for intramolecular reaction are retained.

REFERENCES

1. Treloar, L. R. G., *The Physics of Rubber Elasticity*, 3rd edn, Oxford University Press, London, 1975.
2. Mark, J. E., *Adv. Polymer Sci.*, 1982, **44**, 1.

3. Flory, P. J., *Br. Polymer J.*, 1985, **17**, 96.
4. Erman, B., *Br. Polymer J.*, 1985, **17**, 140.
5. Brotzman, R. W. and Flory, P. J., *Polymer Preprints*, 1985, **26**, 51.
6. Gottlieb, M., Macosko, C. W., Benjamin, G. S., Meyers, K. O. and Merrill, E. W., *Macromolecules*, 1981, **14**, 1039.
7. Batsberg, W. and Kramer, O., in *Elastomers and Rubber Elasticity*, Amer. Chem. Soc. Symposium Series Vol. 193, Mark, J. E. and Lal, J. (Eds), American Chemical Society, Washington, DC, 1982, Chapter 23.
8. Ross-Murphy, S. B. and Stepto, R. F. T., in *Cyclic Polymers*, Semlyen, J. A. (Ed.), Elsevier Applied Science Publishers, London, 1986, p. 349.
9. Stepto, R. F. T., *Polymer*, 1979, **20**, 1324.
10. Fasina, A. B. and Stepto, R. F. T., *Makromol. Chem.*, 1981, **182**, 2479.
11. Stanford, J. L. and Stepto, R. F. T., in *Elastomers and Rubber Elasticity*, Amer. Chem. Soc. Symposium Series Vol. 193, Mark, J. E. and Lal, J. (Eds), American Chemical Society, Washington, DC, 1982, Chapter 20.
12. Stanford, J. L., Stepto, R. F. T. and Still, R. H., in *Reaction Injection Molding and Fast Polymerization Reactions*, Kresta, J. E. (Ed.), Plenum Publishing Corp., New York, 1982, p. 31.
13. Stanford, J. L., Stepto, R. F. T. and Still, R. H., in *Characterization of Highly Crosslinked Polymers*, Amer. Chem. Soc. Symposium Series Vol. 243, Dickie, R. A. and Labana, S. S. (Eds), American Chemical Society, Washington, DC, 1984, Chapter 1.
14. Macosko, C. W. and Saam, J. C., *Polymer Preprints*, 1985, **26**, 48.
15. Stepto, R. F. T., in *Developments in Polymerisation—3*, Haward, R. N. (Ed.), Elsevier Applied Science Publishers, London, 1982, p. 81.
16. Temple, W. B., *Makromol. Chem.*, 1972, **160**, 277.
17. Cawse, J. L., Stanford, J. L. and Stepto, R. F. T., *Proc. 26th IUPAC Int. Symposium on Macromolecules*, Mainz, 1979, p. 693.
18. Lloyd, A. C. and Stepto, R. F. T., *Br. Polymer J.*, 1985, **17**, 190.
19. Shy, L. Y. and Eichinger, B. E., *Br. Polymer J.*, 1985, **17**, 200.
20. Stepto, R. F. T., in *Advances in Elastomers and Rubber Elasticity*, Lal, J. and Mark, J. E. (Eds), Plenum Publishing Corp., New York, 1986, p. 329.
21. Stepto, R. F. T. and Waywell, D. R., *Makromol. Chem.*, 1972, **152**, 263.
22. Stanford, J. L. and Stepto, R. F. T., *Br. Polymer J.*, 1977, **9**, 124.
23. Hopkins, W., Peters, R. H. and Stepto, R. F. T., *Polymer*, 1974, **15**, 315.
24. Smith, R. S. and Stepto, R. F. T., *Makromol. Chem.*, 1974, **175**, 2365.
25. Stepto, R. F. T., *Faraday Disc. Chem. Soc.*, 1974, **57**, 69.
26. Ahmad, Z. and Stepto, R. F. T., *Colloid and Polymer Sci.*, 1980, **258**, 663.
27. Ahmad, Z., Stepto, R. F. T. and Still, R. H., *Br. Polymer J.*, 1985, **17**, 205.
28. Jacobson, H. and Stockmayer, W. H., *J. Chem. Phys.*, 1950, **18**, 1600.
29. Stanford, J. L., Stepto, R. F. T. and Waywell, D. R., *J. Chem. Soc., Faraday Trans I*, 1975, **71**, 1308.
30. Gordon, M. and Temple, W. B., *Makromol. Chem.*, 1972, **152**, 277.
31. Gordon, M. and Temple, W. B., *Makromol. Chem.*, 1972, **160**, 263.
32. Temple, W. B., *Makromol. Chem.*, 1972, **160**, 277.
33. Cawse, J. L., Stanford, J. L. and Stepto, R. F. T., *Proc. 26th IUPAC Int. Symposium on Macromolecules*, Main, 1979, p. 693.
34. Askitopoulos, V., MSc Thesis, University of Manchester, UK, 1981.

35. Lloyd, A. C. and Stepto, R. F. T., *Br. Polymer J.*, 1985, **17**, 190.
36. Shy, L. Y. and Eichinger, B. E., *Br. Polymer J.*, 1985, **17**, 200.
37. Stockmayer, W. H., *J. Polymer Sci.*, 1952, **9**, 69; 1953, **11**, 424.
38. Demirörs, M., PhD Thesis, University of Manchester, UK, 1985.
39. Flory, P. J., *Principles of Polymer Chemistry*, Cornell University Press, Ithaca, New York, 1953.
40. Stepto, R. F. T., unpublished work.

11

THE PHYSICS OF TEMPORARY POLYMER NETWORKS: A COMPARISON OF THEORY AND EXPERIMENT

E. Kröner, D. Chassapis and R. Takserman-Krozer

Institut für Theoretische und Angewandte Physik der Universität Stuttgart und Max-Planck-Institut für Metallforschung, Pfaffenwaldring 57/VI, 7000 Stuttgart 80, FRG

ABSTRACT

Compared with permanent networks, the temporary polymer networks in solution show additional mobility in the form of viscoelasticity. This mobility results from the kinetic processes of decay and formation of junctions. The molecular-statistical theory of Takserman-Krozer and Kröner gives the viscoelastic material functions (of the velocity gradient tensor) within a generalized spring–bead model where the springs represent the network chains and the beads represent the junctions which are not conserved. The (integro-differential) diffusion equation contains the transition probabilities for junction decay and formation. The equations for these are solved in the one-junction approximation, simultaneously with the diffusion equation in the relaxation time approach. The material functions thus obtained are compared with various experiments, above all those on stationary shear flow. Adaptation of the theoretical to the experimental curves occurs only in the Newtonian range so that the non-Newtonian part of the theoretical curves represents a true prediction. Further results concern the mean number of chains per macromolecule (up to 10) and the number of decays per second of a junction (e.g. $10^{-2}/s$ for the fluid at rest and 65 at a velocity gradient of $10^3/s$).

1 INTRODUCTION

The subject of this chapter is the physics of temporary polymer networks, in particular those in solution. Such networks possess junctions which decay and form in certain time intervals. Therefore, as compared with permanent networks, the temporary networks have additional mobility which is macroscopically observed as viscoelasticity. The aim is to derive the pertaining non-Newtonian material functions by means of a molecular-statistical theory which uses the molecular parameters and gives results which depend on the ambient conditions (temperature and concentration). The parameters involved are:

η_s — (Newtonian) viscosity of solution when polymer bonds are removed (no more chains, but monomers still present). η_s is usually negligibly small

l_0 — Length of monomer

l — Elementary bond length

ψ — Bond angle

σ — Structure factor

ε_D — Energy of dissolution of a junction

ε — Energy of *trans* state relative to *gauche* state

$\mathcal{M}$ — Molecular weight of polymer

$\mathcal{M}_{mon}$ — Molecular weight of monomer

For illustration and comparison with experiment we deal with four-functional networks, because these occur frequently. Other networks can be treated analogously. The junctions may represent chemical or physical (e.g. van der Waals) binding. We believe that, after some modification, even networks whose 'junctions' are fluctuating little crystallites can be included. In our comparison with experiments we use networks of the physical binding type.

We mention here a peculiar difficulty which arises if purely four-functional (or s-functional, s being any integer) networks are considered. Real networks are built from molecules which usually possess two open ends, besides consisting of (network) chains connecting two junctions. The open ends do not contribute to the elasticity of the network, but they have a friction effect and also play a role in the formation of the networks. To keep our model as simple as possible without giving up the essential physics, we shall include the friction of the open ends in an effective friction constant ζ, which also contains the friction effect of entanglements possibly present. This should be a reasonable procedure, since we are interested here in the viscoelasticity due to the decay and formation of junctions. Furthermore, we take into account the free ends when the formation of the network is investigated, above all when the number of junctions is calculated. We neglect the existence of the free ends where the elasticity of the network is concerned. Of course, this model can later be improved if so desired. In such an improved model one may also wish to include other topological pecularities such as those classified by Ziabicki.[1] For reasons of simplicity we renounce such a refinement at the present time.

A rigorous statistical treatment of all monomeric motions is beyond present day possibilities. We therefore reduce the number of atomic degrees of freedom of our polymeric system by introducing a spring–bead model, where the beads represent the network junctions and, therefore, are not conserved. The (massless) springs take care of the chain elasticity which varies from one chain to the other, because the chain lengths vary. This model is a direct generalization of the model used by Rouse[2] and Zimm;[3] see also Yamakawa.[4]

The beads are labelled by numbers $i(=1,2\ldots M)$, where M is the total number of junctions in the network system. The chains connecting two junctions i and j are marked by (ij). They are $2M$ in number.

The elastic spring constant k_{ij} of the spring (ij) is taken from the small-scale statistics of Volkenstein[5] and Flory:[6]

$$k_{ij} = \frac{\alpha}{z_{ij}}, \quad \alpha \equiv \frac{3k\theta}{ll_0} \times \frac{1}{\sigma} \times \frac{1-\cos\psi}{1+\cos\psi} \tag{1}$$

where z_{ij} denotes the number of monomers of the chain (ij), i.e. z_{ij} is the chain contour length in units of l_0, k is Boltzmann's constant and θ is the absolute temperature. Equation (1) is considered as a good approximation as long as the chains are not 'almost stretched'. According to eqn (1), the constant α can be considered as given, if the species of polymer is known.

Although the spring–bead model implies a large reduction of the problem, it is still immensely complex and needs further simplifications, which will be explained in the text. Due to the limitation of space and time, equations are often not derived but only explained qualitatively. Much of the quantitative description is found in Refs 7, 8, 9 and 10.

Two very different problems have to be solved. The first problem concerns the partially oscillatory motion of the spring–bead network in solution. It is solved essentially by the transition to normal co-ordinates and requires the diagonalization of the $(M \times M)$ elastic matrix specifying the chain network. The diagonalization can be done only numerically. In fact, the elastic constants are determined by the contour lengths z_{ij} (see eqn (1)) but, unlike in Rouse's and Zimm's theory,[2,3] these vary from one chain to the other which causes the complication.

The necessity to work in normal co-ordinates reflects the fact that this problem is of global nature, i.e. the oscillations comprise the whole network. The decay or formation of a junction, on the other hand, is a rather local process which has to be described in local co-ordinates. Keeping this in mind we develop the following picture of the physical situation.

During the motions of the network, elastic forces are transmitted through the chains and pull at the junctions with fluctuating strength. Thus the stress in the network must be known in order to calculate the required transition probabilities for the decay and formation of the junctions. On the other hand, the stress in the network cannot be calculated without knowing the transition probabilities, because the stress, as a response to the flow of the fluid, clearly depends on the rate of the kinetic processes. Thus we have a strong coupling between the global problem of calculating the stress and the local problem of calculating the transition probabilities. Whereas the best adapted co-ordinates for the first problem are the (global) normal co-ordinates, the second problem is best treated in the (local) position co-ordinates. It is the very coupling between global and local quantities which makes the theory so complex.

2 TEMPORARY POLYMER NETWORK IN EQUILIBRIUM

Let us first specify our thermodynamic-statistical system in the situation of equilibrium. Consider an almost infinitely extended (volume V_∞), macroscopically homogeneous, four-functional polymer network in solution. Imagine a fixed large partial volume V of V_∞ such that $V \ll V_\infty$. The physical system treated in the present theory is the portion of polymer in V. This system is taken as open so that besides solvent molecules also monomers and junctions can pass through the boundary walls of V. The solvent in V and the polymer plus solvent outside V form the heat and particle bath for our open system.

To specify the equilibrium state of the system, we can use the extensive variables S (entropy in V), L (number of monomers in V) and M (number of junctions in V). The pertaining intensive variables are then θ (temperature), λ (chemical potential of a monomer) and μ (chemical potential of a junction, a negative quantity). L can be replaced by the mass concentration c (portion of polymer in g/cm^3), according to the equation

$$c = \bar{L} m_{mon}/V = \bar{L} \mathcal{M}_{mon}/N_A V \tag{2}$$

with N_A the Avogadro–Loschmidt number and m_{mon} the monomer mass. Here and later the overbar denotes ensemble averages. Similarly, $\bar{M}$ can be replaced by $n = \bar{M}/V$, the number density of junctions. Note that eqn (2) is valid if L includes the monomers of the open ends. We shall later also denote by L the number of monomers resulting when the free end chains are neglected. We shall then speak of the 'L-inconsistency'. This should be a reasonable approximation as long as the density of junctions is sufficiently large.

From the postulate of equal *a priori* probabilities we conclude that the networks of the statistical ensemble (all with volume V) which have equal energy (E), number of monomers (L) and junctions (M) are equally probable. We specify the kinetic and potential ($=$elastic) energy by[8]

$$E_{\text{kin}} = \frac{1}{2\alpha m_{\text{mon}}} \sum_{i,j}^{M} \kappa_{ij} \mathbf{p}_i \cdot \mathbf{p}_j, \quad E_{\text{pot}} = \frac{1}{2} \sum_{i,j}^{M} \kappa_{ij} \mathbf{r}_i \cdot \mathbf{r}_j \quad (3)$$

where $\mathbf{p}_i$ are the position vectors of the momentum part and $\mathbf{r}_i$ the position vectors of the configuration part of the phase space. The elasticity matrix (κ_{ij}) is connected with the matrix (k_{ij}) of the spring constants by

$$\kappa_{ij} = \sum_{j}^{M} k_{ij} \quad (i = j), \qquad \kappa_{ij} = -k_{ij} \quad (i \neq j) \quad (4)$$

Thus also κ_{ij} is determined by z_{ij}. Note that L, too, can be replaced, namely by $L = \sum_{(ij)}^{2M} z_{ij}$, where the sum goes over all chains (ij). This replacement holds within the L-inconsistency.

The grand-canonical probability density function describing the statistical state of the ensemble is now easily obtained in the form[8,10]

$$W_{\text{eq}} = W_{\text{eq}}(\mathbf{P}, \mathbf{R}, Z, M), \quad \mathbf{P} \equiv \{\mathbf{p}_i\}, \quad \mathbf{R} \equiv \{\mathbf{r}_i\}, \quad Z \equiv \{z_{ij}\} \quad (5)$$

In principle, the grand partition function $C(\theta, \lambda, \mu)$ follows from eqn (5) by integration $(\mathbf{P}, \mathbf{R})$ and summation (Z, M) over the phase space. Whereas the integration and the first summation can be done in closed form, this is not so with the summation concerning M, so that at present $C(\theta, \lambda, \mu)$ is not known analytically. This means that also the grand thermodynamic potential $G(\theta, \lambda, \mu)$, from which the three equations of state follow by

$$S = \partial G/\partial T, \quad L = \partial G/\partial \lambda, \quad M = \partial G/\partial \mu \quad (6)$$

is not known. Nevertheless, it is possible to get closer to two equations of state. In Ref. 7 we have described the decay of a junction as a first-order reaction and the formation as a second-order reaction. Making, then, extensive use of results of Flory and co-workers,[11-14] the density in equilibrium $n = \bar{M}/V$ was calculated and the result rewritten as a quartic equation for L in Ref. 9. Note that with eqn (2) $n = \bar{M} N_A c / \mathcal{M}_{mon} \bar{L}$. Observing that the theory is good only for mean contour lengths $\bar{z} = \bar{L}/2\bar{M} \gg 1$, the quartic can be replaced by a quadratic equation with solution

$$n = n(\theta, c, \mu) \equiv \frac{N_A c}{2 \mathcal{M}(\sqrt{(2c/A(\theta, \mu))} - 1)} \tag{7}$$

This equation can also be derived directly from Ref. 7. The physical content of eqn (7) is the same as that of eqns (42) and (43) of Ref. 9.

Obviously, eqn (7) has the quality of an equation of state. Note that we can give $\bar{M}$ the value of an arbitrary integer, if we adjust the system volume V accordingly.

The function $A(\theta, \mu)$, which does not depend on the concentration c, is found in eqn (25) of Ref. 7 or in eqn (43) of Ref. 9 (see also Ref. 10, app. 7). A contains, besides θ and μ, Flory's lattice coordination number, the effective volume of a junction and the so-called front factor, all quantities which are not well known. We therefore take A from experiment for each species of polymer at given θ and μ (see Section 4). Hence we take A in eqn (7) as given.

Let us now draw some conclusions from eqn (7). Obviously, this equation becomes invalid when c decreases to $A/2$. To see this, let us rewrite eqn (7) in the form

$$\bar{z} = (\mathcal{M}/\mathcal{M}_{mon})(\sqrt{(2c/A(\theta, \mu))} - 1) \tag{8}$$

which follows with eqn (2) and $\bar{z} = \bar{L}/2\bar{M}$. In fact, $\bar{z}$ must be a positive number, but not too small, in order to have sufficiently flexible chains. Note that small $\bar{z}$ implies a large number of chains per macromolecule, say $\bar{j}$. On the other hand, when c increases then $\bar{j}$ decreases, so that finally no coherent network exists anymore. This is a basic, perhaps surprising, result of the theory: the average number of chains per macromolecule decreases with concentration. This prediction is valid in the concentration range to which eqn (7) applies, i.e. in a middle concentration range. Perhaps this situation may be compared with a gas which, when dense, is more ionized.

The second equation of state can be obtained as follows. Suppose that within a finite time interval any chain assumes in succession all contour lengths which at a given time are found in the network. Then the probability to find a contour length z in the interval dz around z, when a chain in the network is picked out at random, is $W_{eq}(z) dz$ and

$$W_{eq}(z) = \tfrac{1}{6} a^4 z^3 \exp\{-az\}, \qquad a \equiv \ln \bar{z} - \frac{\lambda}{k\theta} \tag{9}$$

For a derivation of this formula see Ref. 10 and also Ref. 8. Equation (9) has the formal appearance of Wien's radiation formula. The mean $\bar{z}$ of the contour lengths comes out as $\bar{z} = 4k\theta/a$. After a simple transformation we obtain

$$\lambda = \lambda(\theta, c, n) = k\theta(\ln \bar{z} - 4k\theta/\bar{z}) \tag{10}$$

where we substitute

$$\bar{z} = \bar{L}/2\bar{M} = N_A c/2\mathscr{M}_{mon}n \tag{11}$$

Here again the L-inconsistency comes into play.

With eqns (7) and (10) we have obtained two equations of state, the first one incomplete because the function $A(\theta, \mu)$ is not specified and the second approximate due to the L-inconsistency. The third equation of state which contains the entropy, and can be interpreted as a caloric equation of state, has not yet been derived. We can ignore it as long as we consider only isothermal processes.

3 TEMPORARY POLYMER NETWORK IN FLOW

In the non-equilibrium the probability density function $W = W(\mathbf{P}, \mathbf{R}, Z, M, t)$ for the state $\{\mathbf{P}, \mathbf{R}, Z, M\}$ at time t changes in time, (i) because the network flows and (ii) because the junctions decay and form. To find the governing equation we first need the equations of motion for the beads. These are Langevin equations with neglected inertia forces. This approximation is rather common in the spring–bead model and is not discussed here. It has the consequence that P in the argument of W drops out. Among the forces acting on the beads we include the elastic forces transmitted through the chains, the friction forces between the beads and the surrounding fluid and the stochastic forces of the Brownian motion. We leave for a later treatment the forces of the hydrodynamic interaction and the excluded volume effect. With this understanding the Langevin equation becomes

$$\dot{\mathbf{r}}_i = \mathbf{r}_i \cdot \mathbf{q} - (1/\zeta)\left(\sum_j^M \kappa_{ij}\mathbf{r}_j - k\theta\nabla_i \ln W\right) \tag{12}$$

Here $\dot{\mathbf{r}}_i$ is the velocity of the ith bead, ∇_i the Nabla operator belonging to $\mathbf{r}_i$, $\mathbf{q}$ the velocity gradient tensor, later considered as given, and ζ the effective friction constant, to be taken from experiment.

$\dot{\mathbf{r}}_i$ is now inserted into the equation

$$\frac{\partial W}{\partial t} + \sum_i^M \nabla_i \cdot (\dot{\mathbf{r}}_i W) = \int \int (W'\vec{p} - W\bar{p}) \, d\mathbf{R} \, dZ \qquad (13)$$

Here $W' \equiv W(\mathbf{R}', Z, t)$ and the integration is taken over the whole configuration space, which includes $Z \equiv \{z_{ij}\}$. Equation (13) is valid for fixed M (see below): otherwise a sum over M should occur on the right. The negative of the sum on the left describes the change of W due to the flow. This form is well known. The right-hand side is equal to the change of W by the kinetic processes. This becomes clear when we now define

$$\vec{p} \, dt \equiv p(\mathbf{R}', Z' \to \mathbf{R}, Z) \, dt, \qquad \bar{p} \, dt \equiv p(\mathbf{R}, Z \to \mathbf{R}', Z') \, dt \qquad (14)$$

as the transition probability densities (with respect to the unprimed ($\mathbf{R}$) and primed ($\mathbf{R}'$) variables, respectively) for the indicated transitions in the configuration space. The first integral of eqn (13) implies the rate at which configurations $\mathbf{R}, Z$ are formed and the second integral tells us at which rate such configurations $\mathbf{R}, Z$ decay. The integro-differential equation (13) with eqns (12) and (14) form the diffusion equation for the spring–bead model of the temporary polymer network.

An important feature of this equation is that the transitions described are configurational, thus many-junction processes. This implies a great complication for the diffusion equation. The second difficulty lies in the very occurrence of the integrals. Both difficulties can be overcome only by simplifications. Let us first attack the integrals. They remind us strongly of the integrals in Boltzmann's kinetic equation. There the integrals are removed in the so-called relaxation time approach. We can proceed in an analogous manner by observing that both sides of eqn (13) vanish if W is replaced by W_{eq}, the equilibrium function. Thus eqn (13) remains correct if we substitute $W - W_{eq}$ for W on the right of eqn. (13). Since $\int \int (W' - W'_{eq})\vec{p}$ describes transitions away from equilibrium, which are much more seldom than those towards equilibrium represented by $\int \int (W - W_{eq})\bar{p}$, we neglect the first of these two integrals and obtain instead of eqn (13) the diffusion equation in the relaxation time approach:

$$\frac{\partial W}{\partial t} + \sum_i^M \nabla_i \cdot (\dot{\mathbf{r}}_i W) = -\bar{p}(W - W_{eq}) \qquad (15)$$

Here

$$\bar{p} = \bar{p}(\mathbf{R}, Z) \equiv \int \int p(\mathbf{R}, Z \to \mathbf{R}', Z') \, d\mathbf{R}' \, dZ' \tag{16}$$

is the probability per unit time that the configuration $\mathbf{R}, Z$ will decay into any other configuration. $\tau(\mathbf{R}, Z) \equiv 1/\bar{p}(\mathbf{R}, Z)$ is the pertaining relaxation time spectrum.

An important consequence of the relaxation time approach must be mentioned. It has been proved,[8] that within this approach

$$W(Z) = W_{\text{eq}}(Z) \tag{17}$$

which means that the distribution of contour lengths is the same in equilibrium and in flow. Note that $W(Z)$ is only a small extract of the full probability density function $W(\mathbf{R}, Z, t)$, for which an equation like eqn (17) does not exist. We expect that eqn (17) will be good as long as the velocity gradient is not too high.

$\bar{p}$ still represents configurational transitions and cannot be generally calculated. This difficulty can be overcome by a simplification which has again an analogue in Boltzmann's kinetic theory. Already in the derivation of his kinetic equation the assumption 'only two-particle collisions' is made. The corresponding simplification in our theory is based on the assumption that junctions decay and form so seldom that processes at different junctions are effectively uncorrelated. We assume furthermore that each decay is immediately followed by a formation process which re-establishes the junction in a somewhat different local configuration. This implies also that M is fixed. Finally, we assume that the time interval between decay and formation is so short that no energy is dissipated in the interval. This we call the 'adiabatic approximation'.[15]

Under the listed assumptions the transition spectrum is calculated[9,10] as

$$\bar{p}(\mathbf{R}, Z) = \sum_{i}^{M} B_i \exp\left\{\frac{1}{2k\theta} \sum_{(i\alpha)}^{4} k_{i\alpha} \mathbf{h}_{i\alpha}^2\right\} \tag{18}$$

where the chain end-to-end vector $\mathbf{h}_{i\alpha} \equiv \mathbf{r}_\alpha - \mathbf{r}_i$,

$$B_i \equiv v^* \left(\frac{k\theta}{\alpha l l_0}\right)^3 (z_{ia} + z_{ib})^{5/2} (z_{ic} + z_{id})^{5/2} \tag{19}$$

and v^* (dimension time^{-1}) is to be taken from the experiment (see below). Equation (18) gives the transition probability as an 'either–or' probability that the transition occurs at the 1st, 2nd or 3rd, etc., junction. At each junction, it is the elastic energy $kh^2/2$ compared with the thermal energy $k\theta$ that determines the transition probability. In fact, the summation in the exponent of eqn (18) goes over the four chains that are labelled $(i\alpha)$ and end in the junction $i(\alpha = 1, 2, 3, 4)$. In eqn (19) the four chains $(i\alpha)$ are marked by a, b, c, d such that the chains a, b belong to one macromolecule and c, d to the other of the two molecules participating in the particular process. Herewith, we take into account that the bonds forming the junction are much weaker than those forming the polymer.

Note that eqn (18) contains the stress implicitly, via the elastic energy. Equation (18) with eqn (19) solves the problem of calculating the transition probabilities when the stress is given. On the other hand, eqn (15) with eqn (12) leads to the stress when $\bar{p}$ is given. Equation (12), however, can only be handled if besides the given velocity gradient q we know also the effective friction constant ζ and the elastic matrix (κ_{ij}). The problem of ζ will be discussed in the next section. As already mentioned, the (stochastic) matrix (κ_{ij}) can be found with the help of eqn (9), provided the quantities $\bar{z}$ and λ are known. Given the concentration c, $\bar{z}$ follows from eqn (8) by adaptation to the experiment, as described in the next section. λ can then be taken from eqn (10). These results are derived for equilibrium, i.e. fluid at rest, but can be used for non-equilibrium, too, thanks to eqn (17).

4 ADAPTATION TO EXPERIMENT OF THE PARAMETERS v^*, ζ AND $A(\theta, \mu)$

The predictive power of a theory is all the greater the less quantities need be adapted to experiment. At the present state of our theory we have no means to calculate v^*, ζ and $A(\theta, \mu)$. Thus we adapt them to the experiment. First we consider simple shear flow in the x_3-direction, such that only the component $q \equiv q^{13}$ of the tensor $\mathbf{q}$ is different from zero. The viscoelastic material functions for this flow (viscometric functions) are

$$\eta(q) = \eta_s + \hbar^{13}/q, \qquad \xi(q) = \frac{\hbar^{11} - \hbar^{22}}{q^2} \qquad (20)$$

Here η_s is the Newtonian viscosity of the solution when all polymer bonds are removed so that no more chains, but only monomers and solvent molecules, are present. The stress tensor of this particular fluid when added to the stress tensor $\hbar$ of the last equations gives the total stress tensor prevailing in our polymer solution. The non-Newtonian flow depends only on $\hbar$. η_s can often be found in the literature. Usually it is so small compared with η that it can be neglected.

Since the hydrodynamic interactions and the excluded volume effect are disregarded, there is no second normal-stress difference, so that the pertaining material function vanishes. For the simple elongational flow in the x_3-direction, considered later, we have only one material function, which we define to be

$$\eta''(q) = (\hbar^{33} - \hbar^{11})/q^{33} \tag{21}$$

For $q \to 0$, η and ξ have limit values η_0 and ξ_0 that can be taken from the Newtonian part of the $\eta(q)$ and $\xi(q)$ curves. Now, the equations of the last section allow us to calculate in the normal way the stress tensor $\hbar$ needed in eqns (20) and (21). The so-obtained expressions for η and ξ contain the parameters ζ and v^*. If we insert the values of η_0 and ξ_0, as taken from experiment, in those expressions then we can calculate ζ and v^*. The procedure is somewhat cumbersome but accessible to the computer. The obtained values of ζ and v^* are valid only for the particular temperature and concentration for which they have been calculated. Although derived in the shear experiment, they can also be utilized for the elongational flow.

As discussed at the end of the last section, the construction of the $(M \times M)$ elastic matrix requires the adaptation of $\bar{z}$. On the other hand, M is disposable as remarked after eqn (7). It is taken as 10 for the calculations which are compared with the experiment in Section 5. $\bar{z}$, or $\bar{j} = (\mathcal{M}/\mathcal{M}_{mon})/\bar{z}$, has been derived from experiment as follows.[10]

We adopt certain plausible values of A at a given θ and μ. For μ we have taken the negative of ε_D, the energy of dissolution of a junction. This is perhaps not quite satisfactory and should later be improved. With these values of A, the curves $\eta(q)$ and $\xi(q)$ can be calculated and plotted. $\bar{j}$ is then chosen such that the Newtonian part of the theoretical curves $\eta(q)$ and $\xi(q)$ coincides as far as possible with the Newtonian range of the experimental viscoelasticity. Practically, this comes to the determination of that point on the experimental curves where the Newtonian behaviour ends and the non-Newtonian flow starts. So far, this adaptation has been done by looking at the curve or manually, which is not completely satisfactory. Fortunately, the curves are not very sensitive to the choice of A or the choice of $\bar{j}$, so that for the moment one may be content. Since A does not depend on the concentration, the adaptation needs be made for only one concentration. If A, ζ and v^* are gained, as described, from the shear experiment, a new adaptation for the elongational experiment is not required.

We see a success of the theory in the fact that the whole adaptation relates to the Newtonian range of the flow. Nothing is adjusted in the non-Newtonian domain. This means that the deviations between theoretical and experimental curves in the non-Newtonian domain say something about the goodness of the theory.

5 COMPARISON OF THEORY AND EXPERIMENT

The eqns (7) and (10), (15) with (12), and (18) with (19) form a complete system from which the stress tensor can be derived. For reasons of space, details of this calculation which are routine are not given. They are found in Refs 8, 9 and 10. We rather have restricted ourselves to give comments which classify the physical situation.

Of course, all details are contained in the computer program involved, but not involved excessively. In fact, the average computer time (language Fortran 77, using the Cyber 835 of the University of Stuttgart computer centre) for producing a curve was about 12 s. Most of this time was consumed calculating the eigenvalues and eigenvectors of the elasticity matrix, and determining the parameters v^* and ζ.

Most comparisons refer to the shear flow where more and clearer experimental results are available. Figure 1(a) and (b) illustrates the adaptation of A, which leads to the average number $\bar{j}$ of chains per macromolecule. Four trial values of $\bar{j}$ were chosen, and with each of them the curves $\eta(q)$ and $\xi(q)$ were drawn. Inspecting both figures gives the impression that the curve belonging to $\bar{j} = 4.52$ should be chosen, if the Newtonian range of the experimental curve would extend further to the right. Similarly, though less distinct, would one choose $\bar{j} = 3.25$ if the experimental Newtonian range would end more to the left. $\bar{j} = 3.50$ seems to be a reasonable compromise. Of course, this is not an accurately determined figure. However, the course of the curves does not depend sensitively on the value of $\bar{j}$.

In the same way all mean chain numbers $\bar{j}$ on our graphs were determined. Figure 2(a) and (b) shows the results on polystyrene solution in decalin at two different concentrations. Higher concentration leads to a lower number of chains per macromolecule and also to higher viscosity. Figure 3(a) and (b) shows the results on polyacrylamide in water for two different molecular weights $\mathcal{M}$. In accordance with the theory, $\bar{j}$ does not depend on $\mathcal{M}$. Higher molecular weight increases the viscosity.

Figure 4(a), (b) and (c) displays another example for the concentration dependence of $\bar{j}$. The low chain number 1·18 indicates that the pertaining

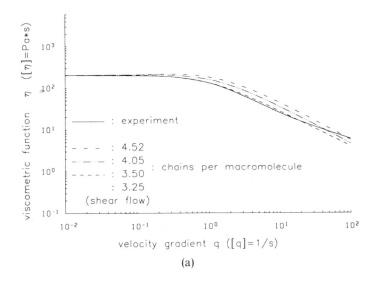

(a)

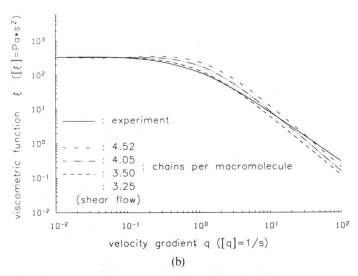

(b)

Fig. 1. On the adaptation of the mean number of chains per macromolecule (polystyrene solution in decalin). Experiments after Krozer and Gruber,[16] Test conditions: $c = 0.125 \, g/cm^3$; $T = 293 \, K$; $\mathcal{M} = 2\,000\,000$; $\mathcal{M}_{mon} = 104$.

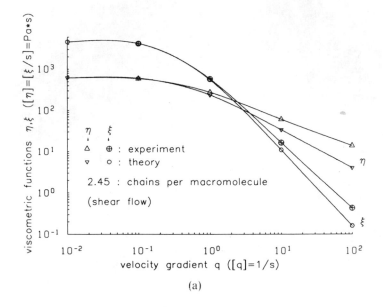

(a)

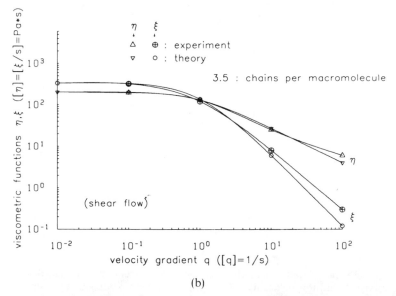

(b)

Fig. 2. Calculated and experimental viscometric functions of polystyrene in decalin. Experiments after Krozer and Gruber.[16] Test conditions as for Fig. 1, except $c = 0.150\,g/cm^3$ in (a).

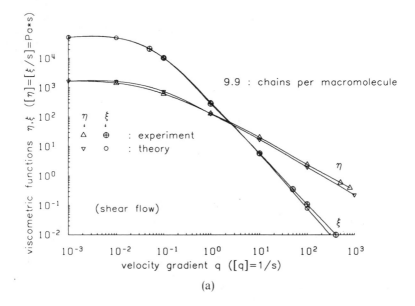

(a)

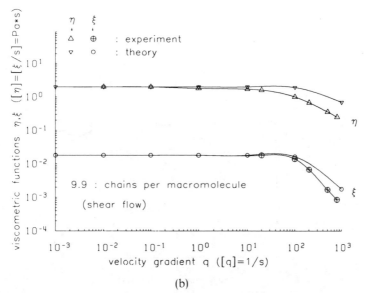

(b)

Fig. 3. Calculated and experimental viscometric functions of polyacrylamide in water. Experiments after Klein and Kulicke.[17] (a) Test conditions: $c = 0.05\,\text{g/cm}^3$; $T = 298\,\text{K}$; $\mathscr{M} = 5\,300\,000$; $\mathscr{M}_{\text{mon}} = 71$. (b) Test conditions as for (a), except $\mathscr{M} = 1\,000\,000$.

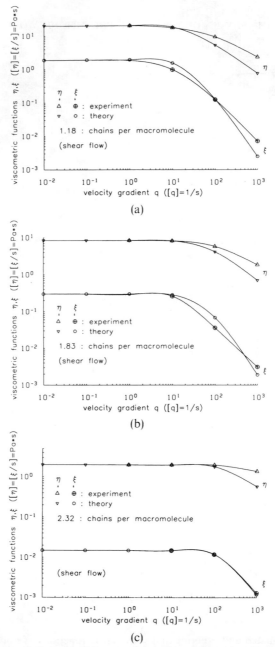

Fig. 4. Calculated and experimental viscometric functions of polystyrene in toluol. Experiments after Krozer *et al.*[18] (a) Test conditions: $c = 0.250 \, g/cm^3$; $T = 298 \, K$; $\mathcal{M} = 670\,000$; $\mathcal{M}_{mon} = 104$. (b) Test conditions as for (a), but $c = 0.200 \, g/cm^3$. (c) Test conditions as for (a), but $c = 0.150 \, g/cm^3$.

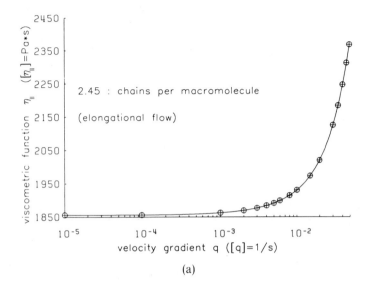

(a)

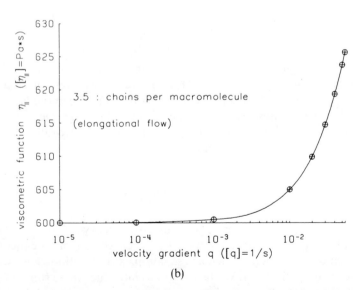

(b)

Fig. 5. Calculated material function for elongational flow of polystyrene in decalin. Values of parameters as in Fig. 2(a) and (b).

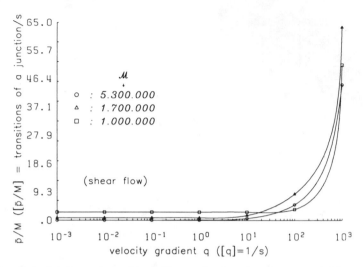

Fig. 6. Calculated number of transitions of a junction per second (shear flow, polyacrylamide in water). By adaptation from experiments of Klein and Kulicke.[17] Test conditions; $c = 0.05 \, g/cm^3$; $T = 298 \, K$; $\mathcal{M}_{mon} = 71$.

concentration $0.25 \, g/cm^3$ is already of a magnitude where networks cease to exist. Figure 5(a) and (b) shows an example for longitudinal flow. Note that the Newtonian domain is very short. The viscosity increases dramatically with concentration. This is an indication of the complex physical situation at higher velocity gradients in elongational flow.

Figure 6, finally, shows us typical examples of decay rates. The number of decays per second is constant and very low (e.g. $10^{-2}/s$) in the Newtonian domain, but increases faster with increasing velocity gradient. It can be anticipated that at still higher velocity gradients the decay rate becomes so high that the network is torn into pieces.

6 CONCLUSION

Comparison with experiment shows that the theory lies in the realm of physical reality. Better agreement cannot be expected from the theoretical approach to phenomena which are so complex that nothing goes without significant simplifications. We have tried to introduce these simplifications in such a way that the main physical effects are not eliminated. Let us now give a short summary of the main simplifications.

The most decisive simplification is certainly the reduction of degrees of freedom which is achieved by the introduction of the spring–bead model of the polymer network, with beads that can decay and form. We have chosen a four-functional network and discussed a certain inconsistency connected herewith, which, however, seems tolerable, at least in the present stage. This so-called 'L-inconsistency' implies that we do not distinguish clearly between the monomer numbers of a network with open ends and without open ends. Removing this inconsistency would mean distinguishing already in the macrodescription between chains with open ends and chains connecting two junctions. We leave this for later investigations, in which other restrictions can also be removed, e.g. that the model contains no branching and only one molecular weight, $\mathcal{M}$, which in our case represents a mean molecular weight of the polymer system.

The model just characterized is governed by the (integro-differential) diffusion equation which is reduced to a differential equation by introducing the relaxation time approach. The most visible physical effect is that the distribution of contour lengths becomes identical in the resting and moving fluid. It is clear that this approximation has its limits, namely it is viable in the limits of high velocity gradients, and it is difficult to say how high these limits are. The similarity to Boltzmann's kinetic theory suggests that this problem is analogous to that of deviations from Maxwell's distribution in non-equilibrium, where these are found small even far away from equilibrium. Regarding the enormous simplification achieved by the relaxation time idea, this approach appears more than justified.

The diffusion equation should contain all forces acting on the beads. We have neglected inertia forces (unimportant) and the forces of hydrodynamic interaction and excluded volume (serious). At present we are working on the latter two effects, but it is troublesome, because both effects are of the many-particle type.

The diffusion equation can be handled only if the transition probabilities are known. These have been calculated in the one-junction approximation, which appears reasonable for not too high velocity gradients. Finally, the restriction to 10 of the number of junctions in our system also reduces the accuracy of our results because the statistics are not very good when based on the number 10. However, an improvement is easily possible if instead of only one ensemble a number, N, of ensembles are constructed so that N different elastic matrices arise which, however, all are compatible with the distribution $W(z)$ of contour lengths. In our computations $N = 10$ was assumed. For the future it is desirable to increase M and N with the help of larger computers.

Several times we have mentioned the L-inconsistency, which makes L, the number of monomers in the system volume V, a poorly defined quantity. This leads to the result that also $\bar{j}$, the mean number of chains per macromolecule, is not a good number. As far as the viscoelastic material functions, the main results of our theory, are concerned, the uncertainty of $\bar{j}$ is not serious, since $\bar{j}$ is adapted to the experiment. The adaptation, however, is done in such a way that the material functions should be correct.

In conclusion, we emphasize that the theory presented is a first step towards a molecular-statistical theory of temporary polymer networks. Many improvements are possible and desirable. We believe, however, that the approach described here already reflects important physical features of the temporary networks.

ACKNOWLEDGEMENT

The authors would like to thank the Deutsche Forschungsgemeinschaft for financial support.

REFERENCES

1. Ziabicki, A., Molecular mechanism of non-linear rubber elasticity, in *Trends in Applications of Pure Mathematics to Mechanics*, Lecture Notes in Physics 249, Kröner, E. and Kirchgässner, K. (Eds), Springer-Verlag, Heidelberg, 1986, pp. 384–408.
2. Rouse, R. E., A theory of the linear viscoelastic properties of dilute solutions of coiling polymers, *J. Chem. Phys.*, 1953, **21**, 1272–80.
3. Zimm, B. H., Dynamics of polymer molecules in dilute solution: Viscoelasticity, flow birefringence and dielectric loss, *J. Chem. Phys.*, 1956, **24**, 269–78.
4. Yamakawa, H., *Modern Theory of Polymer Solutions*, Harper and Row, New York, 1971.
5. Volkenstein, M. W., *Configurational Statistics of Polymer Chains*, Interscience, New York, 1963.
6. Flory, P. J., *Statistical Mechanics of Chain Molecules*, Interscience, New York, 1969.
7. Takserman-Krozer, R., Krozer, S. and Kröner, E., On the kinetics of polymer networks with temporary junctions, *Colloid & Polymer Sci.*, 1979, **257**, 1033–41.
8. Takserman-Krozer, R. and Kröner, E., Statistical mechanics of temporary polymer networks I. The equilibrium theory, *Rheol. Acta*, 1984, **23**, 1–9.
9. Kröner, E. and Takserman-Krozer, R., Statistical mechanics of temporary polymer networks II. The non-equilibrium theory, *Rheol. Acta*, 1984, **23**, 139–50.

10. Chassapis, D., Zur molekular-statistischen Theorie temporärer Polymer-netzwerke—Theoretische Untersuchungen und Vergleich mit Experimenten, Dissertation, Fakultät für Physik, Universität Stuttgart, 1986.
11. Flory, P. J., Thermodynamics of high polymer solutions, *J. Chem. Phys.*, 1942, **10**, 51–61.
12. Flory, P. J., Statistical mechanics of swelling of network structures, *J. Chem. Phys.*, 1950, **18**, 108–11.
13. Flory, P. J., Statistical mechanics of semiflexible chain molecules, *Proc. Roy. Soc.*, 1956, **A234**, 60–73.
14. Flory, P. J., Hoeve, C. A. J. and Ciferri, A., Influence of bond angle restrictions on polymer elasticity, *J. Polymer Sci.*, 1959, **34**, 337–47.
15. Kröner, E. and Takserman-Krozer, R., Molecular descriptions of temporary polymer networks, in *Continuum Models of Discrete Systems 4*, Brulin, O. and Hsieh, R. K. T. (Eds), North Holland, Amsterdam, 1981, pp. 297–310.
16. Krozer, S. and Gruber, E., Zur Temperaturabhängigkeit der viskoelastischen Eigenschaften konzentrierter Polystyrol-Lösungen in Dekalin, *Rheol. Acta*, 1979, **18**, 86.
17. Klein, J. and Kulicke, W.-M., Rheologische Untersuchungen zur Struktur hochmolekularer Polyacrylamide in wäßrigen und nichtwäßrigen Lösungen, *Rheol. Acta*, 1976, **15**, 558–76.
18. Krozer, S., Tawadjoh, M. and Gruber, E., Zur Beschreibung der rheologischen Eigenschaften konzentrierter Polystyrol-Lösungen, *Rheol. Acta*, 1977, **16**, 438–43.

12

KINETICS OF RING FORMATION IN POLYMERIZATION REACTIONS

Lu Binglin* and Thor A. Bak

*Department of Chemistry, University of Copenhagen,
Universitetsparken 5, DK-2100 Copenhagen Ø, Denmark*

ABSTRACT

The kinetic equations for polymerization systems with ring formation are established. We solve the equations with arbitrary initial conditions both for the high functionality system and for the general f-functionality system. Ring formation in long chain systems and the general Flory–Stockmayer model with arbitary initial conditions are included as special cases.

1 INTRODUCTION

Intermolecular reactions which lead to ring formation are important phenomena in network formation. Although the equilibrium size distribution of ring molecules has been studied by several authors,[1−5] the kinetic aspects of ring formation so far have largely been neglected.

In this paper, we study the kinetics of ring formation in polymerization reactions for branched chain polymers. Generally, the kinetic equations for polymerization reactions have been established following Flory's two assumptions: (1) the assumption of equal reactivity, i.e. the assumption that the reactivities of the identical chemical functionalities are independent of the size of the polymer and their position in the polymer; (2) intramolecular reactions are neglected. The kinetic equation which one derives from this has the following form

$$\frac{dc_n}{dt} = \frac{1}{2} \sum_{i+j=n} K_{ij} c_i c_j - c_n \sum_{j=1}^{\infty} K_{nj} c_j \qquad (1)$$

* Permanent address: Fushun Petroleum Institute, Liaoning, China.

This equation has been widely used in many fields of physics and chemistry.[6] Especially with reference to its use as a polymerization model for f-functionality units, the details of the mathematics required to obtain the solution were given by Dušek[7] and Ziff and Stell,[8] but their result is limited to a special set of initial conditions, i.e. the monodisperse initial conditions.

A polymerization system which includes ring formation is more complicated than the tree-like polymerization system described by eqn (1). As was established by us in a previous paper,[9] the ring formation is a first-order reaction and the reactions in the whole system can be written as

$$\left. \begin{aligned} X_{im} + X_{jn} &\xrightarrow{K_{im,jn}} X_{i+j,m+n} \\ X_{im} &\xrightarrow{K_{im}} X_{i,m+1} \end{aligned} \right\} \tag{2}$$

where X_{im} represents the molecule which is composed of i units with m rings and $K_{im,jn}$, K_{im} are rate constants. The kinetic equation can be written as

$$\frac{dc_{kl}}{dt} = \frac{1}{2} \sum_{\substack{i+j=k \\ m+n=l}} K_{im,jn} c_{im} c_{jn} - c_{kl} \sum_{p,q} K_{kp,lq} - K_{kl} c_{kl} + K_{k,l-1} c_{k,l-1} \tag{3}$$

Here, of course, l is limited by the total number of free functionalities σ_k: we have $l \le \sigma_k/2 = \frac{1}{2}[k(f-2)+2]$.

Equation (3) is difficult to solve for the general polymerization model. Noting that, in a polymerization system, the fraction of the total number of molecules which contains rings already before gelation is relatively small,[10] we can neglect the coagulation of ring molecules both with ring molecules and non-ring molecules. Writing down the reactions in a more precise way, we have

$$X_i + X_j \xrightarrow{K_{ij}} X_{i+j}$$
$$X_i \xrightarrow{A_{i0}} R_{i,1}$$
$$R_{il} \xrightarrow{A_{il}} R_{i,l+1} \tag{4}$$

where X_i represent non-ring molecules and R_{il} represent a molecule which is composed of i units and contains l rings. The following simplified kinetic equation is thus established:

$$\frac{dc_n}{dt} = \frac{1}{2} \sum_{i+j=n} K_{ij} c_i c_j - c_n \sum_j K_{nj} c_j - A_{n0} c_n$$
$$\frac{dc_n^l}{dt} = A_{n,l-1} c_n^{l-1} - A_{nl} c_n^l \tag{5}$$

where c_n is the concentration of non-ring molecules with size n at time t, and c_n^l is the concentration of molecules with size n and ring number l.

2 RATE CONSTANT AND KINETIC EQUATIONS

Theoretically, the rate constant K_{ij} and A_{il} in eqn (5) can be determined from the special model one uses. Following the assumption of equal reactivities of free functionalities, K_{ij} can easily be determined. The determination of rate constants A_{il} is more complicated. From eqn (5), we can see that if c_n is known, c_n^l can easily be obtained for known A_{nl}. To get the size distribution $c_n(t)$, we must first obtain an expression for A_{n0}. In this paper, we consider two cases: (1) the rate constant A_{n0} is independent of the size of the molecule; (2) the rate constant A_{n0} is proportional to the size of the molecule. For a branched chain polymerization system, the second case sounds more realistic, and the real situation is probably in between these two cases. After determining A_{n0}, we can introduce a new quantity

$$c_{nR} = \sum_{l=1}^{l \le \sigma_n/2} c_n^l \tag{6}$$

which is the total number of ring molecules with size n. Summing the second equation of (5) over all possible l, we get

$$\frac{\mathrm{d}c_{nR}}{\mathrm{d}t} = A_{n0}c_n \tag{7}$$

Noting the fact that the coagulation between ring molecules and non-ring molecules will not change the total number of ring molecules, we can understand that the quantity c_{nR} is a useful quantity in the model.

We consider the ring formation in a high functionality system and in the general f-functionality system. The kinetic equations will be

$$\left.\begin{aligned}
\frac{\mathrm{d}c_n}{\mathrm{d}t} &= \frac{1}{2} \sum_{i+j=n} ijc_ic_j - nc_n \sum_{j=1}^{\infty} jc_j - \lambda c_n \\
\frac{\mathrm{d}c_n^l}{\mathrm{d}t} &= A_{n,l-1}c_n^{l-1} - A_{nl}c_n^l
\end{aligned}\right\} \tag{8}$$

$$\left.\begin{array}{l} \dfrac{dc_n}{dt} = \dfrac{1}{2} \sum_{i+j=n} ijc_ic_j - nc_n \sum_{j=1}^{\infty} jc_j - \lambda nc_n \\[4mm] \dfrac{dc_n^l}{dt} = A_{n,l-1}c_n^{l-1} - A_{nl}c_n^l \end{array}\right\} \qquad (9)$$

$$\left.\begin{array}{l} \dfrac{dc_n}{dt} = \dfrac{1}{2} \sum_{i+j=n} \sigma_i\sigma_jc_ic_j - c_n\sigma_n \sum_{j=1}^{\infty} \sigma_jc_j - \lambda c_n \\[4mm] \dfrac{dc_n^l}{dt} = A_{n,l-1}c_n^{l-1} - A_{nl}c_n^l \end{array}\right\} \qquad (10)$$

$$\left.\begin{array}{l} \dfrac{dc_n}{dt} = \dfrac{1}{2} \sum_{i+j=n} \sigma_i\sigma_jc_ic_j - c_n\sigma_n \sum_{j=1}^{\infty} \sigma_jc_j - \lambda\sigma_nc_n \\[4mm] \dfrac{dc_n^l}{dt} = A_{n,l-1}c_n^{l-1} - A_{nl}c_n^l \end{array}\right\} \qquad (11)$$

In eqns (10) and (11), $\sigma_k = (f-2)k + 2$, is the number of free functionalities of a k-mer. We solve all these four sets of equations with arbitrary initial conditions. The mathematics is quite complicated, and here we just give a review which is kept as simple as possible.

3 SIZE DISTRIBUTION OF NON-RING MOLECULES AND RING MOLECULES IN A HIGH FUNCTIONALITY SYSTEM

First, we consider eqn (8) and solve this equation before 'gelation'. As usual we define the moments as

$$M_n = \sum_{k=1}^{\infty} k^n c_k(t) \qquad (12)$$

From eqn (8), we have

$$\frac{dM_1}{dt} = -\lambda M_1 \qquad (13)$$

The solution of eqn (13) is

$$M_1 = M_1(0)\,e^{-\lambda t} \qquad (14)$$

Set

$$c_n(t) = x_n(t) \exp\left[-\int_0^t (nM_1 + \lambda)\, dt' \right] \tag{15}$$

and

$$\tau = \int_0^t \exp(-\lambda t')\, dt' = \frac{1}{\lambda}[1 - \exp(-\lambda t)] \tag{16}$$

From the kinetic equation (8) we then obtain a kinetic equation for $x_n(t)$:

$$\frac{dx_n}{dt} = \frac{1}{2} \sum_{i+j=n} ij x_i x_j \tag{17}$$

Introducing the generating function

$$G = \sum_{k=1}^{\infty} x_k(\tau)\, e^{kz} \tag{18}$$

and solving the partial differential equation

$$\frac{\partial G}{\partial \tau} = \frac{1}{2}\left(\frac{\partial G}{\partial z}\right)^2 \tag{19}$$

we have

$$\frac{\partial G}{\partial z} = v\left(z + \frac{\partial G}{\partial z}\tau\right) \tag{20}$$

where $v(z) = \sum_{k=1}^{\infty} k\, e^{kz} x_k^{(0)}$ is a function which is determined by the initial condition. Using Lagrange's expansion, we have

$$x_n = (\tau n^2 n!)^{-1} \sum_{\{k_j\}} \frac{n!\, n^m \tau^m}{\prod_j k_j!} \prod_j [jc_j(0)]^{k_j} \tag{21}$$

where the summation goes over all possible sets $\{k_j\}$ which satisfy the condition $\sum_j jk_j = n$ and $\sum_j k_j = m$.

Hence the size distribution of non-ring molecules can be obtained from eqns (21), (15) and (16):

$$c_n = (n^2 n!)^{-1}\, e^{-\lambda t} \exp\left\{ -\frac{n}{\lambda} M_1(0)(1 - e^{-\lambda t}) \right\}$$
$$\times \sum_{\{k_j\}} \frac{n!\, n^m (1 - e^{-\lambda t})^{m-1}}{\lambda^{m-1} \prod_j k_j!} \prod_j [jc_j^{(0)}]^{k_j} \tag{22}$$

For the special case of monodisperse initial conditions, we have

$$c_n = c_1(0) \frac{n^{n-2} e^{-\lambda t}}{n!} \exp\left[-\frac{n}{\lambda}(1 - e^{-\lambda t}) \right] \frac{[c_1(0)(1 - e^{-\lambda t})]^{n-1}}{\lambda^{n-1}} \qquad (23)$$

The total number of ring molecules with size n is

$$c_{nR} = \sum_{\{k_j\}} \frac{\lambda h^{-2} \prod_j [jc_j(0)]^{k_j}}{\prod_j k_j!}$$

$$\times \left\{ \frac{(m-1)!}{[M_1(0)]^m} - \exp\left[-\frac{nM_1(0)}{\lambda}(1 - e^{-\lambda t}) \right] \sum_k \frac{(m-1)!}{k!} \frac{(1 - e^{-\lambda t}/n\lambda)^k}{[M_1(0)]^{m-k}} \right\} \qquad (24)$$

For the case of monodisperse initial conditions, we have

$$c_{nR} = \lambda n^{-3} \left\{ 1 - \exp\left[-\frac{nc_1(0)}{\lambda}(1 - e^{-\lambda t}) \right] \sum_{k=0}^{n-1} \frac{(1 - e^{-\lambda t})^k}{k!(c_1(0)\lambda n)^k} \right\} \qquad (25)$$

The size distribution $c_n^l(t)$ can be obtained by solving the differential equation for c_n, $l = 1, 2, 3, \ldots$

$$c_n^l = \exp(-A_{ml}t) \left[\int_0^t A_{n,l-1} c_n^{l-1}(t') \exp(A_{nl}t') \, dt' \right] \qquad (26)$$

The time at which the divergence of the second moment occurs can be obtained from

$$\frac{dM_2}{dt} = M_2^2 - \lambda M_2 \qquad (27)$$

with the solution

$$M_2 = \frac{\lambda M_2(0) e^{-\lambda t}}{\lambda - M_2(0)[1 - e^{-\lambda t}]} \qquad (28)$$

When $t \to t_c$, $M_2 \to \infty$:

$$t_c = \frac{1}{\lambda} \log \frac{M_2(0)}{M_2(0) - \lambda} \qquad (29)$$

This is the so-called 'gelation time' for a tree-like system. The evolution of the total number of non-ring molecules before 'gelation' can be obtained from the equation

$$\frac{dM_0}{dt} = -\tfrac{1}{2} M_1^2 - \lambda M_0^{(1)} \qquad (30)$$

the solution of which is

$$M_0(t) = e^{-\lambda t} \left[M_0(0) - \frac{[M_1(0)]^2}{2\lambda} (1 - e^{-\lambda t}) \right] \qquad (31)$$

We now consider the situation where A_{n0} is proportional to the size of molecules, i.e. we are dealing with the kinetic equation (9).

Setting

$$c_n(t) = x_n(t) \exp\left[-n \int_0^t (\textstyle\sum jc_j(t') + \lambda) \, dt' \right] \qquad (32)$$

we get

$$c_n(t) = \frac{1}{n! \, n^2 t} \exp\left[-n\phi(t) \right] \left\{ \left(\frac{d}{ds} \right)^n \exp\left[ntv(s) \right] \right\}_{s=0} \qquad (33)$$

where

$$\phi(t) = \int_0^t (\textstyle\sum jc_j(t') + \lambda) \, dt'$$

and

$$v(z) = \sum k x_k(0) z^k$$

The function $M(t) = \sum_j jc_j(t)$ can be obtained from the equation

$$M(t) = v(-\phi(t) + M(t)t) \qquad (34)$$

In the special case of monodisperse initial conditions, we have

$$t = \frac{e^{M/\lambda}}{\lambda} \int_1^M \frac{e^{-M/\lambda}}{M} \, dM \qquad (35)$$

Hence the size distribution of non-ring molecules and ring molecules can be obtained.

4 SOME ANALYTICAL RESULTS FOR THE GENERAL f-FUNCTIONALITY SYSTEM

For the general f-functionality system with ring formation, the kinetic equations are more complicated. We give an expression for the size distribution for non-ring molecules which contains a function μ, determined by an equation. The size distribution of ring molecules can be obtained by integration.

To solve eqn (10), we introduce

$$\mu = \sum_{k=1}^{\infty} \sigma_k c_k(t) \tag{36}$$

It satisfies the equation

$$\dot{\mu} = -\mu^2 - \lambda\mu \tag{37}$$

before 'gelation'.

The solution of eqn (37) is

$$\mu = \frac{\lambda\mu_0 e^{-\lambda t}}{\lambda + \mu_0(1 - e^{-\lambda t})} \tag{38}$$

We set

$$c_n = x_n \exp\left[-\int_0^t \sigma_n \mu(t')\,dt' - \lambda t \right] \tag{39}$$

and

$$\tau = \int_0^t \exp\left\{ -\left[\lambda t' + 2\int_0^{t'} \mu(t'')\,dt'' \right] \right\} dt' \tag{40}$$

The kinetic equation then reduces to

$$\frac{dx_n}{d\tau} = \frac{1}{2} \sum_{i+j=n} \sigma_i \sigma_j x_i x_j \tag{41}$$

Introducing a generating function, solving the partial differential equation for the generating function and using Lagrange's expansion, we have

$$x_n = \frac{1}{\sigma_n!} \sum_{\{k_j\}} \frac{[(f-2)n+m]!}{\prod_j k_j!} \tau^{m-1} \prod_j [\sigma_j c_j(0)]^{k_j} \tag{42}$$

where the summation goes over all possible sets of $\{k_j\}$, which satisfies $\sum_j k_j = n$ and $\sum k_j = m$. We thus obtain the size distribution for non-ring molecules

$$c_n = \frac{1}{\sigma_n!} \sum_{\{k_j\}} \frac{[(f-2)n+m]!}{\prod k_j!}$$

$$\times e^{-\lambda t} \left[\frac{\lambda}{\mu_0(1 - e^{-\lambda t}) + \lambda} \right]^{(f-2)n+m+1} \left(\frac{1 - e^{-\lambda t}}{\lambda} \right)^{m-1} \prod_j [\sigma_j c_j(0)]^{k_j} \tag{43}$$

For monodisperse initial conditions, we have

$$c_n = e^{-\lambda t} \left[\frac{\lambda}{\mu_0(1 - e^{-\lambda t}) + \lambda} \right]^{(f-1)n+1} \left(\frac{1 - e^{-\lambda t}}{\lambda} \right)^{n-1} \frac{[(f-2)n]!}{[(f-2)n+2]!} [fc_1(0)]^n \tag{44}$$

For a general f-functionality system in which the rate of the ring formation is proportional to the size of the molecule, we have the kinetic equation (11). After a calculation, which is nearly the same as the one given above, we obtain

$$c_n = \frac{1}{\sigma_n!} \sum_{\{k_j\}} \frac{[(f-2)n+m]!}{\prod_j k_j!} \tau^{m-1} \exp(-\sigma_n \phi(t)) \prod_j [\sigma_j c_j(0)]^{k_j} \tag{45}$$

where

$$\phi(t) = \int_0^t \left[\sum \sigma_j c_j(t') + \lambda \right] dt' = \int_0^t [\mu(t') + \lambda] dt' \tag{46}$$

It satisfies the equation

$$\mu(t) \exp[\phi(t)] = v \left\{ \exp[-\phi(t)] + \mu(t) \exp \phi(t) \int_0^t \exp - 2\phi(t') \right\} \tag{47}$$

where v is determined by the initial condition

$$v(z) = \sum_{k=1}^{\infty} \sigma_k c_k(0) z^{\sigma_k - 1} \tag{48}$$

In the case of monodisperse initial conditions, we have

$$\dot{\psi} = f e^{(f-2)\lambda t} \left[\psi^{-1} + \dot{\psi} \int_0^t \exp(-2\lambda t') \psi^{-2} dt' \right] \tag{49}$$

where

$$\psi(t) = \exp \left[\int_0^t \mu(t') dt' \right] \tag{50}$$

5 THE CONNECTION BETWEEN RING FORMATION IN A LINEAR CHAIN SYSTEM AND GENERAL F-S THEORY

Linear chain polymerization systems are characterized by the fact that ring molecules cannot react with other molecules. We therefore have the kinetic equation

$$\frac{dc_n}{dt} = \sum_{i+j=n} c_i c_j - 2c_n \sum_{j=1}^{\infty} c_j - kc_n \tag{51}$$

$$\frac{dc_n^{(1)}}{dt} = kc_n$$

For polydisperse initial conditions, we have

$$c_n = \sum_{\{k_j\}} \frac{m!}{\prod_j k_j!} \frac{K^2(1-e^{-Kt})^{m-1}}{[M_0(0)(1-e^{-Kt})+K]^{m+1}} e^{-Kt} \prod_j [c_j(0)]^{k_j} \tag{52}$$

$$c_n^{(1)} = \sum_{\{k_j\}} [M_0(0)]^{-m} \frac{(m-1)!}{\prod_j k_j!} \left[\frac{M_0(0)(1-e^{-Kt})}{M_0(0)(1-e^{-Kt})+K}\right]^m \prod_j [c_j(0)]^{k_j} \tag{53}$$

For monodisperse initial conditions, we have

$$c_n = \frac{c_1(0)K^2 e^{-Kt}[c_1(0)(1-e^{-Kt})]^{n-1}}{[c_1(0)(1-e^{-Kt})+K]^{n+1}} \tag{54}$$

and

$$c_n^{(1)}(t) = \frac{K}{n}\left[\frac{c_1(0)}{c_1(0)+K}\right]^n \tag{55}$$

We can also include the general F-S model as a special case of our model. When the rate constant for ring formation is zero, we have the kinetic equation

$$\frac{dc_n}{dt} = \frac{1}{2}\sum_{j=1}^{n-1} \sigma_{n-j}\sigma_j c_{n-j}c_j - c_n(t) \sum_{j=1}^{\infty} \sigma_n \sigma_j c_j \tag{56}$$

The solution has the following form

$$c_n(t) = \frac{1}{\sigma_n!} \sum_{\{k_j\}} \frac{[(f-2)n+m]!}{\prod_j k_j!} \frac{t^{m-1}}{(1+\mu_0 t)^{(f-2)n+m+1}} \prod_j [\sigma_j c_j(0)]^{k_j} \tag{57}$$

which is also a new result whose probability explanation generalizes Flory's probability argument [11] to polydisperse initial conditions.[12]

6 CONCLUSIONS

The main purpose of this paper has been to show that the very complicated equations which arise, when one includes ring formation in the coagulation equation, can actually be solved. Equation (29) makes precise the, not unexpected, result that ring formation postpones gelation and may indeed suppress it completely. In Section 5 we finally show how our results can be considered a generalization of the probability arguments originally presented by Flory and Stockmayer.

REFERENCES

1. Jacobsen, H. and Stockmayer, W. H., *J. Chem. Phys.*, 1950, **18**, 1600.
2. Harris, Frank E., *J. Chem. Phys.*, 1955, **23**, 1518.
3. Spouge, J. E., *Can. J. Chem.*, 1984, **62**, 1262.
4. Whitle, P., *Adv. Appl. Prob.*, 1980, **12**, 94.
5. Gordon, M. and Scantleburg, G. R., *Proc. Roy. Soc.* (*London*), 1965, **A292**, 380.
6. Drake, R. L., in *Topics in Current Aerosol Research*, Hidy, G. M. and Brock, J. R. (Eds), Pergamon Press, Oxford, 1972, p. 180.
7. Dušek, K., *Polym. Bull.*, 1979, **1**, 523.
8. Ziff, R. M. and Stell, G., *J. Chem. Phys.*, 1980, **73**, 3492.
9. Bak, Thor A. and Lu, Binglin, *Lectures in Applied Mathematics AMS*, Vol. 24, American Mathematics Society, Providence, 1986.
10. Falk, M. and Thomas, R. E., *Can. J. Chem.*, 1974, **52**, 3285.
11. Flory, P. G., *Principles of Polymer Chemistry*, Cornell University Press, Ithaca, New York, 1953.
12. Bak, Thor A. and Lu, Binglin, *Chem. Phys.*, 1986, **112**, 189.

13

CALCULATION OF AVERAGE NETWORK PARAMETERS USING COMBINED KINETIC AND MARKOVIAN ANALYSIS

Douglas R. Miller

*Department of Operations Research, George Washington University,
Washington, DC 20052, USA*

and

Christopher W. Macosko

*Department of Chemical Engineering and Materials Science,
University of Minnesota, Minneapolis, Minnesota 55455, USA*

ABSTRACT

Molecular and network parameters of branched polymer systems can be computed using various theoretical models and analysis techniques. Kinetic analysis and Markov modeling are two general approaches. For certain complex polymer systems the best approach may be a hybrid approach which combines kinetic and Markovian analysis. This hybrid approach is described in general, and several specific examples and applications are given.

INTRODUCTION

Theoretical models are useful for calculating various average parameters during network formation, such as weight-average molecular weight before gelation, the gel point and weight fraction solubles, and concentration of pendant chains and elastically effective junctions after gelation. These models and their methods of solution may take various forms: the most common are kinetic models and Markovian models. The kinetic approach applies to a larger class of systems; however, it may become intractable, especially for branching polymerizations, because a large number of differential equations must be solved. The Markovian approach (sometimes referred to as the statistical or cascade approach) is attractive because it permits more direct calculation of average properties by exploiting the recursive structure of the material. An optimal approach would combine the two methods. This paper surveys some systems which can be efficiently analyzed with a combined methodology.

The kinetic approach and the Markovian approach can be contrasted by considering the gel point of an ideal A_f stepwise homopolymerization. Before gelation, the kinetic approach models the concentration of each size of molecule. If P_i represents a polymer consisting of i A_f monomers, then

$$P_i + P_j \rightarrow P_{i+j} \tag{1}$$

at a rate proportional, ideally, to $[(fi - 2i + 2)(fj - 2j + 2)]$ for $i, j \geq 1$. Differential equations are solved to find the concentration of different molecular sizes as a function of time or conversion. Analytical solution of these equations is quite difficult for even the simplest systems,[1-3] and appears to be impossible for complex systems, and thus the equations must be solved numerically. From the concentration of different species, the weight-average molecular weight is computed. Gelation occurs when the weight-average molecular weight diverges.

The Markovian approach is based on the fact that the A_f monomers combine randomly in an ideal homopolymerization, i.e. all intermolecular AA bonds are equally likely to form. This results in a branching structure which has a recursive structure with lack of memory, which can be exploited by various analysis methods.[4-7] For illustration we use the 'in–out' recursion: let EW_A^{out} equal the expected weight seen looking out from a randomly chosen A site (directions 1 in Fig. 1). At conversion α:

$$EW_A^{out} = \alpha EW_A^{in} + (1 - \alpha)0 \tag{2}$$

where EW_A^{in} is the expected weight looking into a random A-site (directions 2 in Fig. 1):

$$EW_A^{in} = M_A + (f - 1)EW_A^{out} \tag{3}$$

where M_A is the weight of an A_f monomer. Finally, the weight-average molecular weight is

$$M_W = M_A + fEW_A^{out} \tag{4}$$

Solving the above three equations gives

$$M_W = M_A(1 + \alpha)/[1 - \alpha(f - 1)] \tag{5}$$

and gelation occurs at $\alpha_c = 1/(f-1)$, which is identical to the answer computed using kinetic analysis.

The Markovian analysis assumes that the species combine in a totally random way. There are many systems for which this is not true. In these cases we recommend combining the two methods. This idea is not new: several authors have used it implicitly and Dusek[8] has explicitly presented it as a formal approach to analyzing different systems.

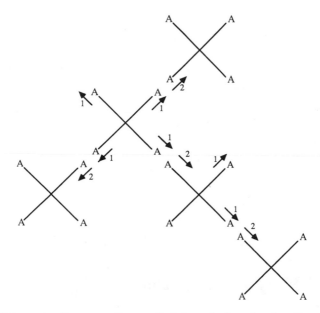

Fig. 1. Schematic diagram of a typical branched molecule after degree of polymerization α.

In a combined analysis the aspects of the reaction which are not totally random are treated kinetically to obtain 'superspecies', which are combined randomly and thus can be treated using a Markovian analysis:

$$\text{Initial Species} \xrightarrow[\substack{\text{non-random} \\ \text{combination}}]{} \text{Superspecies} \xrightarrow[\substack{\text{random} \\ \text{combination}}]{} \text{Product}$$

The two steps in this model may either correspond to two distinct sequential steps or they may be occurring simultaneously, in which case the construct is merely a convenient way to think about the process. This paper discusses several different systems whose average properties can be computed efficiently using a combined kinetic–Markovian approach.

COPOLYMERIZATION WITH SUBSTITUTION EFFECT

Consider an A_3 plus B_2 copolymerization for which there is a first-shell substitution effect on the A_3 monomers. This system has been analyzed by numerous authors, including Miller and Macosko.[9] By explicitly recognizing the combined kinetic–Markovian approach some clarity is revealed: it is a good example for illustrating the concepts.

We define seven superspecies: four types of A-monomer and three types of B-monomer. Let

$A_3^i =$ A-monomer with i reacted sites, $i = 0, 1, 2, 3$
$B_2^i =$ B-monomer with i reacted sites, $i = 0, 1, 2$
$k_i =$ rate constant for reaction of A_3^i functionality with a B functionality, $i = 0, 1, 2$.

From a kinetic analysis we can find the concentration of each of the seven superspecies by solving the following system of kinetic equations:

$$\frac{d}{dt} A_3^0(t) = -3k_0 A_3^0(t)(2B_2^0(t) + B_2^1(t)) \tag{6}$$

$$\frac{d}{dt} A_3^1(t) = (3k_0 A_3^0(t) - 2k_1 A_3^1(t))(2B_2^0(t) + B_2^1(t)) \tag{7}$$

$$\frac{d}{dt} A_3^2(t) = (2k_1 A_3^1(t) - k_2 A_3^2(t))(2B_2^0(t) + B_2^1(t)) \tag{8}$$

$$\frac{d}{dt} A_3^3(t) = k_2 A_3^2(t)(2B_2^0(t) + B_2^1(t)) \tag{9}$$

$$\frac{d}{dt} B_2^0(t) = -2B_2^0(t)(3k_0 A_3^0(t) + 2k_1 A_3^1(t) + k_2 A_3^2(t)) \tag{10}$$

$$\frac{d}{dt} B_2^1(t) = (2B_2^0(t) - B_2^1(t))(3k_0 A_3^0(t) + 2k_1 A_3^1(t) + k_2 A_3^2(t)) \tag{11}$$

$$\frac{d}{dt} B_2^2(t) = B_2^1(t)(3k_0 A_3^0(t) + 2k_1 A_3^1(t) + k_2 A_3^2(t)) \tag{12}$$

The seven superspecies are shown in Fig. 2. The combination is random: any pair of reacted A-site and reacted B-site are equally likely to combine. Thus instead of thinking of the system as two species, A_3 and B_2, combining together in a rather complicated, not completely random way (substitution effect), we think of seven superspecies which combine in a completely

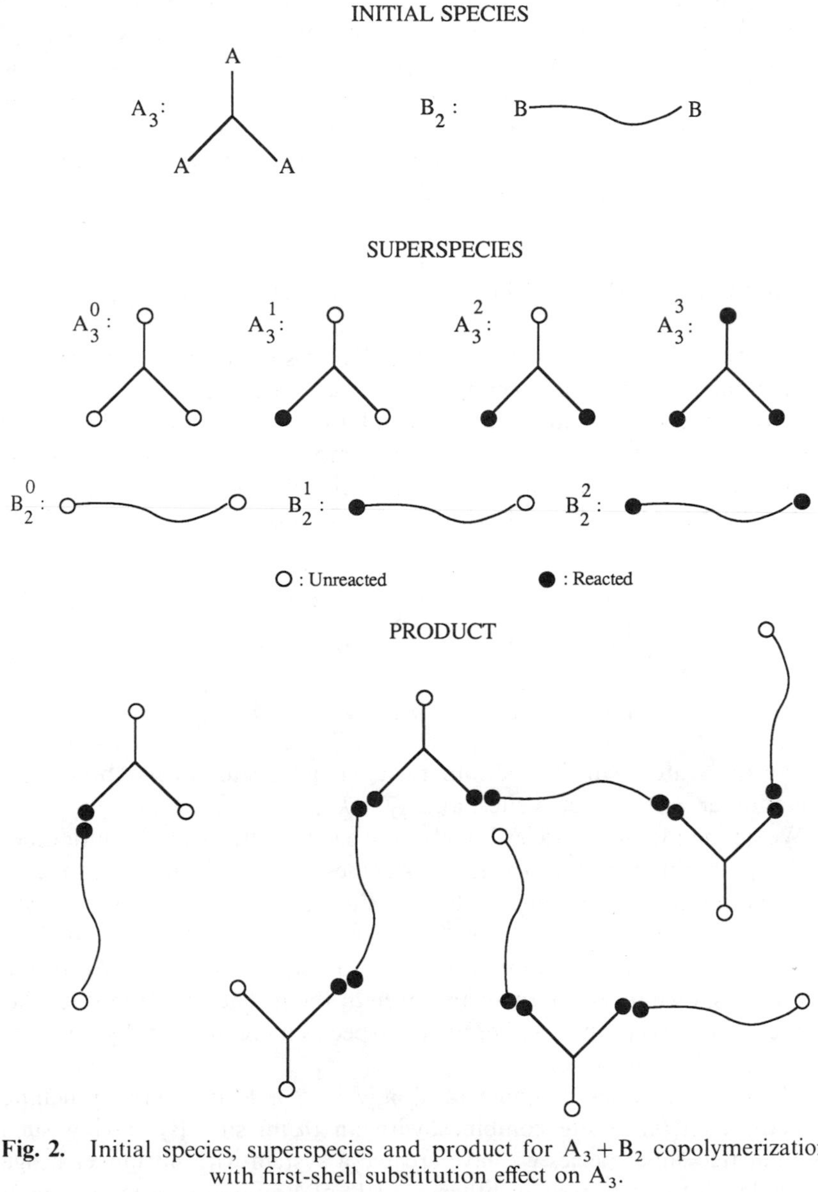

Fig. 2. Initial species, superspecies and product for $A_3 + B_2$ copolymerization with first-shell substitution effect on A_3.

random way. Furthermore, the seven superspecies go to complete reaction. The resulting structure is a branching process with seven types of components: it can be analyzed using any of several methods (the pgf approach of Gordon,[5] the recursive approach of Macosko and Miller,[6] the analysis of generation sizes as mentioned by Burchard,[7] or even the Flory–Stockmayer combinatorial approach[4]). The point is that by the construct of superspecies it is possible to identify a Markovian branching process which can then be analyzed using existing techniques.

HOMOPOLYMERIZATION WITH SUBSTITUTION EFFECT

Consider an A_f homopolymerization with first-shell substitution effect. At first glance this might be considered as a special case of the above copolymerization example, but in general it is more complicated. For this case it is necessary to define more complicated superspecies of A-monomers, if our goal is a Markovian branching structure.

For a general first-shell substitution effect homopolymerization, we have a matrix of kinetic rates

$$
\mathbf{K} = \begin{bmatrix}
k_{00} & k_{01} & k_{02} & \cdots & k_{0,f-1} \\
k_{10} & k_{11} & k_{12} & \cdots & k_{1,f-1} \\
\vdots & & & & \\
k_{f-1,0} & k_{f-1,1} & k_{f-1,2} & \cdots & k_{f-1,f-1}
\end{bmatrix}
\tag{13}
$$

where $k_{i,j} =$ rate at which a monomer with i reacted sites forms a bond with a monomer with j reacted sites, and $k_{i,j} = k_{j,i}$.

We define the superspecies by labelling each reacted site of a monomer with a pair of integers (m, n), where m signifies that it is the mth site to react on the monomer and n signifies that the site with which it reacts is the nth site to react on the monomer to which it belongs, $1 \le m, n \le f$. The superspecies for an A_2 homopolymerization with first-shell substitution effect are shown in Fig. 3: there are seven of them. The situation is similar for general f, except the number of superspecies increases rapidly: for $f = 3$ there are 40 superspecies.

These superspecies combine randomly into a Markovian branching structure: an (m, n) site combines with an (n, m) site. By having such complicated superspecies we have created a system without longer range dependencies which exist in other models of homopolymerization with general first-shell effect.

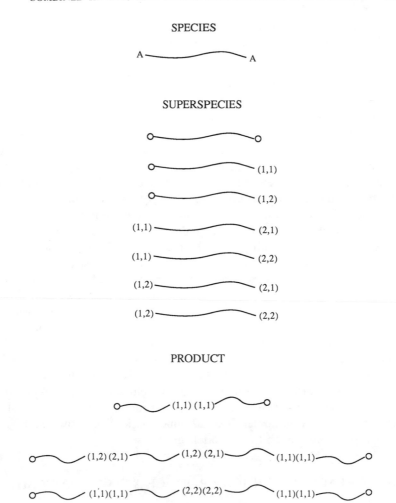

Fig. 3. Species, superspecies and product for A_2 homopolymerization with first-shell substitution effect.

CROSSLINKING OF PRE-MADE CHAINS

Consider a system which forms by crosslinking of pre-made chains which have arbitrary distributions of mass and functionality. In this case the chains play the role of superspecies so the two stages are distinct: manufacture of the functional polymer (typically kinetically controlled) and then random crosslinking using radiation or a small monomer (typically Markovian). This case is treated by Miller and Macosko.[10,11]

COPOLYMERIZATION WITH CYCLES

Consider an $A_3 + B_2$ copolymerization in which intramolecular bonds are allowed to form. Sarmoria et al.[12] have analyzed a special case of this system in which the only allowable intramolecular bonds are those which close a cycle consisting of a single B_2-monomer forming two bonds with a single A_3-monomer. They define 13 superspecies which correspond to the possible configurations within one shell of A-monomers and four superspecies corresponding to B-monomers. The concentrations of the various superspecies are kinetically modelled assuming a Gaussian end-to-end distance and competing ring–chain reactions. The superspecies are then combined randomly into a Markovian branching structure.

Sarmoria et al.[13] have extended the above model to include large cycles. Cycles involving a single B_2 are modeled as before, but larger cycles are modeled using the spanning-tree approximation of Gordon and Scantlebury.[14] In this case, as the polymerization is proceeding, intermolecular bonds are labeled with α's, intramolecular bonds are labeled with σ's and the unreacted sites with ω's. The intramolecular bonds on cycles with more than one B_2 are then broken (each site retaining its σ label). This results in a tree structure made up of 22 different superspecies. As before, the superspecies reflect all possible configurations within the first shell surrounding an A-monomer. The 22 superspecies are shown in Fig. 4. Numerous variations on these models are possible.

STEP REACTIONS WHICH PRODUCE A SECONDARY SITE

Consider an $A_f + B_g$ copolymerization such as an f-functional epoxy and a g-functional cresol: such a system is used for transistor encapsulation. In these systems bond formation between two sites creates a new site. One of the initial reactants can react with this site, often in a chainwise manner. The block of connected sites generated from one initial reaction is one of the superspecies, although in this context it might be more appropriate to call them 'superbonds'. Figure 5 shows the situation.

Bokare and Gandhi[15] and Dusek[16] have analyzed the reaction of epoxy and diamine with polyetherification using a combined kinetic–Markovian approach. Riccardi and Williams[17,18] have also studied these systems.

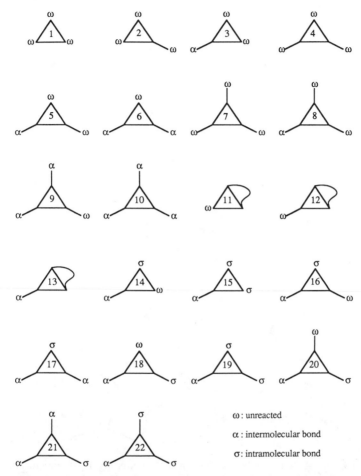

Fig. 4. Superspecies for $A_3 + B_2$ copolymerization with cycles.

Other such systems include isocyanate and hydroxyl with allophanate formation.

ANIONIC COPOLYMERIZATION OF VINYL WITH DIVINYL MONOMERS

Worsfold[19] and Lutz et al.[20] describe anionic copolymerizations. Lutz et al. consider α-methylstyrene (α-MS) plus 1,3-diisopropenylbenzene (1,3-DIB). This system is depicted in Fig. 6: the chemical structure is shown only

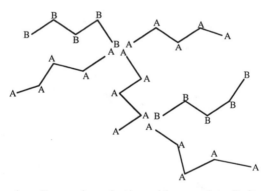

Fig. 5. $A_f + B_g$ copolymerization with secondary B-site produced.

in unreacted species. The superspecies are the living polymer chains; when a 1,3-DIB monomer is part of a chain its other group may be unreacted (labelled with ω in Fig. 6) or reacted, in which case the monomer acts as a crosslink (and the other group is labelled with an α). The relevant properties of the superspecies are computed from a kinetic analysis: these properties are very similar to those needed in the analysis of the crosslinking of ready-made chains studied by Macosko and Miller.[10,11] The superspecies are then randomly combined together into a Markovian structure and the product analyzed by branching theory. Dusek and Somvársky[21] have also studied similar systems using a combined kinetic–Markovian analysis.

UNREACTED SPECIES

α-MS

1,3-DIB

SUPERSPECIES

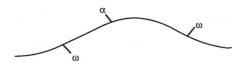

PRODUCT WITH LABELS

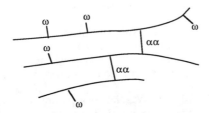

PRODUCT

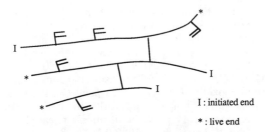

I : initiated end

* : live end

Fig. 6. α-MS + 1,3-DIB copolymerization.

ACKNOWLEDGEMENTS

This work was supported by a grant from the US Army Research Office. The authors thank Neil Dotson for helpful discussions and preparation of the manuscript.

REFERENCES

1. Dusek, K., Formation–structure relationship in polymer networks, *Brit. Polym. J.*, 1985, **17**, 185–89.
2. Ziff, R. M., Kinetics of polymerization, *J. Stat. Phys.*, 1980, **23**, 241–63.
3. Ziff, R. M. and Stell, G., Kinetics of polymer gelation, *J. Chem. Phys.*, 1980, **73**, 3492–9.
4. Flory, P. J., *Principles of Polymer Chemistry*, Cornell University Press, Ithaca, New York, 1953.
5. Gordon, M., Good's theory of cascade processes applied to the statistics of polymer distributions, *Proc. Roy. Soc. London*, 1962, **A268**, 240–59.
6. Macosko, C. W. and Miller, D. R., A new derivation of average molecular weights of nonlinear polymers, *Macromolecules*, 1976, **9**, 199–206.
7. Burchard, W., Static and dynamic light scattering from branched polymers and biopolymers, *Adv. Polym Sci.*, 1983, **48**, 1–124.
8. Dusek, K., Network formation in curing of epoxy resins, *Adv. Polym. Sci.*, 1986, **78**, 1–58.
9. Miller, D. R. and Macosko, C. W., Substitution effects in property relations for stepwise polyfunctional polymerization, *Macromolecules*, 1980, **13**, 1063–89.
10. Miller, D. R. and Macosko, C. W., Molecular weight relations for crosslinking of chains with length and site distribution, *J. Polym. Sci., Phys. Ed.*, 1987, **25**, 2441–69.
11. Miller, D. R. and Macosko, C. W., Network parameters for crosslinking of chains with length and site distribution, *J. Polym. Sci., Phys. Ed.*, 1988, **26**, 1–54.
12. Sarmoria, C., Valles, E. M. and Miller, D. R., Ring–chain competition kinetic models for linear and nonlinear step-reaction copolymerizations, *Macromol. Chem., Macromol. Symp.*, 1986, **2**, 69–87.
13. Sarmoria, C., Valles, E. M. and Miller, D. R., Models for nonlinear step-reaction polymerisations with intramolecular reactions, 1988, in press.
14. Gordon, M. and Scantlebury, G. R., Theory of ring–chain equilibria in branched non-random polycondensation system, with applications to $POCl_3/P_2O_5$, *Proc. Roy. Soc. London*, 1966, **A292**, 380.
15. Bokare, U. M. and Gandhi, K. S., Effect of simultaneous polyaddition reaction on the curing of epoxides, *J. Polym. Sci., Chem. Ed.*, 1980, **18**, 857–70.
16. Dusek, K., Build-up of polymer networks by initiated polyreactions 2. Theoretical treatment of polyetherification released by polyamine-polyepoxide addition, *Polym. Bull.*, 1985, **13**, 321–8.
17. Riccardi, C. C. and Williams, R. J. J., Statistical structural model for the build-up of epoxy–amine networks with simultaneous etherification, *Polymer*, 1986, **27**, 913–20.

18. Riccardi, C. C. and Williams, R. J. J., Kinetic and statistics of the formation of epoxy–amine networks with simultaneous etherification, 1987, in press.
19. Worsfold, D. J., Anionic copolymerization of styrene and *p*-divinylbenzene, *Macromolecules*, 1970, **3**, 514–17.
20. Lutz, P., Beinert, G. and Rempp, P., Anionic polymerization and copolymerization of 1,3- and 1,4-diisopropenylbenzene, *Makromol. Chem.*, 1982, **183**, 2787–97.
21. Dusek, K. and Somvársky, J., Build-up of polymer networks by initiated polyreactions 1. Comparison of kinetic and statistical approaches to the living polymerization type of build-up, *Polym. Bull.*, 1985, **13**, 313–19.

14

EFFECT OF DILUTION DURING NETWORK FORMATION ON CYCLIZATION AND TOPOLOGICAL CONSTRAINTS IN POLYURETHANE NETWORKS

KAREL DUŠEK and MICHAL ILAVSKÝ

*Institute of Macromolecular Chemistry, Czechoslovak Academy of Sciences,
162 06 Prague 6, Czechoslovakia*

ABSTRACT

The sol fraction and equilibrium modulus of networks, prepared from poly(oxypropylene) triols and 4,4'-diphenylmethane diisocyanate in the presence of 0–60 vol. % of xylene (diluent), were compared with theoretical predictions of the branching theory based on the spanning-tree approximation to ring formation. For dilutions ranging from 0 to 60%, the fraction of bonds lost in elastically inactive cycles increases from 2·5% to 5–7·5%, depending on the initial molar ratio of OH to NCO groups. The equilibrium modulus falls very steeply with increasing dilution, much more than would correspond to the increase of the extent of cyclization. By comparing the effect of dilution on the sol fraction, one could conclude that the theory somewhat overestimates this effect, but the differences are within the variations of about 1% in the conversion of the minority (isocyanate) groups. The steep decrease in modulus with dilution during network formation can be explained by a decrease in the number of permanent interchain constraints. The concept of trapped entanglement seems to be a reasonable approximation.

1 INTRODUCTION

Dilution during network formation is known to promote cyclization. In the pre-gel stage, the ring formation does not increase the degree of polymerization. The gel point is shifted to higher conversion. Beyond the gel point, the sol fraction increases but the effect on the equilibrium

233

elasticity is more complex. The addition of a diluent increases the population of such cycles that are not active or are partially active in the equilibrium rubber elasticity. Also, it is known that the addition of a diluent to linear polymers weakens interchain interactions and affects viscosity, particularly the position of the crossover from the unentangled to the entangled règime. The diluent thus decreases the concentration of transient entanglements. If in networks there are permanent topological constraints (permanent entanglements), these would also be affected by dilution.

The aim of this chapter is to analyze the effect of a diluent on the sol fraction and equilibrium elasticity of polyurethane networks in terms of the existing theories of network formation and rubber elasticity. The polyurethane networks were prepared from a poly(oxypropylene) triol and 4,4'-diphenylmethane diisocyanate. The experimental details and main results were published elsewhere.[1]

2 THEORETICAL TREATMENT OF NETWORK FORMATION INCLUDING CYCLIZATION

The existing theoretical approaches to network formation that take into account formation of cycles are based either on essentially tree-like models perturbed by the formation of cycles, or on simulation in n-dimensional space. In the case of the perturbed tree-like models, either a statistical build-up of structures from units or structural fragments is used, or cyclization can be taken into account in the kinetic (coagulation) generation of structures (cf., for example, Ref. 2). The extent of cyclization in networks obtained by the end-linking of reactive oligomers by step reactions is relatively weak,[3,4] so that there is a fair chance for success in applying the perturbed tree-like theories. In the pre-gel stage and at the gel point, the results of these theories are in a good agreement with those of the off-lattice simulations by Leung and Eichinger.[5] The perturbed statistical theories are, however, able to supply information on details of the gel structure.

In the statistical theories, the branching probability is decreased by cyclization, and this effect is either averaged over all possible sizes of the rings,[6] or only the smallest possible ring is considered.[4,7] The latter approximation is reasonable only if the smallest rings prevail in the population of cyclic structures.

The approach, in which the formation of rings of all sizes is considered, is based on the so-called spanning-tree approximation. In this approach, the cyclization probability is proportional to the probability that any pair of unreacted groups within the same molecule can meet in the reaction volume. The branched and crosslinked structures are built up from units (in this particular case from poly(oxypropylene) triol (PPT) and diisocyanate units (DI)) differing in the number of unreacted groups and groups reacted intra- or intermolecularly. This distribution is represented by p_{ijk}, i.e. the molar fraction of units having i groups reacted intermolecularly, j groups reacted intramolecularly and k unreacted groups. The physically possible distribution of units is shown in Fig. 1(a). This distribution is obtained by solving a set of seven and four kinetic differential equations.[6] The procedure can be improved if the distribution is amended by the fragment with the smallest cycle, as shown in Fig. 1(b).

It is worth noting that the distribution of units obtained by solution of the 11 differential equations is not random, i.e.

$$p_{ijk} \neq \frac{f!}{i!\,j!\,k!}\, \alpha^i \sigma^j \omega^k \qquad (1)$$

where $f = i + j + k$, and α, σ and ω ($\alpha + \sigma + \omega = 1$), respectively, are fractions of groups reacted intermolecularly and intramolecularly, and of the unreacted groups. The random distribution has been assumed in a number of gelation theories.

The definition of a ring for the pre-gel stage is no more applicable in the post-gel stage because of extensive ring (circuit) formation in the gel. In the gel, one has to distinguish between ring structure with elastically active chains and ring structures elastically inactive, e.g. rings within dangling chains or rings residing on elastically active network chains. In addition, a ring activation process is to be considered in which cycles bearing an unreacted group can become attached to the gel structure. Then, the ring chains may become elastically active.

The ring formation and ring activation processes are taken into account in the differential equations for p_{ijk}, which describe the transformation

$$p_{ijk} \xrightarrow{\text{intermolecular reaction}} p_{i+1,j,k-1}$$

$$p_{ijk} \xrightarrow{\text{ring formation}} p_{i,j+1,k-1}$$

$$p_{ijk} \xrightarrow{\text{ring activation}} p_{i+1,j-1,k}$$

The results of the application of this theory[6] to the above-mentioned polyurethane system[1] are shown in Fig. 2 as a function of dilution; $s_g = \sigma/(\sigma + \omega)$ is the fraction of bonds wasted in elastically inactive cycles (EIC) in the gel.

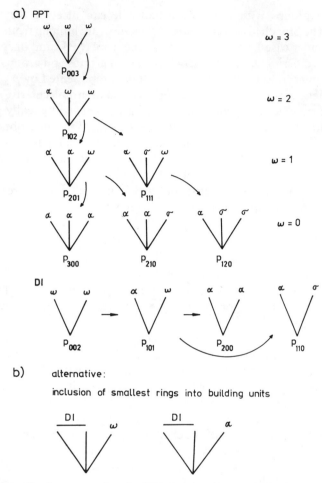

Fig. 1. Distributions p_{ijk} of units differing in the number of reacted groups engaged in inter- and intramolecular bonds and of unreacted groups: PPT, poly(oxypropylene) triol units; DI, diisocyanate units. (a) Spanning-tree approximation; (b) units containing the smallest ring.

The calculation of p_{ijk} and σ respects differences in the composition of the sol, elastically active network chains (EANC) and dangling chains, but eventually a single set of p_{ijk} is used for the build-up of the network, irrespective of whether the ring forming bonds are a part of the sol or gel. The network is then generated by routine methods of the branching theory in which only the intermolecular bonds (α) are active in branching. This

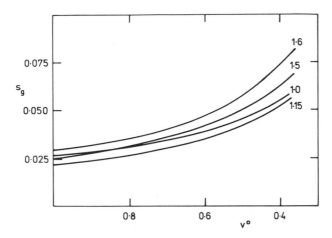

Fig. 2. Dependence of the fraction of bonds wasted in elastically inactive cycles in the gel, s_g, on the volume fraction of polymer at network formation v°. Numbers on the curves denote $r_H = [OH]/[NCO]$.

randomization is evidently the most serious approximation used in the theory. Cyclization within the sol molecules is assumed to be lower compared with the whole system because the sol molecules are less branched and their molecular weight gradually decreases. Therefore, it is expected that the theory suggested by Dušek and Vojta[6] may somewhat overestimate cyclization in the sol.

3 EXPERIMENTAL RESULTS AND DISCUSSION

Networks were prepared from the Union Carbide poly(oxypropylene) triol NIAX LHT-240 (number-average molecular weight $M_n = 708$, number-average functionality $f_n = 2.89$). The networks were prepared using varying molar ratios of OH to NCO groups, $r_H = [OH]/[NCO]$ in the range of $r_H = 1.0–1.6$, and varying amounts of the diluent xylene (volume fractions of polymer at network formation, v°, were in the range $v^\circ = 1–0.4$).

3.1 Sol Fraction
Figure 3 shows the weight fraction of the gel, $w_g(= 1 - w_s$, where w_s is the weight fraction of the soil), as a function of the molar ratio r_H and dilution expressed by the volume fraction v°.[1] The experimental data are compared with the theoretical dependence, using the theory of Dušek and Vojta[6] and

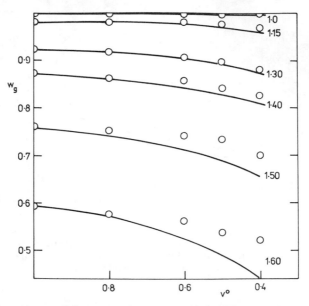

Fig. 3.   Dependence of the weight fraction of the gel, w_g, on the volume fraction of polymer at network formation, $v°$. Numbers on the curves denote $r_H = [OH]/[NCO]$. ○, experiment; ——, theory (Ref. 6, eqn (20)) for conversions found for networks with $v° = 1$.

supplements given in papers by Ilavský and Dušek.[1,8] The value of the cyclization parameter was $\Lambda = (1/N_A)[3(13 + 24\cdot4)/2\pi(10 \times 13 \times 0\cdot15^2 + 4\cdot8 \times 24\cdot4 \times 0\cdot15^2)]^{3/2}$ mol/nm³, which satisfied the gelation experiments by Matějka and Dušek.[9]

Inspection of Fig. 3 reveals that the calculated values of w_g become lower than the experimental ones when dilution increases. This result would mean that the theory overestimates the effect of cyclization on the sol fraction. However, a closer analysis of the results using the branching theory shows that the value of w_g is extremely sensitive to the final conversion of isocyanate groups. The theoretical dependences in Fig. 3 were calculated assuming that the conversion of isocyanate groups was the same for all samples containing a diluent and equal to that for bulk samples. For these the final conversion was calculated from the sol fraction and varied between 0·986 and 0·997. To make the theoretical curves coincide with experimental data, an increase in conversion by 0·008–0·014 is sufficient. Such a small change in conversion is within the limits of precision and accuracy of analytical methods available for the determination of NCO groups. Therefore, it is not excluded that the final conversion of the NCO groups may increase somewhat with increasing dilution. The conclusion made before that the theory somewhat overestimates the effect of dilution on the sol fraction cannot be considered as undoubtedly proved by the experiments.

3.2 Equilibrium Elasticity

Small-strain equilibrium moduli of samples containing the sol and the diluent, G_p, and of dry samples after evaporation of the diluent but still containing the sol, G_d, were measured. According to the rubber elasticity theory, the reduced equilibrium modulus, G_r, can be expressed as

$$G_r = A v_{eg}(\overline{r_{nf}^1}/\overline{r_0^2}) = G_p/w_g v^\circ RT_1 = G_d/w_g(v^\circ)^{2/3} RT_2 \tag{2}$$

where A is the front factor equal to $A_{ph} = (f_e - 2)/f_e$ (f_e is the effective functionality of an elastically active crosslink) for a phantom network and $A > A_{ph}$ for a real network; v_{eg} is the concentration of elastically active network chains in the gel (EANC); v° is the volume fraction of the polymer in the mixture with the diluent; $\overline{r_{nf}^2}$ is the mean-square end-to-end distance of EANCs at network formation and $\overline{r_0^2}$ is the corresponding value for the reference state (it is assumed that $\overline{r_{nf}^2} = \overline{r_0^2}$); R is the gas constant; and $T_1 = 298$ K and $T_2 = 333$ K are the temperatures of measurement of diluted and dry samples, respectively. It was found that the measurements of both dry samples and samples with the diluent obey eqn (2). This means that also the value of the front factor A is the same for dry and swollen samples.

The question is whether the decrease of the equilibrium modulus with dilution can be explained solely by cyclization, or whether a weakening of permanent topological constraints also comes into play. Figure 4 shows that the experimental decrease of the modulus is much higher than the theory predicts for a constant value of the front factor. Two possibilities exist for an explanation of this fact: (a) much more intensive formation of elastically inactive cycles in the gel than predicted by the theory; and (b) a decrease in G_r, i.e. weakening of the interchain constraints by dilution.

Let us first examine the alternative (a). In order to explain the observed decrease in the modulus solely by cyclization at a constant value of the front factor, the concentration of elastically inactive cycles should be much higher than predicted by the theory. For instance, for $r_H = 1.5$ the decrease in G_r caused by cyclization and by dilution according to eqn (2) when passing from $v^\circ = 1$ to $v^\circ = 0.4$ should amount to a factor of lower than 3, but in reality this factor amounts to almost 6. There is no theoretical reason for such increased formation of EIC. The simulation by Leung and

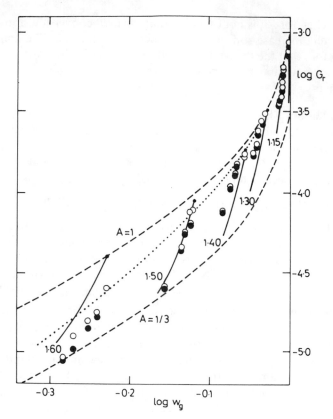

Fig. 4. Dependence of the reduced modulus G_r (mol/cm³) on the weight fraction of the gel, w_g (the numbers on the curves denote r_H values): ○, dry samples; ●, diluted samples; ....., experimental dependence found earlier[8] with undiluted networks; - - - -, theory without entanglement contribution with denoted values of the factor A; ——, theory (eqn (3)) with $A = 1/3$ and entanglement contribution εT_{eg}.

Eichinger[5] gives an even lower value for the concentration of EIC than does the spanning-tree approximation by Dušek and Vojta.[6]

Therefore, a weakening of topological constraints by dilution at network formation is likely. The values of the moduli lie within the limits corresponding to the front factor $A = 1/3$ and $A = 1$ of the Flory theory.[10] In this theory, however, the topological constraints are considered to be of transient nature and should disappear with increasing swelling and deformation. A contribution of permanent constraints is not considered. In view of the good agreement between data on swollen and dry samples (eqn

(2)), it would be difficult to explain the observed dependence of the modulus on dilution merely by the existence of transient constraints.

The other possibility is to explain the rubber elasticity data by a contribution of permanent topological constraints, e.g. in the form of trapped entanglements (cf. the theory of Langley[11] extended by Ilavský and Dušek[1,8]), to systems discussed here. According to this theory,

$$G_r = A v_{eg} + \varepsilon T_{eg} \tag{3}$$

where T_{eg} is the trapping factor calculated using the branching theory and ε is a proportionality constant. The trapping factor is the probability of contacts between segments of the EANCs and is proportional to $(v°)^2$. The factor T_{eg} thus decreases for two reasons: (a) because the number of segments in EANCs is lowered due to cyclization; and (b) because of the proportionality to $(v°)^2$. The dependence on $v°$ has a decisive effect on the dependence of the entanglement contribution to the gel fraction.

While for samples prepared at constant dilution or in bulk the effect of cyclization or incomplete conversion can largely be compensated in the G_r vs. w_g plot, it is not so when the trapped entanglement contribution comes into play, mainly due to its direct $v°$ dependence. Figure 4 shows the experimental data and the theoretical curves calculated using eqn (3) for $A = 1/3$ and $\varepsilon = 5 \times 10^{-4} \, \text{mol/cm}^3$. The agreement is relatively good. It gives strong evidence that permanent topological constraints imposed on network chains are weakened by dilution during network formation and that the theory of trapped entanglement offers a reasonable approximation to the dilution effect.

REFERENCES

1. Ilavský, M. and Dušek, K., The structure and elasticity of polyurethane networks. 5. The effect of diluent in the formation of model networks from poly(oxypropylene) triols and 4,4'-diisocyanatodiphenylmethane, *Macromolecules*, 1986, **19**, 2093.
2. Dušek, K., Formation–structure relationships in polymer networks, *Brit. Polym. J.*, 1985, **17**, 185.
3. Dušek, K., Formation and structure of end-linked elastomer networks, *Rubber. Chem. Technol.*, 1982, **55**, 1.
4. Stepto, R. F. T., Intra-molecular reaction in condensation or random polymerisation, in *Developments in Polymerisation—3*, Haward, R. N. (Ed.), Elsevier Applied Science Publishers, London, 1982, p. 81.
5. Leung, Y.-K. and Eichinger, B. E., Computer simulation of end-linked elastomers. I. Trifunctional networks cured in the bulk, *J. Chem. Phys.*, 1984, **80**, 3878.

6. Dušek, K. and Vojta, V., Concentration of elastically active networks chains and cyclization in networks obtained by alternating stepwise polyaddition, *Brit. Polym. J.*, 1977, **9**, 164.
7. Stepto, R. F. T., Intramolecular reaction and network formation and properties, *this volume*, p. 153.
8. Ilavský, M. and Dušek, K., The structure and elasticity of polyurethane networks. 1. Model networks from poly(oxypropylene) triol and diisocyanate, *Polymer*, 1983, **24**, 981.
9. Matějka, L. and Dušek, K., Formation of polyurethane networks studied by the gel point method, *Polym. Bull.*, 1980, **3**, 489.
10. Flory, P. J., Theory of elasticity of polymer networks. The effect of local constraints on junctions, *J. Chem. Phys.*, 1977, **66**, 5720.
11. Langley, N. R., Elastically effective strand density in polymer networks, *Macromolecules*, 1968, **1**, 348.

15

^{13}C-NMR ANALYSIS OF CROSSLINKING SITES IN BRANCHED POLYESTERS

SØREN HVILSTED

Polymer Group, Department of Chemistry, Risø National Laboratory, DK-4000 Roskilde, Denmark

ABSTRACT

A detailed structural analysis of a four-component polyester based on adipic acid, isophthalic acid, neopentyl glycol (NPG) and trimethylol propane (TMP) is performed by ^{13}C-NMR spectroscopy. The ^{13}C resonances of the quaternary carbons of NPG and TMP are sensitive to all the possible different dicarboxylic acid combinations. All nine possible TMP polyester units are mimicked by synthesis of models with short well-defined adipic and/or isophthalic ester branches. The ^{13}C resonance of the quaternary TMP carbon in each model corresponds exactly to one of the nine different resonances in the quaternary TMP polyester multiplet. This enables an unequivocal identification of three fundamental kinds of TMP structures in the four-component polyester: the four different crosslinking sites, the three linear segments and the two end groups. The four ^{13}C resonances of the quaternary carbons of the crosslinking sites are evenly spaced by 0·4 ppm as a result of the stronger deshielding power of an isophthalate moiety when replacing the adipate analogue.

INTRODUCTION

Polyesters comprising many different constituents are extensively used by the surface coatings industry as binders. A variety of aromatic as well as aliphatic acids with mono-, di- and trifunctionality in conjunction with different alcohols, primarily glycols and multifunctional alcohols (polyols), are employed. The use of polyols invariably results in branched polyesters.

243

Comprehensive analysis of such branched polyesters by gas chromatography of volatile derivatives results in overall determination of
constituents,[1] but provides *no* sequencial monomer distribution and thus
no structural information due to the inherently destructive derivations.
Proton (^{1}H) NMR provides a non-destructive tool which normally allows
quantifications of carboxylic acids.[2] Glycol and polyol proton resonances,
on the other hand, often overlap in polyesters, frequently even preventing
reliable polyol identifications by ^{1}H-NMR.

We have recently demonstrated[3] that a large number of glycols
condensed with a dicarboxylic acid resulting in structurally simple
polyesters are characterized by sets of conclusive carbon-13 (^{13}C)
resonances, allowing unequivocal identifications of the glycols. Moreover,
the class of constituting carboxylic acid in the polyester influences the
glycol ^{13}C chemical shift. Consequently, it is normally possible to perform
a ^{13}C-NMR identification of the five different glycol structural units
resulting from a glycol randomly copolymerized with an aliphatic and an
aromatic dicarboxylic acid.[4] These structural units comprise three glycol
triads and two methylol end groups.

STRUCTURAL COMPLEXITY

Polycondensation of a trifunctional alcohol such as trimethylol propane
(TMP) with two dicarboxylic acids is accordingly expected to accomplish
increased structural complexity, including branching. The predicted
structural complexity of a polyester based on TMP, neopentyl glycol
(NPG), adipic acid and isophthalic acid is reflected in the ^{13}C-NMR
spectrum depicted in Fig. 1. The spectral assignments also displayed in Fig.
1(a) have previously been thoroughly discussed.[4,5] It is evident that all
compositional parts of the polyester are characterized by particular [13]C
resonances. A striking feature, however, is the apparent difference between
the carboxylic single peak resonance feature and most of the TMP and
NPG resonances, which appear as highly split multiplets. A likely
explanation is the lack of carboxylic acid ^{13}C spectral sensitivity to
sequential monomer influence,[4] in marked contrast to the behaviour of the
alcohols. The apparent lack of spectral sensitivity could partly result from
the applied ^{13}C spectroscopic conditions, since the recording frequency is
only 15 MHz with 8 K data points available.

The sequential influence of the two carboxylic acids on NPG in this

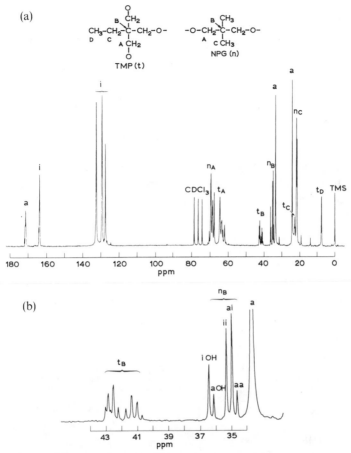

Fig. 1. ¹³C-NMR spectrum in CDCl₃ of polyester based on TMP, NPG, adipic (a), and isophthalic (i) acid: (a) Total spectrum; (b) Expansion of quaternary carbon region.

polyester is persuasively visible in the expanded ¹³C polyester spectrum in Fig. 1(b). The five clearly separated peaks all originate from the quaternary carbon (n_B) of NPG in different structural units. The peaks represent, from right to left, the three sequences where NPG is flanked by adipic (aa), mixed (ai) and isophthalic (ii) acids, followed by the two NPG end groups next to adipic (a, OH) and isophthalic (i, OH) acids, respectively.[4,5] The other multiplet (t_B) in the expanded ¹³C polyester spectrum originates from the quaternary carbon of TMP.

MODEL APPROACH AND EVALUATION

The lack of appropriate TMP ^{13}C chemical shift information necessitated an analytical model approach, initially in line with the approach applied to solve the NPG situation, in order to assign the peaks in the TMP quaternary carbon multiplet. The initial strategy aimed at preparation and ^{13}C analysis of TMP ester models, each containing only one TMP unit, in chemical environments mimicking the structural units anticipated in the four-component polyester. Accordingly, the models will additionally require adipate and/or isophthalate moieties in different ratios.

Condensation of TMP with less than three equivalents of methyl 5-(chloroformyl)-valerate (Cl-Adip) is expected to result in a mixture of TMP esters containing one, two and three adipate chains terminated with a methyl ester, respectively:

$$\text{TMP} + \underset{\text{Cl-Adip}}{\text{ClC}(CH_2)_4 \text{COCH}_3} \xrightarrow{\text{pyridine}} \underset{n=1,2,3}{\text{TMP}\!-\!(C(CH_2)_4COCH_3)_n}$$

The acid chloride condensation with TMP in pyridine[6] was deliberately chosen in order to avoid ester interchange on the adipate chains. The

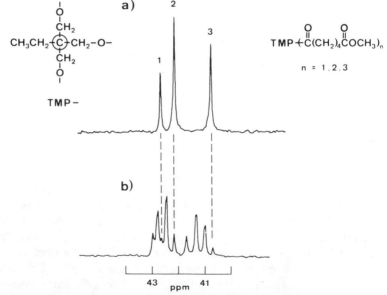

Fig. 2. Expansions of quaternary carbon region of ^{13}C-NMR spectra of TMP esters: (a) model mixture with short adipate chains; (b) four-component polyester.

quaternary carbon region of the ^{13}C spectrum of the model mixture resulting from the condensation of TMP with two equivalents of Cl-Adip is shown in Fig. 2(a). This model mixture evidently contains three different TMP esters. More importantly, however, the ^{13}C chemical shifts of these three model peaks correspond exactly to those of three of the peaks in the polyester multiplet, as indicated in Fig. 2b. The assignments of the individual adipate-containing TMP models/structural units nevertheless still remain.

One possibility is separation of the adipate model mixture by preparative-scale HPLC, followed by ^{13}C-NMR characterization of the isolated models. Alternatively, at least two of the three adipate models can be selectively prepared. In fact the model with three adipate chains is actually exclusively formed when three equivalents of Cl-Adip are employed in the condensation with TMP.

Selective preparation of the model with only one short adipate chain (TMP-Adip$_1$) is possible,[6] utilizing a ketal protecting group in the following way:

$$
\begin{array}{ccc}
\text{CH}_2\text{OH} & & \text{CH}_2\text{OH} \\
| & & | \\
\text{CH}_3\text{CH}_2{-}\text{C}{-}\text{CH}_2\text{OH} + (\text{CH}_3)_2\text{CO} & \xrightarrow[\text{toluene}]{p\text{-TSA}} & \text{CH}_3\text{CH}_2{-}\text{C}{-}\text{CH}_2 \\
| & & | \quad \diagdown \\
\text{CH}_2\text{OH} & & \text{CH}_2 \quad \text{O} \\
& & \diagdown \quad | \\
& & \text{O}{-}\text{C}{-}\text{CH}_3 \\
& & | \\
\text{TMP} & & \text{CH}_3
\end{array}
$$

$$
\text{pyridine} \Big\downarrow \quad \underset{\text{ClC(CH}_2)_4\text{COCH}_3}{\overset{\text{O} \quad\quad \text{O}}{\overset{\|\quad\quad\|}{}}}
$$

$$
\begin{array}{ccc}
\quad\quad \overset{\text{O}\quad\quad\text{O}}{\overset{\|\quad\quad\|}{\text{CH}_2\text{O}{-}\text{C(CH}_2)_4\text{COCH}_3}} & & \quad\quad \overset{\text{O}\quad\quad\text{O}}{\overset{\|\quad\quad\|}{\text{CH}_2\text{O}{-}\text{C(CH}_2)_4\text{COCH}_3}} \\
\diagdown & & \diagdown \\
\text{CH}_3\text{CH}_2{-}\text{C}{-}\text{CH}_2\text{OH} & \xleftarrow{\;\;H^+\;\;} & \text{CH}_3\text{CH}_2{-}\text{C}{-}\text{CH}_2 \\
| & & | \quad\quad \diagdown \\
\text{CH}_2\text{OH} & & \text{CH}_2 \quad\quad \text{O} \\
& & \diagdown \quad | \\
\text{TMP-Adip}_1 & & \text{O}{-}\text{C}{-}\text{CH}_3 \\
& & | \\
& & \text{CH}_3
\end{array}
$$

The advantage of the ketal group temporarily making TMP strictly monofunctional is its resistance to the basic reaction media (pyridine), in

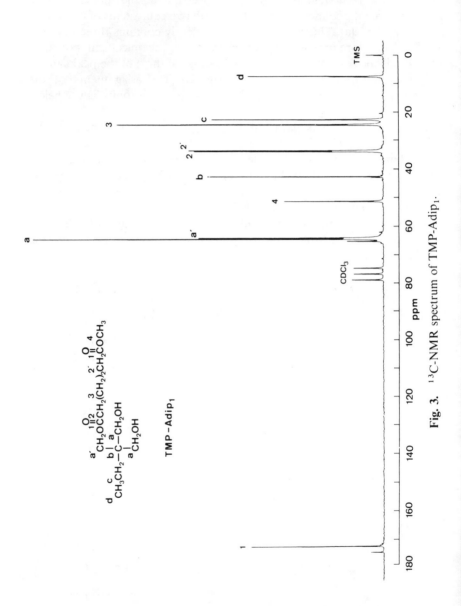

Fig. 3. ^{13}C-NMR spectrum of TMP-Adip$_1$.

combination with its subsequent easy removal under weakly acidic conditions performed with 10% aqueous acetic acid.[7]

The entire ^{13}C spectrum of TMP-Adip$_1$ in Fig. 3 demonstrates that all non-symmetrical carbons have characteristic resonances in this structure model. All carbons in both the TMP and the adipate part of the model are assigned, as seen in Fig. 3. The oxymethylene carbons (a, a') are discriminated by use of the almost 2:1 intensity ratio of the two resonance peaks. Even the resonance of the adipate methylene carbons (2, 2') next to carbonyl is split due to the proximity of different ester groups, TMP and methyl. Most important analytically in the present context, however, is the

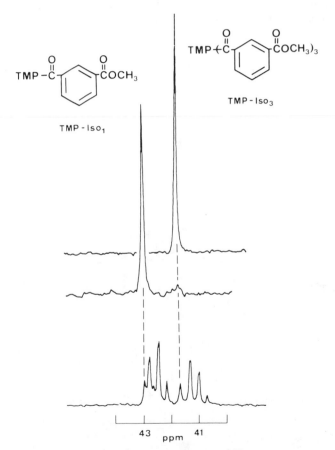

Fig. 4. Expansions of quaternary carbon region of ^{13}C-NMR spectra of TMP-Iso$_1$, TMP-Iso$_3$ and four-component polyester.

[13]C chemical shift of the quaternary carbon (b), which corresponds exactly to that of peak 1 from the model mixture in Fig. 2(a). A similar exact correspondence is confirmed between peak 3 of the mixture and the quaternary carbon of the model with three adipate chains. This model is likewise characterized by specific ^{13}C resonances. The selective preparation of models and subsequent ^{13}C analysis of these thus unequivocally identifies the compounds with one and three adipate chains, respectively, in the adipate model mixture. Accordingly, peak 2 of Fig. 2(a) originates from the model with two adipate chains. The outlined adipate model strategy positively enables the identification of the TMP moieties linked through adipate chains exclusively in the four-component polyester.

A similar selective route was designed for the isophthalate analogues.

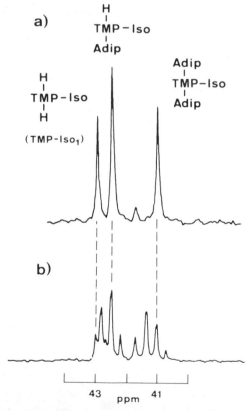

Fig. 5. Expansions of quaternary carbon region of ^{13}C-NMR spectra of TMP esters: (a) models with mixed ester chains; (b) four-component polyester.

The necessary precursor, isophthalic acid chloride methyl ester (Cl-Iso), was prepared from dimethyl isophthalate via the mono-potassium salt and the mono-acid, finally reacting the latter with thionyl chloride.[6] The TMP ester with one short isophthalate chain (TMP-Iso$_1$) was recovered after condensation of the ketal-blocked TMP and Cl-Iso catalysed by 4-dimethylaminopyridine and subsequent deblocking. The TMP model with three short isophthalate chains (TMP-Iso$_3$) was prepared by a similar condensation of TMP with the stoichiometric amount of Cl-Iso. Both models again reveal specific ^{13}C-NMR resonances. The quaternary peaks of these models are shown in Fig. 4. The positioning of the peaks in comparison with the respective peaks of the multiplet from the four-component polyester is also indicated in Fig. 4.

Preparation of models with mixed ester chains was accomplished utilizing the selectively prepared models with one chain as precursor, as illustrated with TMP-Iso$_1$:

When Cl-Adip is employed in a less than stoichiometric ratio, the above-indicated component mixture is expected.

Figure 5 confirms the expectation in as much as the depicted three quaternary carbon peaks are found in the ^{13}C spectrum of the mixture. Although the two models with mixed ester chains are not yet separated and individually characterized, the ^{13}C spectral assignment is based on the previously observed combined deshielding effects on the quaternary carbon exerted by β and higher positioned (γ, δ, etc.) neighbour groups. A hydroxyl group in the β position deshields the ^{13}C resonance more than the combined effects of the replacing ester group. Accordingly, the high-field peak of the mixed models in Fig. 5 is assigned to the quaternary carbon from the three-chain model. The corresponding peaks in the four-component polyester are again indicated in Fig. 5.

Again, a small modification of the reaction mixture to a stoichiometric ratio will allow this mixed three-chain model (TMP-Adip$_2$-Iso$_1$) to be produced exclusively. In fact, the remaining three-chain model with two isophthalate arms can likewise be selectively prepared starting from TMP-Adip$_1$.

TMP STRUCTURAL UNITS IN THE BRANCHED POLYESTER

The outlined strategy based on preparation of TMP esters with definite structural composition and subsequent ^{13}C analysis has revealed an exact correspondence in pairs between the resonances of the quaternary TMP carbons in the multiplet from the four-component polyester and in each of the nine TMP models. The nine possible TMP structural units resulting from the co-condensation with both adipic and isophthalic acid are shown in Fig. 6. The spectral assignments also depicted here clearly demonstrate how the ^{13}C resonances are grouped in three fundamental TMP structures: crosslinking sites (three branches), linear segments or triads (two branches + one methylol) and end groups (one branch + two methylol).

The primary reason for the ^{13}C resonance grouping is the deshielding power of the hydroxyl group in the β position to the quaternary carbon, which is superior to the effect of either adipate or isophthalate. The same effect was observed on the quaternary carbon of NPG as shown in Fig. 1(b). Also, Newmark et al.[8] observe hydroxyl deshielding on the quaternary carbon of pentaerythritol (PA, tetrafunctional) in the simple two-component polyester formed when caprolactone polymerization is initiated by PA. The quaternary PA carbon peaks differed by more than 0·5 ppm and were assigned to the various branched structures containing three, two, one and no methylol groups, respectively.

Within each structural group a consistent periodicity is recognized. This is most easily observed in the group with the crosslinking sites (Fig. 6). But even within the two other structural groups the peaks are equally spaced. An isophthalate moiety deshields the quaternary carbon of TMP by

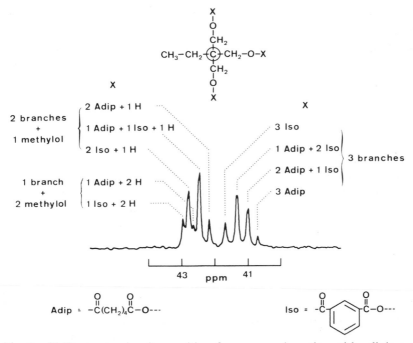

Fig. 6. TMP structural units resulting from co-condensation with adipic and isophthalic acid and assignments in expansion of quaternary TMP carbon region of ^{13}C-NMR spectrum of four-component polyester.

0·4 ppm more than the adipate analogue. Furthermore, a simple additivity of this effect is experienced. The combination of the ^{13}C resonance sensitivity to long-range substitutional effects and the additivity of the effects is, thus, the reason for the substantiated ^{13}C-NMR structural sensitivity in this TMP-based branched four-component polyester.

FUTURE PROSPECTS

The completed ^{13}C-NMR structural analysis of this four-component polyester—revealing all 14 possible polyol structural units, five involving NPG and nine including TMP—is to our best knowledge the most detailed structural analysis ever performed on branched polyesters. The potential identification of the different crosslinking sites appears especially as an exciting new possibility. The approach presented is suggested as a universal structural tool for investigation of other branched polyesters or complex polycondensation systems before gelling.

Still remaining, however, is the attractive quantification of the structural polyol units including the crosslinking sites, which could be used for determination of the relative reactivities of the acids and of the possible substitution effect in the polyol. This information in combination with the branching theory could describe gelation and network formation in terms of viscosity changes, sol fraction and crosslinking density. Consequently, separation and ^{13}C-NMR analysis of all the TMP ester models are currently in progress in order to evaluate parameters influencing the quantitative use of ^{13}C-NMR, especially spin–lattice relaxation times, T_1. The other parameter, the Nuclear Overhauser enhancement (NOE), is not expected to complicate the quantifications. All TMP structural units will be quantified by use of the elucidated specific quaternary ^{13}C resonances, which all have the same four α-methylene substituents, one in an ethyl group and three in methoxy groups, and thus the same NOE.

ACKNOWLEDGEMENT

The assistance of Birgit Svendsen and Ulrik Jørgensen in recording NMR spectra is gratefully acknowledged. The author additionally thanks the Danish Technical Research Council for partial financial support.

REFERENCES

1. IUPAC, Recommended methods for the analysis of alkyd resins, *Pure Appl. Chem.*, 1973, **33**, 413–35.
2. Marshall, M., NMR analysis of paint media, *J. Oil Col. Chem. Assos.*, 1983, **66**, 285–93.
3. Hvilsted, S. and Jørgensen, N. U., Polyol structural elucidation in binder polyesters. 1. ^{13}C NMR study of fundamental polyols in aliphatic and aromatic polyesters, *Polym. Bull.*, 1983, **10**, 236–43.
4. Hvilsted, S., Structure elucidation of polyester binders for coatings, in *Organic Coatings—Science and Technology, Vol. 8*, Parfitt, G. B. and Patsis, A. V. (Eds), Marcel Dekker Inc., New York, 1986, pp. 79–108.
5. Hvilsted, S., Potential structural elucidation in binder polyesters, *XVIIth FATIPEC Congress*, Vol. I, Lugano, 1984, pp. 347–61.
6. Hvilsted, S., Experimental details, in preparation.
7. Schöllner, R. and Löhnert, P., Darstellung einiger Adipinsäureester des Glycerins und des Trimethylolpropans, *J. prakt. Chem.*, [4], 1968, **38**, 162–7.
8. Newmark, R. A., Runge, M. L. and Chermack, J. A., ^{13}C NMR sequence analysis of polyesters from pentaerythritol and caprolactone, *J. Polym. Sci., Polym. Chem. Ed.*, 1981, **19**, 1329–36.

16

FLUOROELASTOMERS: REACTION PRODUCTS IN EARLY STAGES OF NETWORK FORMATION

Gianna Cirillo, Graziella Chiodini, Natalino Del Fanti, Giovanni Moggi

Montefluos/CRS, Via S. Pietro 50, 20021 Bollate, Milan, Italy

and

Febo Severini

Department of Chemical Industry, The Polytechnic, Piazza Leonardo da Vinci 32, 20133 Milan, Italy

ABSTRACT

Reactions occurring in the first steps of crosslinking of vinylidene fluoride (VDF)–hexafluoropropene (HFP) elastomer copolymers (molar ratio 4/1) are investigated operating in homogeneous solution, phase transfer catalysis conditions and bulk. Two kinds of principal unsaturation, monitored by FT-IR and ^{1}H-NMR spectrometry, —CH=CF— and (CF$_3$)C=CH—, have been detected. The former seems to be the curing site in the nucleophilic crosslinking of the fluoroelastomer. Polymer chain scission is also observed and has to be related to the HFP units as well as to the base content in the system.

INTRODUCTION

The crosslinking reaction of vinylidene fluoride (VDF)–hexafluoropropene (HFP) copolymers can be summarized in a three-step process:

—base-induced dehydrofluorination producing polymer chain unsaturation;
—primary network formation by reaction of unsaturated chains with a
   bisnucleophilic agent;
—ultimate network formation after heating.

The dehydrofluorination reactions have been investigated in great detail by ^{19}F-NMR in dimethylacetamide (DMAC) solution.[1] It has been found that the sequence HFP-VDF-HFP, or

$$-CF_2-\underset{\underset{CF_3}{|}}{CF}-CH_2-CF_2-CF_2-\underset{\underset{CF_3}{|}}{CF}-$$

is the only selective base-sensitive site leading to the diene group

$$-CF{=}\underset{\underset{CF_3}{|}}{C}-CF{=}CF-CF_2-\underset{\underset{CF_3}{|}}{CF}-$$

while about two equivalents of OH^- are consumed per equivalent of detected F^-. It was suggested that the diene group is the only cure site.[1]

The aim of this work is to study the first and second steps of this process. The dehydrofluorination reaction has been investigated operating with homogeneous and heterogeneous systems. Reactions with the unsaturated polymer (after the first step) have also been studied.

EXPERIMENTAL

Homogeneous Phase Dehydrofluorination
Dehydrofluorination was investigated by dissolving the copolymer in suitable solvents, such as acetone, THF, DMAc, and DMSO in the presence of alcoholic KOH. The reaction was monitored by analysing the F^- ions in the solution. It proceeded to completion within a few minutes and, at the same base concentration, its rate was found to depend on the nature of the solvent. Homogeneous phase dehydrofluorination has been studied in more detail using THF as a solvent.

Phase Transfer Catalysis (PTC) Dehydrofluorination
The reaction was carried out by dissolving the copolymer in a water insoluble solvent such as methyl-t-butyl ether, and treating the stirred solution in the presence of a PTC agent, usually an onium salt such as $[C_{16}H_{33}P(C_4H_9)_3Br]$.

Bulk Dehydrofluorination
Copolymers containing 6% $Ca(OH)_2$ and 0.5% onium salt were mixed on a cold lab mill and then heated to 150°C for 30 min. Only the MEK-soluble fraction has been used for the analysis.

Techniques

FT-IR spectra were recorded on a Nicolet 20 SXB. Films of the copolymer were obtained from a THF solution. The ^{1}H- and ^{19}F-NMR spectra were recorded on a Varian XL-200 spectrometer, working in FT at 200 MHz for ^{1}H and 188 MHz for ^{19}F. Number-average molecular weights ($\bar{M}_n$) were measured with a Knauer membrane osmometer at 30°C, solvent MEK.

RESULTS AND DISCUSSION

Figure 1 shows the relationship between KOH concentration and unsaturation as well as molecular weight ($\bar{M}_n$) operating in a THF solution of VDF/HFP copolymer (molar ratio close to 4/1) at 25°C (homogeneous phase).

While unsaturation is proportional to the KOH content, $\bar{M}_n$ decreases until an asymptotic value is reached. The addition of further KOH induces only new unsaturation but no more polymer chain scission.

The FT-IR spectra of the VDF/HFP copolymer before and after alkaline treatment in the homogeneous phase, as reported above, show important differences. Concerning C–H stretching absorption, we notice after

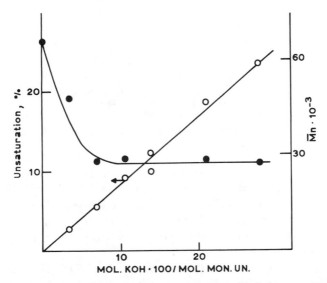

Fig. 1. Dependence of added base on unsaturation level and on average number molecular weight $\bar{M}_n$ of the VDF/HFP copolymer.

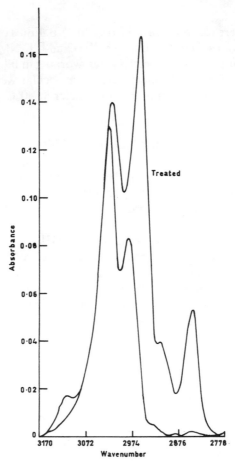

Fig. 2. FT-IR spectrum (C–H stretching region) of the indicated and alkaline-
treated VDF/HFP copolymer.

treatment the appearance of some new signals different from the symmetric
and asymmetric stretching of —CH_2— due to the VDF units (Fig. 2).

Two strong absorptions (Fig. 3) are observed in the double bond region
at 1720 cm^{-1} and 1684 cm^{-1}, related, respectively, to —CH=CF— and to
—$C(CF_3)$=CH—. The 1638 cm^{-1} band is probably related to conjugated
double bonds.

Figure 4 shows the FT-IR spectra of the VDF/HFP copolymer
dehydrofluorinated with different techniques. The dotted line represents
homogeneous treatment, with alcoholic KOH, of the copolymer dissolved

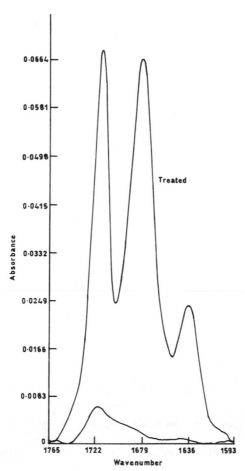

Fig. 3. FT-IR spectrum (double bond region) of the untreated and alkaline-treated VDF/HFP copolymer.

in THF, and shows the formation of three bands. The dashed line refers to the heterogeneous PTC dehydrofluorination. The signal at $1720\,cm^{-1}$ is also strong here, while that at $1684\,cm^{-1}$ is weaker. This suggests that PTC treatment seems to attack VDF units having no HFP adjacent groups more easily than HFP-VDF diads having the CF_3- side bulk substituent. This fact can be related to steric reasons, since the PTC agent is a large molecule $([C_{16}H_{33}P(C_4H_9)_3]^+OH^-)$. The thick line in Fig. 4 is very similar to the dashed one, suggesting that the heterogeneous bulk treatment shows a feature similar to PTC dehydrofluorination.

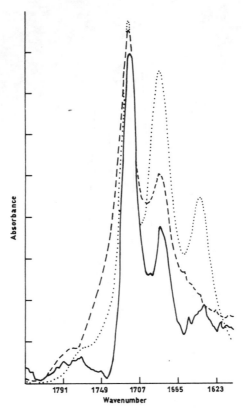

Fig. 4. FT-IR spectra (double bond region) of VDF/HFP copolymers after alkaline treatments: ·····, homogeneous phase dehydrofluorination; -----, PTC dehydrofluorination; ———, bulk dehydrofluorination.

Figure 5 shows the ^{1}H-NMR spectrum of a VDF/HFP copolymer after alkaline treatment. Signals at 6·3 and 5·3 ppm have been attributed to the proton in the —$\underline{H}$C=CF— and —C(CF$_3$)=C$\underline{H}$— groups, respectively.

In Fig. 6 the intensity of the 6·3 ppm ^{1}H-NMR signals has been plotted against the optical density of the IR 1720 cm^{-1} band. The observed linear relationship indicates a good agreement between the results of these techniques.

Differences between ^{19}F-NMR spectra of VDF/HFP copolymers before (Fig. 7) and after (Fig. 8) alkaline treatment are shown by arrows. The decreasing intensity of B″, G, F, and K$_f$ signals demonstrates that monomer sequences HFP-VDF-HFP and VDF-VDF-HFP are important sites of

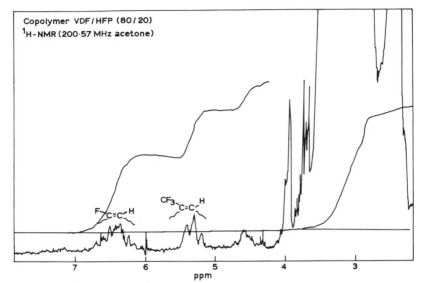

Fig. 5. 200 MHz ^{1}H-NMR spectrum of the VDF/HFP copolymer after homogeneous alkaline treatment.

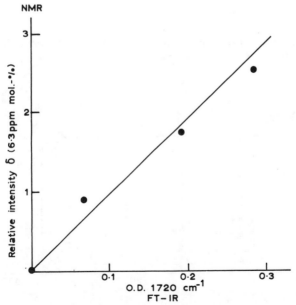

Fig. 6. Intensity of —CH=CF NMR signal vs. optical density of the IR 1720 cm^{-1} band.

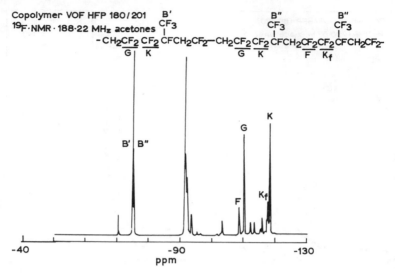

Fig. 7. 188 MHz ¹⁹F-NMR spectrum of the VDF/HFP copolymer before alkaline treatment.

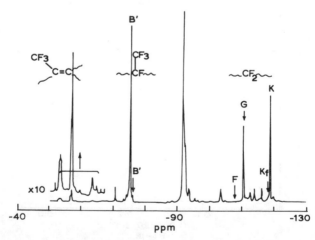

Fig. 8. 188 MHz ¹⁹F-NMR spectrum of the VDF/HFP copolymer after homogeneous alkaline treatment.

reaction, while VDF inversions are unaffected by alkaline treatment. The new signals appearing in the range -50 to -70 ppm are attributed to the different kinds of $-C(CF_3)\!\!=\!\!C-$ groups. There is a linear relationship between the intensity of the changing signals and the amount of alkali added to the copolymer.

Figure 9 shows the FT-IR spectra of dehydrofluorinated VDF/HFP (80/20) copolymer before and after reaction with sodium p-chlorophenate. The appearance of the 1591 cm^{-1} band due to the aromatic ring, the

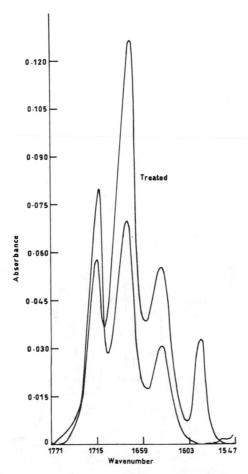

Fig. 9. FT-IR spectra (double bond region) of dehydrofluorinated VDF/HFP copolymer before and after treatment with sodium p-chlorophenate.

decrease of the $1720 \, \text{cm}^{-1}$ band, and the increase of the $c. 1680$ and $1640 \, \text{cm}^{-1}$ bands are evidenced. The decrease of the $1720 \, \text{cm}^{-1}$ band suggests that the —CF=CH— reacts with the p-chlorophenate ion, while the increase of the other band is likely to be due to further dehydro-fluorination.

Figure 10 shows the FT-IR spectra of dehydrofluorinated VDF/HFP (80/20) copolymer before and after reaction with bromine in ionic conditions. The intensities of the 1720 and $1640 \, \text{cm}^{-1}$ bands do not change,

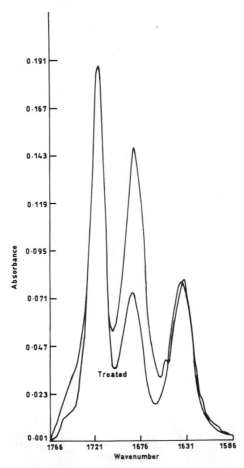

Fig. 10. FT-IR spectra (double bond region) of dehydrofluorinated VDF/HFP copolymer before and after treatment with bromine.

while the $1680\,cm^{-1}$ band decreases. This indicates that bromine reacts with —C(CF$_3$)=CH—, as proposed by Blagodatova.[2]

$$
\underset{\displaystyle \overset{|}{\underset{}{}}}{\overset{\displaystyle CF_3}{\underset{\displaystyle -C=CH-}{|}}} + Br_2 \longrightarrow \underset{\displaystyle Br\ \ Br}{\overset{\displaystyle CF_3}{\underset{\displaystyle -C-CH-}{|\ \ \ \ |}}}
$$

CONCLUSIONS

Alkaline treatment of VDF/HFP copolymers (molar ratio 4/1) in homogeneous, PTC and bulk systems induces in all cases at least two kinds of unsaturation. FT-IR spectra show absorptions (proportional to the added base) at:

(1) $1720\,cm^{-1}$ due to —CF=CH— groups. This absorption is also the only one detected after alkaline treatment of chlorotrifluoro-ethylene–vinylidene fluoride and tetrafluoroethylene–vinylidene fluoride copolymers.

(2) $1680\,cm^{-1}$ related to —(CF$_3$)C=CH—. Its relative intensity increases with the HFP content in the copolymer.

Both these kinds of unsaturation are also detected by ^{1}H-NMR. The —CF=CH— signals appear at 6·3 ppm. Moreover, ^{19}F-NMR shows a decrease of —CF$_2$—C(C$\underline{F}_3$)F— signals and the appearance of —C(C$\underline{F}_3$)=C— at 53·5 and 56·8 ppm, which are different results from those reported in the literature.[1]

A signal at $1640\,cm^{-1}$ is present only in the copolymers treated in alkaline homogeneous solution, which is probably related to conjugated double bonds. This band seems to be peculiar to this treatment, as it does not appear in the copolymers treated in the heterogeneous systems.

Polymer chain scission is also observed; this has to be related to the HFP units. No degradation is in fact observed in similar experimental conditions with poly(vinylidene fluoride) and tetrafluoroethylene–vinylidene fluoride copolymers. FT-IR spectra of the copolymer treated with sodium p-chlorophenate suggest that —CF=CH— should be the curing site in the bisnucleophile crosslinks of the fluoroelastomers.

REFERENCES

1. Schmiegel, W. W., *Kautsch Gummi Kunstst.*, 1978, **31**, 137; *Angew. Makromol. Chem.*, 1979, **76/77**, 39.
2. Blagodatova, O. V., Zeperalova, T. B. and Kozlova, O. V., *Soedin. Ser. B*, 1985, **27**, 362.

17

A MODEL FOR INHOMOGENEOUS NETWORK FORMATION BY CHAIN-REACTION POLYMERIZATION

H. M. J. Boots

Philips Research Laboratories, PO Box 80000, 5600 JA Eindhoven, The Netherlands

'Though this be madness, there is method in't.'
William Shakespeare, *Hamlet*

ABSTRACT

Network formation by chain polymerization is an intrinsically inhomogeneous process. The Kinetic Gelation Model has been shown to produce such inhomogeneities in a natural way. These results are extended in several ways. The chance of radical trapping and of cage recombination is calculated and the effect of chain stiffness on the extent of crosslinking is determined. Moreover, the extent of crosslinking of material formed during a specific conversion interval is discussed in connection with recent neutron scattering experiments by Sperling and co-workers.

INTRODUCTION

Strong indications for inhomogeneity during the formation of polymer networks by chain reactions have been found by many authors.[1-9] Here we want to mention the permeability,[2] the early strong crosslinking[3-5] and radical trapping and various scattering results.[1,6] However, the usual models of network formation,[7] which are very satisfactory in many situations, cannot be used to describe inhomogeneity, because they disregard any spatial dependence. The only model for free-radical crosslinking polymerization which does include spatial dependence is the

267

simulation model given by Manneville and De Sèze,[8] which is known as the Kinetic Gelation Model. In general[8,9] it is not used for studying inhomogeneity. Our aim in using the model for this purpose[5,10] is to describe the essential features and trends of the process rather than to obtain quantitative agreement by the introduction of adjustable parameters or specific interactions. 'Snapshots' taken during a simulation of polymerization in two dimensions showed an inhomogeneous spatial distribution of polymer. We also found that the model yielded the early strong crosslinking which was observed experimentally.

In the present paper the model is extended to include chain stiffness. The early trapping of radicals and the results of recent neutron scattering work by Sperling and coworkers[6] are discussed in the framework of the model.

THE MODEL

The chain reactions are modeled as random walks on a simple cubic lattice. Each lattice site contains a monomer unit. If the unit is bifunctional (monovinyl) it may be visited not more than once by a random walk. A tetrafunctional site may be visited not more than twice; if it has been visited once, we speak of a 'pendant double bond', if twice, it constitutes a crosslink between two chains or between two parts of one chain. (Here no distinction is made between crosslinking and cyclization.)

Initiation occurs by creating radicals on free monomer sites at a constant rate. One may either form single radicals or pairs of radicals on neighboring sites. To obtain the lowest possible initiation rate, we also studied the situation where one radical is active at a time; if it is terminated, a new one is created. On a lattice of one million sites such a low rate is typical of many experiments.

A propagation step consists of randomly choosing a radical unit and one of its neighbors; if the neighbor has not fully reacted a bond is formed and the radical character is transferred to the neighbor site. Usually it is assumed that stepping backward is forbidden for steric reasons.

Termination occurs if the two sites chosen for a propagation step are both radicals (termination by encounter) or if a radical is trapped between fully reacted sites. A pair of radicals that has just been formed may recombine; then the situation before the formation is restored. The enhancement of recombination when many neighbors of the pair of radicals have fully reacted is called cage recombination.

Chain transfer to monomer or polymer is not considered.

INHOMOGENEITY

The inhomogeneity of the process is illustrated in Fig. 1(a) by snapshots of a two-dimensional version of the model. Comparison between Fig. 1(a), (b) and (c) clearly illustrates the experimental observation by Silberberg[2] that shorter kinetic chain lengths lead to more homogeneous systems; in fact,

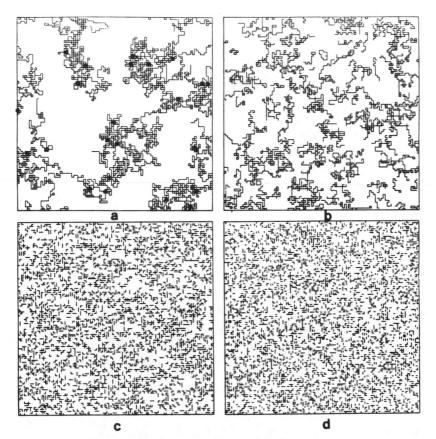

Fig. 1.   Snapshots of polymerization of tetrafunctional monomer on a square 100 × 100 lattice at a bond conversion of 25%. Bonds between units are indicated, units are not. Boundary conditions are periodic. Part 1(a) illustrates a chain reaction of one active radical at a time; when it is trapped (dark squares) a new one starts. To show the increase in inhomogeneity on decreasing kinetic chain length, the maximum kinetic chain length is limited to 20 steps and to two steps in Parts 1(b) and 1(c). Part (c) is as homogeneous as Part (d) for a step polymerization after conversion of 25% of the reactive groups.

chain processes at unrealistically short kinetic chain lengths (Fig. 1(c)) lead to a homogeneity comparable to that of step processes (Fig. 1(d)).

Towards the end of the reaction the system becomes more and more homogeneous. By creating single radicals on free monomer sites, we can reach complete monomer conversion: pair creation of radicals leads to almost complete monomer conversion. Thus in the end, (almost) all sites of the lattice are polymer sites. Experimentally, inhomogeneity at the end of the reaction is possible due to polymerization shrinkage and vitrification. These phenomena are beyond the scope of the model.

PENDANT DOUBLE BONDS

In Fig. 2 the fraction of pendant double bonds, i.e. the fraction of polymer units that are not crosslinks, is plotted against monomer conversion. The experimentally[3-5] observed early strong crosslinking (in comparison with the result of a 'classical' approach based on average concentrations of reactants) is well described by the simulation model. This early strong

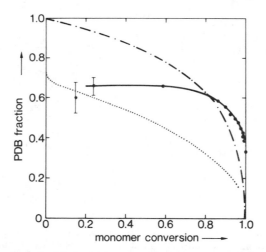

Fig. 2. The pendant double-bond fraction (the fraction of polymer units carrying an unreacted vinyl group) as a function of monomer conversion. The dashed-dotted line (—·—·) results from the usual mean-field approach, the drawn curve (——) guides the eye through the data from Kloosterboer et al.[5] (●) on hexane diol diacrylate (HDDA), and the dotted curve (·····) represents the model results for divinyl monomer on a $60 \times 60 \times 60$ lattice.

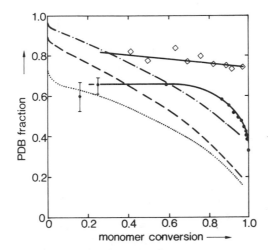

Fig. 3. The effect of stiffness on the crosslink probability: ·····, propagation of a radical in all six directions allowed; − − −, stepping backward is forbidden; − · − · −, stepping backward is forbidden and only after two steps in the same direction a radical function is allowed to change direction. (●) flexible (HDDA) and (◇) stiffer (ethylene glycol diacrylate) monomer from Kloosterboer et al.[5,11] (solid lines are to guide the eye).

crosslinking has been interpreted before as an indication of inhomogeneity;[4] the simulation model supports this interpretation.

At high conversion there is a large discrepancy between the experiment and the simulation model. This discrepancy has been ascribed to the fact that the model overestimates the accessibility of pendant double bonds at high conversion.[5]

In Fig. 3 we show the effect of chain stiffness. Increase of chain stiffness diminishes the effect of early strong crosslinking. This is in qualitative agreement with experiments by Kloosterboer.[5,11] However, the experimental plateau is not predicted by the model. This serious discrepancy is of a similar origin as the discrepancy discussed in connection with Fig. 2.

Figure 4 is meant to illustrate an effect which has been used by Sperling and co-workers to explain their neutron scattering results. During a copolymerization of styrene with 1% divinyl benzene, they used deuterated styrene only in a small interval of the polymerization. They observed that the molecular weight of deuterated polymer was higher, if the deuteration interval started at higher conversion. This 'non-randomness' is explained by the fact that deuterated parts of chains grown in that interval are restricted to the holes in the inhomogeneous polymer structure built up

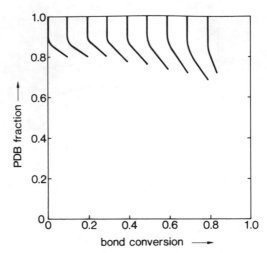

Fig. 4. Pendant double-bond fraction of material formed in intervals of 10% bond conversion for nine consecutive intervals as a function of bond conversion.

before deuteration. As the chains do not easily enter into the highly reacted protonated polymer, the number of crosslinks in deuterated material is enhanced.

In Fig. 4 the amount of crosslinking in polymer produced in a limited conversion interval is shown for bulk divinyl polymerization. For consecutive intervals of 10% bond conversion, we plot the pendant double-bond fraction of polymer units converted to polymer in the course of the interval under consideration. It is seen that there is a tendency towards stronger crosslinking if the interval is taken at a higher conversion, in qualitative agreement with the results of Sperling and co-workers.

THE FATE OF RADICALS

In electron spin resonance experiments, trapped radicals are observed from 25% monomer conversion (50% bond conversion) onwards.[2] The early occurrence of trapping is considered to be an indication of inhomogeneity. Extraction of initiator after polymerization shows that cage recombination of primary radicals is very strong at high conversion.[13]

Due to the absence of spatial dependence, one cannot apply the usual models for network formation to the problem of radical trapping and cage recombination. The predictions of the present model for the fate of radicals are presented in Fig. 5 for two initiation rates. The lowest rate is still high in

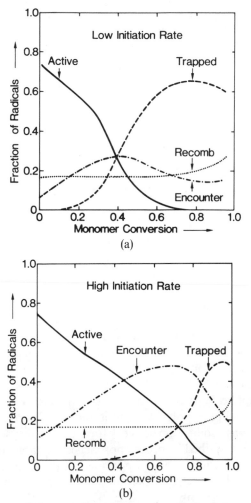

Fig. 5. Fractions of the number of formed radicals that are active, that are trapped, that have been terminated by encounter, and that have recombined before propagation, as a function of monomer conversion on a $60 \times 60 \times 60$ lattice for initiation rates of one pair creation every (a) 1000 and (b) 10 steps.

comparison with the usual experimental rates. At low conversion the model predicts a chance of 1/6 for recombination of a pair of primary radicals. This number is rather arbitrary: it results from the use of a cubic lattice and equal probabilities for all possible reactions. At high conversions the recombination rate of primary radicals increases due to cage effects.

Termination by encounter (combination or disproportionation) is largest at rather low conversions; in this region it is certainly underestimated due to the absence of segmental diffusion in the model. Trapping of radicals increases strongly in the range from 30 to 50% monomer conversion in Fig. 5(a). It must be kept in mind that such numbers cannot be related directly to experiment. However, the model result confirms the experimental finding that trapping is important long before the free monomer is exhausted. In the final reaction stage the relative importance of trapping decreases somewhat due to the increasing probability of cage recombination.

DISCUSSION

The Kinetic Gelation Model is a kinetic model in the special sense that relaxation to equilibrium conformations is supposed to be very slow compared with growth. Therefore effects like termination by segmental diffusion, the Trommsdorf effect, polymerization shrinkage and phase separation by specific interactions between the units are not included in the model. Extension of the model to include dynamics is problematic for network formation: a lattice model can neither be used for the dynamics of crosslink units nor for shrinkage on chemical reaction. Off-lattice models for crosslinking polymerization and dynamics have been found to require a prohibitive amount of computer time due to the fast increase of relaxation times during reaction.

Though the applicability of the method is limited to systems which soon form strongly crosslinked polymer regions and to properties which do not depend strongly on mobility, the network formation processes to which it is limited cannot be described by other approaches available to date. For the largely kinetic effects which are of importance for the formation of densely crosslinked networks by chain reactions, the model is very useful. The model is conceptually simple and may be used to obtain a qualitative insight into general features of networks formed by chain reactions.

The model predicts inhomogeneity in such networks and it demonstrates clearly that this inhomogeneity is an inherent property of the chain-reaction crosslinking process: no specific interactions are assumed. Moreover, a similar model for step reactions does lead to homogeneous results. The higher the initiation rate one uses, the more homogeneous is the system, in accordance with experiment. The model supports the interpretation of early strong crosslinking and radical trapping as resulting from inhomogeneity. It predicts that more stiffness leads to less crosslinking. The model shows some effects of inhomogeneity which have recently been found in neutron scattering experiments.

The absence of a plateau in the plot for the pendant double-bond fraction and the small cage effect for the recombination of radicals both indicate an underestimation of the shielding in the model. This may be explained as follows. In the model, units are points and pendant double bonds cannot be screened from further reaction by other parts of the same unit. In the experimental situation, however, this shielding effect is expected to become important when the mobility of the polymer units is low. Since the monomer mobility is much higher, the formation of a pendant double bond is then preferred to the formation of a crosslink in a dense polymer region. The simplest way to avoid the assumption of point-like monomer would be to model divinyl units as dimers on the lattice. This will be studied in the near future.

The foregoing discussion is of importance for the interpretation of the results on labeled polymer in Fig. 4 as an increased tendency for crosslinking of polymer formed in the holes between dense polymer regions. In these holes the polymer density is low and crosslinking will not be suppressed; in fact we find an increasing tendency for crosslinking inside the holes during conversion, as growing chains are confined to smaller and smaller holes. In the experiment, it will be even more difficult for a growing chain to enter the dense regions than in the model, so that the confinement to the holes will be better. The confinement in the experiment, however, will not be effective anymore when the pendant double-bond fraction has reached the plateau value, i.e. when the hole has been 'filled'. In the model the shielding effect is smaller; the confinement will be less effective but it will also take longer before a hole has been 'filled'.

ACKNOWLEDGEMENT

I am indebted to R. B. Pandey, D. Stauffer, A. Baumgartner, J. G. Kloosterboer and M. F. H. Schuurmans for stimulating discussions.

REFERENCES

1. Dušek, K., In *Developments in Polymerization—3*, Haward, R. N. (Ed.), Elsevier Applied Science Publishers, London, 1982 (and references therein).
2. Silberberg, A., *Networks '86—8th Polymer Networks Group Meeting,* abstracts.
3. Aso, C., *J. Polymer Sci.,* 1959, **39**, 475.

4. Malinsky, J., Klabán, J. and Dušek, K., *J. Macromol. Sci. Chem.*, 1971, **A5**, 1071.
5. Kloosterboer, J. G., van de Hei, G. M. M. and Boots, H. M. J., *Polym. Commun.*, 1984, **25**, 354; Boots, H. M. J., Kloosterboer, J. G., van de Hei, G. M. M. and Pandey, R. B., *Brit. Polymer J.*, 1985, **17**, 219.
6. Fernandez, A. M., Widmaier, J. M., Sperling, L. H. and Wignall, G. D., *Polymer*, 1984, **25**, 1718; Weismann, J. G. and Sperling, L. H., *Macromolecules*, 1985, **18**, 1720.
7. Dušek, K., *Brit. Polymer J.*, 1985, **17**, 185 (for a review).
8. Manneville, P. and de Sèze, L., in *Numerical Methods in the Study of Critical Phenomena*, della Dora, I., Demongeot, J. and Lacolle, B. (Eds), Springer, Berlin, 1981.
9. See, for example, Herrmann, H. J., Stauffer, D. and Landau, D. P., *J. Phys.*, 1983, **A16**, 1221; Pandey, R. B., *J. Stat. Phys.*, 1984, **34**, 163.
10. Boots, H. M. J. and Pandey, R. B., *Polymer Bulletin*, 1984, **11**, 415; Boots, H. M. J., in *Integration of Fundamental Polymer Science and Technology*, Kleintjens, L. A. and Lemstra, P. J. (Eds), Elsevier Applied Science Publishers Ltd, London, 1986.
11. Kloosterboer, J. G. and Lijten, G. F. C. M., in *Chemistry, Properties and Applications of Crosslinking Systems*, Dickie, R. A., Bauer, R. S. and Labana, S. (Eds), ACS Symp. Ser., 1988.
12. Kloosterboer, J. G. and Lippits, G. J. M., *J. Imaging Sci.*, 1986, **30**, 177.
13. Kloosterboer, J. G., unpublished results.

18

SOL–GEL TRANSITION INDUCED BY FRIEDEL–CRAFTS CROSSLINKING OF POLYSTYRENE SOLUTIONS

C. COLLETTE,* F. LAFUMA, R. AUDEBERT
and L. LEIBLER

Laboratoire de Physico-Chimie Macromoléculaire de l'Université Pierre et Marie Curie, CNRS UA 278, ESPCI, 10 rue Vauquelin, 75231 Paris Cedex 05, France

ABSTRACT

Friedel–Crafts crosslinking of polystyrene has been performed with 9,10-dichloromethylanthracene (A(MC)$_2$) in 1,2-dichloroethane solution. The extent of the reaction is followed by UV spectroscopy. The number of crosslinks is a strongly non-linear function of the reaction time. We found that the variation of properties of the sol molecule distribution such as weight, average molecular weight and average radius of gyration and the viscosity near the gel point depend on kinetics. We performed a computer simulation of the kinetics of the reaction which may explain this observation.

The number of crosslinks necessary to form a gel depends on the concentration and molecular weight of precursor chains. Beyond the gel point the existence of intramolecular crosslinks has been demonstrated by analyses of elastic moduli and swelling equilibrium concentrations.

1 INTRODUCTION

A considerable effort has been devoted to studies of the formation of gels by reversible or irreversible crosslinking of solutions of polymer chains. The understanding of phase equilibria, of critical properties near the gelation threshold and of swelling and elastic properties of well-formed networks

* Present address: CAL, ATOCHEM, 92303 Levallois Perret, France

requires a good knowledge of the number of crosslinks formed in the system. For instance, according to statistical theories of gelation such as classical Flory–Stockmayer theory or percolation theory, the number of crosslinks (conversion factor) alone determines the state of the given system near the gelation threshold.[1] It has been pointed out recently that for some systems kinetic effects may play an important role and influence the properties of both sol and gel phases.[2,3] Unfortunately, for many gelation processes (e.g. polycondensation reactions or random crosslinking induced by γ-irradiation) the difficulties of controlling the reaction kinetics, and in particular the distance from the gelation threshold, prevent precise comparisons with theoretical models.

The present contribution deals with a new Friedel–Crafts crosslinking reaction of semidilute polystyrene (PS) solutions. The reaction is achieved in a good solvent, dichloroethane (DCE), with 9,10-dichloromethyl-anthracene (DCMA) as crosslinking agent, in the presence of anhydrous $SnCl_4$ as catalyst, according to the scheme[4]

The major advantage of this system for the study of gelation transition is the possibility of following the reaction kinetics by the accurate determination of the number of grafted DCMA units by UV spectroscopy. Besides, the reaction can be carried out at room temperature. The aim of this work was to study some critical properties both in the sol and the gel phase and to examine the importance of kinetic effects.

2 KINETICS OF CROSSLINKING

The crosslinking reaction can be stopped at any moment by destroying catalyst activity through precipitation and washing of macromolecular species in ethanol or acetone. The advancement of the reaction was followed by sampling small fractions of the homogeneous reaction bath and precipitating them at once. The number of grafted DCMA in the recovered polymer was obtained with good accuracy from the light spectrophotometric measurements at the characteristic absorption band at 403 nm.[5]

The reaction kinetics are controlled by the initial concentrations of polystyrene c, of DCMA molecules $[A]_0$ and of catalyst $[Sn]_0$, but are independent of the molecular weight M_w of the precursor chains.[5] Figure 1 shows typical examples of the variation of the number of grafted DCMA molecules, N, as a function of time for different initial concentrations $[A]_0$ of the crosslinking agent. The most striking effect, rich in interesting

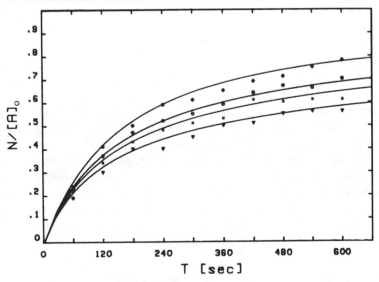

Fig. 1. Variation of the normalized concentration of grafted DCMA, $N/[A]_0$ as a function of time for four different initial DCMA concentrations $[A]_0$: $\diamondsuit$, $0\cdot875 \times 10^{-3}$ mol/litre; $\bigcirc$, $1\cdot45 \times 10^{-3}$ mol/litre; $\times$, $1\cdot75 \times 10^{-3}$ mol/litre; $\bigtriangledown$, $2\cdot31 \times 10^{-3}$ mol/litre. $M_w \simeq 2\cdot5 \times 10^5$, $c = 50$ g/litre, $[SnCl_4]_0 = 4\cdot3 \times 10^{-3}$ mol/litre for all reaction kinetics. The full lines represent the variations predicted by the kinetic model with $K = 12$, $K_r = 0\cdot3$, $L = 0\cdot1$, $K_c = 0\cdot2$ and $K_d = 25$ litres mol/s.

consequences for physical properties of the system, is the strongly non-linear character of the reaction kinetics. Most DCMA molecules are grafted to polystyrene chains at the initial stages of the reaction. Then the reaction becomes much slower and quasi-linear.

Because of these very non-linear kinetics and their complex dependence on initial concentration of reagents, it seemed to us very important and challenging to model the reaction path in order to be able to control and predict the reaction kinetics for any initial conditions. The Friedel–Crafts reaction is known to proceed via an activated complex of the halogenated species.[6] Thus the simplest model for the grafting of one DCMA (A) on a styrene unit (St) would be a two-step path:

$$A + SnCl_4 \underset{K_r}{\overset{K}{\rightleftharpoons}} A^* \tag{1}$$

$$A^* + St \overset{L}{\longrightarrow} X \tag{2}$$

with A^* denoting the activated complex of DCMA and X the grafted species titrated by spectrophotometry. Unfortunately, whatever the values for the reaction rates K, K_r and L are, this model cannot fit the data. In particular, such a model predicts that if the number of grafted units $N = [X]$ is normalized by the initial concentration of the crosslinks $[A]_0$, all the kinetics for different $[A]_0$ should be identical. Figure 1 clearly shows that this is not the case. A possible explanation of this effect could be deactivation of the catalyst, which can occur through several mechanisms, e.g. complexation with traces of water, with evolved HCl and/or with disubstituted aromatic rings of the crosslinks. These three mechanisms can be easily introduced to the kinetic equations scheme.[4,5]

It turns out that they fail to reproduce the quasi-stationary reaction rate at late stages and only predict an end of the crosslinking reaction when enough catalyst has been consumed. In order to explain the data it has been necessary to consider a more complex scheme, namely a possible self-condensation of the crosslinking agent, e.g. by the following path:

$$A^* + A \overset{K_d}{\longrightarrow} AA \tag{3}$$

$$AA + Sn \overset{K_c}{\longrightarrow} AA^* \tag{4}$$

$$AA^* + St \overset{L}{\longrightarrow} Y \tag{5}$$

where Y denotes a grafted unit (which we call a 'double' crosslink). Actually, the 1, 4, 5 and 8 CH groups on the anthracene ring of DCMA are especially reactive, and we have observed the self-condensation when $SnCl_4$ was added to a pure DCMA solution in DCE. Although reaction (3) represents only the first step in polycondensation of DCMA, the above reaction scheme ((1)–(5)) allowed a very good fit of all experimental results (for very different initial concentrations c, $[A]_0$ and $[SnCl_4]$) with a *single set* of reaction rates (cf. Fig. 1).

It should be noted that some grafted units X or Y might not be crosslinks but may contain pendent chloromethyl groups. Their number seems to be, however, negligible since we could not find any direct evidence of their existence from elementary analysis of chloride or IR spectrum or by the NMR spin–echo technique.

Hence the complexity of the reaction kinetics seems to originate from the formation of intermediate charge-transfer complexes as well as from polycondensation of crosslinking agents. The number of grafted DCMA units, N, as measured by visible spectrophotometry is not equal to the number of effective crosslinks, because some crosslinks contain two (or more) anthracene rings. The fraction of double crosslinks may be quite important and even reach about 30% at the gel point.

3 GEL POINT

Very important from a practical point of view and for a quantitative comparison with theory is the determination of the time and of the number of crosslinks where gelation takes place. For this purpose we followed in a low shear rheometer (Contraves LS 30) the increase in viscosity as a function of time and of the number of grafted DCMA molecules, N. We found that the gel point determined by such a viscosimetric method coincides well with the value obtained from the solubility measurements.

The statistical Flory–Stockmayer theory[7,8] and simple kinetic models based on Smoluchowski equations[7] predict that X_g, the *minimum* fraction of crosslinks (with respect to the total number of monomers) necessary to reach the gel point, varies as $1/M_w$, where M_w is the molecular weight of precursor chains. Actually, for our system we do observe such a simple dependence (Fig. 2). It is interesting to remark that, due to the very non-linear kinetics of the crosslinking reaction discussed above, the gel time t_g is shorter for bigger precursor chains than that predicted by the simple $1/M_w$ law (Fig. 3).

Figure 4 illustrates the dependence of the percentage of grafted units at the gel point X_g on the concentration of polystyrene in solution. For

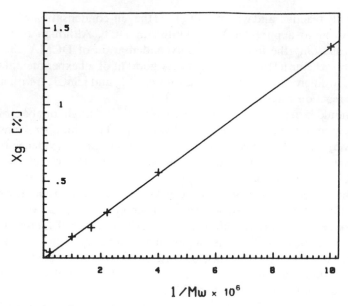

Fig. 2. Variation of the minimum fraction of grafted DCMA necessary to reach the gel point X_g (%) as a function of the inverse of the average molecular weight M_w for stoichiometric kinetics ($c = 30$ g/litre, $[SnCl_4]_0 = 4.27 \times 10^{-3}$ mol/litre).

concentrations higher than the overlap concentration c^*, X_g seems to be proportional to $1/c$. Such a strong variation can be explained in terms of the analogy, first introduced by Daoud,[9] between the random crosslinking and the polycondensation of units with functionality f equal to the number of blobs in a chain. This analogy suggests that the fraction of crosslinks at the gel point should scale like $1/f$ or $1/M_w C^{5/4}$). Perhaps a more intuitive explanation would be to invoke the existence of intrachain crosslinks. In fact, the probability that a given crosslink links two different chains is proportional to c or $c^{5/4}$ when the excluded volume effects are important. The existence of such intrachain crosslinks has interesting consequences in swelling and elastic properties of gels far from the gelation threshold. Indeed we observed that our gels swelled much more and had much smaller elastic moduli than would be predicted from a simple estimation of mesh size from stoichiometry considerations.[4,5]

We would like to stress that the situation is more complex than the above statistical models might suggest, and that the fraction of the grafted units at the gel point does depend on the reaction kinetics, i.e. on the initial concentration of the crosslinking units $[A]_0$ (Fig. 4). This is partially due to

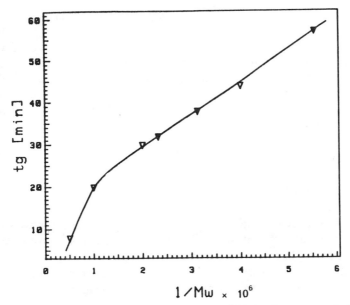

Fig. 3. Variation of the gel time t_g as a function of $1/M_w$ of precursor chains ($c = 30$ g/litre, $[SnCl_4]_0 = 4 \cdot 3 \times 10^{-3}$ mol/litre, $[A]_0 = 2 \cdot 3 \times 10^{-3}$ mol/litre).

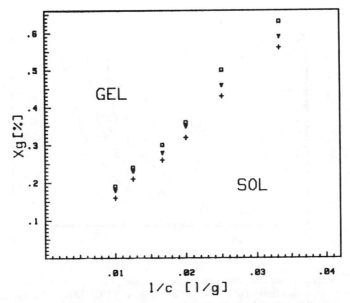

Fig. 4. Phase diagram showing the sol–gel transition line for different reaction kinetics: $+$, stoichiometric kinetics; ∇, $[A]_0 = 2 \cdot 31 \times 10^{-3}$ mol/litre; $\square$, $[A]_0 = 3 \times 10^{-3}$ mol/litre ($M_w = 2 \cdot 5 \times 10^5$, $[SnCl_4]_0 = 4 \cdot 3 \times 10^{-3}$ mol/litre).

the polycondensation of the crosslinking agents. However, even if we take into account the existence of double crosslinks (using the kinetic model), we find that the more rapid the reaction, the more crosslinks are needed to attain the gel point. The minimum number of crosslinks needed to form an insoluble macroscopic cluster for a given M_w and c is obtained for the stoichiometry reaction kinetics. It thus seems that the number of intrachain crosslinks (small loops) might depend on the reaction kinetics.

4 SOL–GEL TRANSITION

In order to study the critical properties of gelation transition we stopped the reaction at different stages, carefully extracted sol molecules and separated soluble and insoluble fractions for samples above the gel point. The extracted sol molecules from pre-gel samples were then diluted in a good solvent, and their weight-average molecular weight M_w and z-average

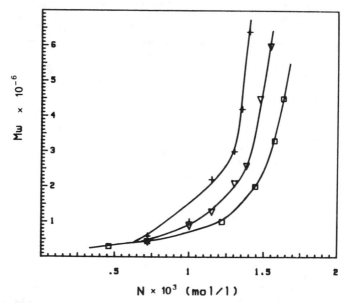

Fig. 5. Variation of the average molecular weight M_w of the extracted sol molecules as a function of N, the number of grafted DCMA units, for different initial concentrations $[A]_0$ of DCMA: $\triangledown$, $2\cdot3 \times 10^{-3}$ mol/litre; $\square$, 3×10^{-3} mol/litre; $+$, stoichiometric kinetics ($c = 30$ g/litre, $[SnCl_4]_0 = 4\cdot3 \times 10^{-3}$ mol/litre, $M_w(0) = 2\cdot5 \times 10^5$).

radius of gyration were measured by static light scattering. We used a modified Fica 40 000 photogoniometer with red light (6328 Å) and scattering angles from 30° to 150°. We also determined the intrinsic viscosity of diluted sol molecules using a Ubbelohde capillary viscometer (Fica).

The very fact that we could follow the number of grafted DCMA molecules, N, by spectrophotometry enabled us to realize that the molecular weight distribution function depends on the reaction kinetics (Fig. 5). This dependence is not simply a result of the influence of kinetic effects on the number of crosslinks necessary to reach the gel point N_g discussed in Section 3. In fact, if we plot M_w as a function of the relative distance from the gel point, defined as $\varepsilon = (N_g - N)/N_g$, we find a power-law dependence; but the exponent and the amplitude seem to depend on the kinetics, at least for the investigated range of ε. At first view this might seem very troublesome, since according to both statistical and kinetic theories the critical exponents near the gelation threshold should be universal, i.e. independent of the reaction kinetics. Actually, it is very probable that the contradiction is only apparent. Figure 6 shows the plot of the average

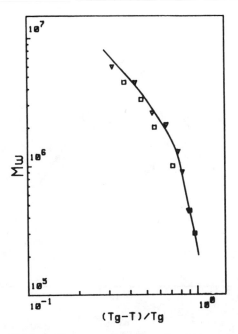

Fig. 6. Variation of M_w of the extracted sol molecules as a function of relative distance from the gel point $\varepsilon_t = (t_g - t)/t_g$ (symbols as in Fig. 5).

molecular weight M_w versus the relative distance from the gel point simply measured by $\varepsilon_t = (t_g - t)/t_g$. All different kinetics seem to tend to the same asymptotic scaling law. The observed curvature makes it difficult to determine the exponent γ defined as $M_w \simeq \varepsilon_t^{-\gamma}$ in the critical regime near the gelation threshold.

The apparent kinetic effect seems to be essentially due to the polycondensation of DCMA and formation of multiple crosslinks. In fact, if we calculate the number of effective crosslinks N_{eff} using the kinetic model, we find a very satisfactory data collapse for different reaction kinetics (Fig. 7) and a simple power law for a large range of ε_{eff}. Moreover, the critical exponent thus determined, $\gamma \simeq 1.65 \pm 0.1$, is very similar to that which may be obtained asymptotically from Fig. 6. This value of γ is close to that predicted by the percolation model ($\gamma \simeq 1.7$) and rather different from the classical value $\gamma = 1$.

The above analysis of the role of kinetic effects is well confirmed if we try to overcome the problem of determining the actual advancement of the

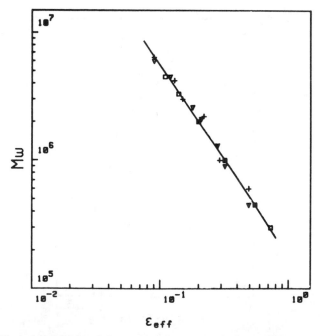

Fig. 7. Variation of M_w of the extracted sol molecules as a function of relative distance from the gel point ε_{eff} (symbols as in Fig. 5).

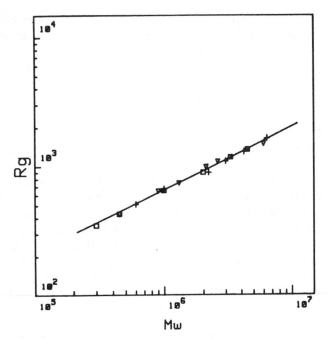

Fig. 8. The variation of the z-average radius of gyration $\langle R_g^2 \rangle_z^{1/2}$ as a function of the weight-average molecular weight M_w for diluted sol molecules (symbols as in Fig. 5).

reaction in a given sample by considering the measurements of two different types of averages (e.g. M_w and $\langle R_g^2 \rangle_z$ and M_w and $\langle [\eta] \rangle$) performed on the same sample (Fig. 8). Once again we obtain a nice data collapse for different reaction kinetics. We find that the effective exponent ν_{eff}, defined by $\langle R_g^2 \rangle_z \simeq M_w^{2\nu_{\text{eff}}}$, equals $\nu_{\text{eff}} \simeq 0.55 \pm 0.05$. Similarly, we find that $\langle [\eta] \rangle \simeq |M_w^a$ with the effective exponent $a \simeq 0.35 \pm 0.05$.

The interpretation of the light scattering and viscometry results obtained for diluted sol requires a clear distinction between two stages of the experiment.[10,11] First, during the reaction chains crosslink and form randomly branched molecules. The molecular weight distribution of sol molecules thus obtained depends on the advancement of the reaction and possibly on the kinetics of the reaction. Secondly, the sol is diluted and the molecules swell. The dilution does not modify the weight distribution function and its moments (e.g. M_w), but changes considerably the conformation of molecules. In particular, the excluded volume interactions are partially screened out in the reaction bath. However, in a dilute solution such a screening is absent and randomly branched molecules swell more. For a branched molecule of weight M the radius $R(M)$ is expected to scale like M with the swelling exponent $v = 0.5$.[10] Hence, the average conformation as measured by the z-average radius of gyration and hydrodynamic properties (intrinsic viscosity) of diluted sol depends both on the distribution function and on these swelling effects, which complicate the quantitative analysis of the results by introducing the new swelling exponent v.[11,12] If we assume that $\langle R_g^2 \rangle_z$ and $\langle [\eta] \rangle$ are dominated by branched molecules, we find simple relations, $v_{eff} = v/(3 - \tau)$ and $a = (3v - \tau + 1)/(3 - \tau)$, between effective exponents and the polydispersity exponent τ and swelling exponent v.[11] The polydispersity exponent τ is equal to about 2·2 and 2 in classical and percolation theories, respectively.

The influence of reaction kinetics on the physical properties of the system complicates the interpretation of stoichiometric experiments, in which a prescribed concentration of crosslinking agent (DCMA) is added and the reaction is left to continue until all DCMA is consumed. In each sample the reaction kinetics are different and so is the number of multiple crosslinks. Hence the number of effective crosslinks is not only given by the initial concentration of DCMA even if the reaction was complete, and the data

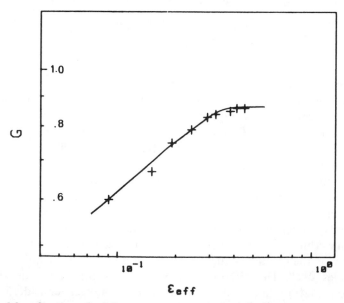

Fig. 9.   Mass fraction of gel G versus the relative distance ε_{eff} from the gel point ($c = 30$ g/litre, $[SnCl_4]_0 = 4.27 \times 10^{-3}$ mol/litre, $M_w = 2.5 \times 10^5$).

must be corrected for the polycondensation of the crosslinking agent, e.g. using the kinetic model. In Fig. 9 the mass fraction of gel, G, after sol–gel transition is plotted as a function of the effective distance from the gel point. In the critical region (which could be studied only for small precursor polymer concentrations), we find that the gel mass fraction scales like $G \simeq \varepsilon_{eff}^{\beta}$ with $\beta \simeq 0.3 \pm 0.05$. This value seems to be close to that of the percolation model $\beta \simeq 0.34$, and much smaller than the classical value $\beta = 1$.

5 CONCLUSIONS

The present work on the Friedel–Crafts crosslinking reaction seems to suggest some interesting qualitative features of gelation transition in polymer solutions which may be also more general and present in other systems. Our results seem to confirm that the *minimum* number of crosslinks necessary to reach the gel point is inversely proportional to the molecular weight of precursor chains. However, kinetic effects may still be important and influence, for instance, the concentration dependence of the gel point and the phase diagram. Kinetic effects may also complicate considerably the analysis of the critical properties of the gelation transition. Due to the non-linearity of reaction kinetics, the reaction time is not a good measure of the advancement of the reaction, and only in very close vicinity of the gelation threshold (difficult to explore experimentally) may the critical exponents be extracted from time dependence of measured quantities. The situation is even further complicated when self-condensation of the crosslinking agent is possible. Then several crosslinking molecules may serve to make only one effective crosslink, and the number of actual crosslinks will be smaller than the number of reacted crosslinking agents. For our systems we found a kinetic model which fits very well the observed reaction kinetics and which enables us to estimate the importance of this self-condensation effect. Then it seems that the critical behaviour near the gelation threshold is universal, i.e. independent of the reaction kinetics. The critical exponents γ and β that we find seem to be close to the percolation exponent. Unfortunately, by spectrophotometry we cannot distinguish between simple and multiple crosslinks and obviously the necessity of introducing the kinetic model to interpret the results reduces considerably the accuracy of determination of the exponents. This source of systematic error is absent when we measure simultaneously two different averages of the distribution. It thus seems that the effective exponents for dependence of the z-average radius of gyration and the intrinsic viscosity as a function of weight-average molecular weight are compatible with percolation exponents for the distribution function and the swelling exponent $v = 0.5$, predicted for randomly branched molecules swollen in a good solvent. Hence, all results seem to be more consistent with percolation model exponents rather than classical exponents. This might be expected for gelation in semi-dilute solutions, for which steric hindrances are important and the critical region is expected to be large.[9]

In this context, it would be extremely interesting to measure the conformation of fractionated randomly branched molecules and find directly the exponent v and also measure directly the distribution function (cf. Ref. 13), in order to be (or not to be) able to confirm the validity of the scaling hypothesis used implicitly in the above interpretations.

REFERENCES

1. Stauffer, D., Coniglio, A. and Adam, M., *Adv. Polym. Sci.*, 1982, **44**, 103; Burchard, W., ibid., 1983, **45**, 1 and references cited therein.
2. Manneville, P. and de Sèze, L., in *Numerical Methods in the Study of Critical Phenomena*, Della Dora, J., Demongeot, J. and Lacolle, B. (Eds), Springer, Berlin, 1981.
3. Herrmann, H., Landau, D. P. and Stauffer, D., *Phys. Rev. Lett.*, 1983, **5**, 412.
4. Collette, C., Thèse de Doctorat, Université Paris VI, 1986 (unpublished).
5. Collette, C., Lafuma, F., Audebert, R. and Leibler, L., to be published.
6. Grassie, N. and Meldrum, I. G., *Eur. Polym. J.*, 1968, **4**, 571; ibid., 1969, **5**, 195; ibid., 1970, **6**, 499, 513; ibid., 1971, **7**, 17, 613, 629, 645, 1253.
7. Stockmayer, W. H., *J. Chem. Phys.*, 1944, **12**, 125.
8. Flory, P. J., *Principles of Polymer Chemistry*, Cornell University Press, Ithaca, New York, 1953.
9. Daoud, M., *J. Phys. Lettres*, 1979, **40**, L201.
10. Isaacson, J. and Lubensky, T. C., *J. Phys. Lettres*, 1980, **41**, L469.
11. Daoud, M., Family, F. and Jannink, G., *J. Phys. Lettres*, 1984, **45**, L199.
12. Schosseler, F. and Leibler, L., *Macromolecules*, 1985, **18**, 398.
13. Leibler, L. and Schosseler, F., *Phys. Rev. Lett.*, 1985, **55**, 1110.

19

PHYSICO-CHEMISTRY OF THE HYDROLYSED POLYACRYLAMIDE–CHROMIUM III INTERACTION IN RELATION TO RHEOLOGICAL PROPERTIES

C. ALLAIN and L. SALOMÉ

Laboratoire d'Hydrodynamique et Mécanique Physique, ESPCI, 10 rue Vauquelin, 75231 Paris Cedex 05, France

ABSTRACT

This paper deals with the behaviour of hydrolysed polyacrylamide solutions in the presence of chromium III. First, using potentiometry and spectroscopy, we study the complexation between chromium III and sodium acetate used as a low molecular weight model for the polymer. Then the same methods are applied to investigate the polymer–Cr(III) complexation and the results are discussed in conjunction with the nature of the physical states of the systems. Finally, we perform rheological measurements in the sol–gel transition regime in a range where the number of crosslinking points can be calculated unambiguously.

1 INTRODUCTION

Chemically crosslinked gels can be obtained either by polymerization of multifunctional monomers, or by crosslinking pre-existent polymer chains. We are interested in this second class of gels, and more specifically in the case of polyelectrolytes crosslinked by plurivalent ions. Depending on the physico-chemical properties of the bonds, the crosslinks in the gels formed by ion addition can be very labile or quasi-irreversible. Experimental measurements have to be performed with a system, the crosslinks of which can be considered as permanent, to allow the comparison with the statistical models of gelation. We have chosen the hydrolysed polyacryl-amide + Cr(III) system. It presents two main advantages: on one hand the bonds between the metal ion and the polymer charged groups are strong; on the other hand their formation is very slow.

In the preceding paper [1] we have compared our rheological experiments with the results predicted by the classical theory and the percolation model: we have found a good agreement with the percolation model. However, our interpretation was based on the hypothesis that the experimental parameter, the time of reaction, was proportional to the theoretical parameter, the crosslink concentration. This assumption is not obvious. As a consequence, we have developed a physico-chemical study of the complexation between the chromium ions and the carboxylic groups of the polymer in order to determine the concentration in crosslinks without any ambiguity.

In the first part we describe the system under study and the various states of the system observed when the concentrations of reagents are varied. The second part deals with the complexation of chromium III with sodium acetate used as a low molecular weight model for the polymer. Finally, in the third part, the complexation with the polymer is discussed and the results of rheological measurements are presented.

2 PRESENTATION OF THE SYSTEM

2.1 Reagents

The hydrolysed polyacrylamide $[CH_2-CHCONH_2]_x-[CH_2CHCOONa]_y$ (Rhône Poulenc, France) used in this study is obtained by copolymerization of acrylamide and sodium acrylate. Its structural and conformational properties have been studied by Truong et al.[2] They found that the degree of hydrolysis, $\tau = y/(x + y)$, is 24% and that the hydrolysed groups follow a random distribution: the weight-average molecular weight is about 6×10^6.

Chromium III is a transition metal ion which is known to form inert complexes [3] because of its electronic configuration (the lowest energy orbitals, t_{2g}, in the third orbital d, are half filled). The ligands are arranged on an octahedron around the central ion. $CrCl_3(H_2O)_6$ salt (Merck quality ultrapure) was used without further purification.

2.2 Preparation of the Samples

The addition of Cr(III) to the polymer was performed by isovolumic mixing of chromium and polymer solutions. The solvent was highly purified water (Milli-Q) containing 0.34 mol/litre NaCl to keep the ionic strength constant: it is a good solvent for the polymer.[4] The polymer in powder form was dissolved gradually in the solvent under gentle stirring which was maintained for 24 h: then the solutions were kept in the refrigerator. Chromium III solutions were prepared by dissolving the $CrCl_3(H_2O)_6$ salt in the solvent: they were kept at room temperature in darkness.

2.3 Concentration Range

The polymer–Cr(III) interaction was studied in concentration ranges where different states were obtained. Once prepared by mixing the reagents, the sample tubes were kept at constant temperature (30°C). The phase diagram (Fig. 1) is established by direct observation of the final stable states of the solutions, 1 day or more after their preparation. Figure 1 shows that sols, gels, precipitates or heterogeneous gels can be obtained when the concentrations of the polymer and Cr(III) are varied. Samples where aggregates lie in the bottom of the tube were called precipitates, while turbid gels or collapsed gels which expelled solvent were called heterogeneous gels.

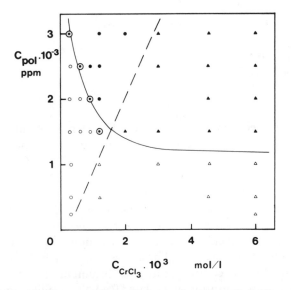

Fig. 1. Nature of the states formed as a function of $CrCl_3$ and polymer concentrations: ○, homogeneous sol; ●, homogeneous gel; △, precipitate; ▲, heterogeneous gel; ⊙, difficult to classify, in the region of sol–gel transition.

3 PHYSICO-CHEMICAL STUDY OF THE DIFFERENT SPECIES IN SOLUTION

3.1 Interaction between Cr^{3+} and Cl^- Ions

Since our solutions are prepared by dissolution of $CrCl_3$ in aqueous solutions of NaCl, we have first studied the chloride complexes. A number of studies have dealt with the properties of $CrCl^{2+}$ and $CrCl_2^+$ ions; in particular, the complexation equilibrium constants (see Table 1) have been found equal to 0·2 and 0·1, respectively.[5] The wavelengths and the molar extinction coefficients corresponding to the peaks of the two d–d absorption bands have been reported also.[6–8]

TABLE 1

Equilibria	Constants
$CrCl_3 \rightarrow CrCl_2^+ + Cl^-$	Complete
$CrCl_2^+ \rightleftharpoons CrCl^{2+} + Cl^-$	$0·1^a$
$CrCl^{2+} \rightleftharpoons Cr^{3+} + Cl^-$	$0·2^a$
$Cr(H_2O)_6^{3+} \rightleftharpoons Cr(H_2O)_5OH^{2+} + H^+$	$7·9 \times 10^{-5b}$
$Cr(H_2O)_5OH^{2+} \rightleftharpoons Cr(H_2O)_4(OH)_2^+ + H^+$	$2·5 \times 10^{-6b}$
$Cr(H_2O)_6^{3+} + CH_3COO^- \rightleftharpoons Cr(H_2O)_5OOCCH_3^{2+} + H_2O$	$1·6 \times 10^{-2c}$
$Cr(H_2O)_5OOCCH_3^{2+} + CH_3COO^- \rightleftharpoons Cr(H_2O)_4(OOCCH_3)_2^+ + H_2O$	$7·9 \times 10^{-2c}$

$^a K = [ML_x][L]/[ML_{x+1}]$.
$^b K_{Ax} = [ML_x][H^+]/[ML_{x-1}]$.
$^c K_x = [ML_x]/[ML_{x-1}][L]$.

The visible spectrum of a 10^{-2} mol/litre $CrCl_3$ solution recorded immediately after dissolution is presented in Fig. 2. The positions of the peaks (627 nm and 437 nm) indicate a large proportion of $CrCl_2^+$ complex. One day later, an important shift toward shorter wavelengths was observed (see Fig. 2) and the spectrum became identical to that measured for a chromium solution of the same concentration prepared in pure water (without NaCl). The wavelengths of the peaks (414 nm and 580 nm) and their absorbances coincide with the values found in the literature for aquochromium complexes.[6,9] These results, which indicate a low proportion of chloride complexes at equilibrium, are in good agreement with the relative concentrations calculated from the equilibrium constants[5] for solutions containing 0·34 mol/litre NaCl: $[Cr^{3+}] = 92\%$, $[CrCl^{2+}] = 6·2\%$, $[CrCl_2^+] = 1·8\%$.

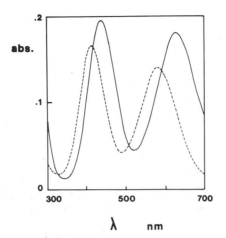

Fig. 2.   Visible spectrum of 10^{-2} mol/litre $CrCl_3$ solution with 0·34 mol/litre NaCl: ——, just after dissolution; - - - - -, 24 h later.

3.2 Acid–Base Equilibrium

Hexaaquochromium $Cr(H_2O)_6^{3+}$ undergoes the first ionization leading to the formation of $Cr(H_2O)_5OH^{2+}$ and the second one leading to $Cr(H_2O)_4(OH)_2^+$ (see Table 1). The equilibrium constant of the first reaction, K_{A1}, has been found ranging from $1·3 \times 10^{-4}$ to 4×10^{-5}, depending on the experimental conditions, while that of the second reaction, K_{A2}, is about 10^{-6}.[10] In the case of our solutions, for which the ionic strength is equal to 0·34 mol/litre, we expect these two equilibrium constants to be equal to $7·9 \times 10^{-5}$ and $2·5 \times 10^{-6}$, respectively.[11]

First, we have investigated the pH of $CrCl_3$ solutions with 0, 0·034 and 0·34 mol/litre NaCl. For chromium concentrations ranging from $2·5 \times 10^{-4}$ to 10^{-2} mol/litre, a good agreement has been found with the calculated values, taking K_{A1} and K_{A2} equal to $7·9 \times 10^{-5}$ and $2·5 \times 10^{-6}$, respectively, and assuming that all complexes are equivalent with respect to the acid–base equilibria. Note that this assumption would fail if the $CrCl_2^+$ concentration was not negligible, since for this species K_{A1} has been found equal to 2×10^{-6}.[5]

We have also investigated the UV–visible light spectrum of chromium solutions in the same range of concentration. We have been especially interested in the absorbance at 250 nm, A_{250}, since acetate complexes are discernible at this wavelength. The variation of A_{250} as a function of the chromium concentration is reported in Fig. 3. The $Cr(OH)_2^+$ concentration is negligible since the pH is always lower than 4·2, and the absorbance is

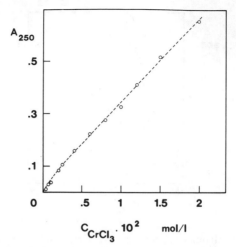

Fig. 3. Experimental (◯) and theoretical (– – –) variations of the absorbance at 250 nm (A_{250}) versus the $CrCl_3$ concentration.

simply related to the Cr^{3+} and $CrOH^{2+}$ concentrations and to their molar extinction coefficients. The preceding values of K_{A1} and K_{A2} were used to fit the experimental points plotted in Fig. 3 and to determine the molar extinction coefficients which have been found equal to 28 and 110 for Cr^{3+} and $CrOH^{2+}$, respectively. The value corresponding to Cr^{3+} at 250 nm is larger than that reported by Deutsch and Taube,[12] but there is a good agreement with the values of the molar extinction coefficient of Cr^{3+} at 410 nm (16·8) and at 570 nm (14·4) which have been calculated in the same way. For $CrOH^{2+}$, we have found 13·5 and 6·2, respectively. The discrepancy observed at 250 nm can be explained by the presence of $CrCl^{2+}$ ions and by the fact that absorbances in the UV range are sensitive to the chemical conditions. Finally, let us note that, after the period of 1 or 2 days needed for chloride complexes to reach equilibrium, the chromium solutions have never been used more than 2 weeks after their preparation. So, aging processes such as olation, that leads to the formation of polynuclear complexes,[13] can be neglected.

3.3 Interaction between Cr^{3+} and CH_3COONa

The equilibrium concentrations of the different species depend on the acid–base reactions of acetate ($K_A = 1·6 \times 10^{-5}$) and chromium, and on the complexation reactions whose equilibrium constants K_1 and K_2 have been found equal to $1·6 \times 10^{-2}$ and $7·9 \times 10^{-2}$, respectively[14] (see Table 1). First, we have measured the pH of the solutions. The acid–base reactions being much more rapid than the complexation reactions which involve a ligand substitution mechanism, we have measured the pH of our samples just after the chromium and acetate solutions have been mixed, and we have measured its variation over several hours. For instance, for concentrations equal to 3×10^{-3} mol/litre $CrCl_3$ and 3×10^{-3} mol/litre CH_3COONa, pH quickly reaches the value 4·3, in good agreement with that calculated taking into account only the acid–base equilibria, and decreases to 4·2 during the following 15 h. For all the concentrations investigated, the same behaviour is observed, showing that the pH variations due to the complexation process are very small.

Secondly, we have studied the complexation reactions by UV–visible light spectroscopy and interpreted the results, assuming that the pH is constant. Figure 4 represents the spectra recorded at different times after the sample preparation. Modifications of the absorbance are observed on the peaks in the visible and in the UV light range. The most important variations being observed at 250 nm, we have studied the difference ΔA between the absorbance measured at the end of the complexation reaction and the absorbance measured two minutes after mixing when only acid–base reactions have taken place. In Fig. 5, ΔA is plotted versus the CH_3COONa concentration for samples where the $CrCl_3$ concentration is equal to 6×10^{-3} mol/litre. For acetate concentrations ranging from 5×10^{-4} to 5×10^{-2} mol/litre, an increase in ΔA by more than a factor of

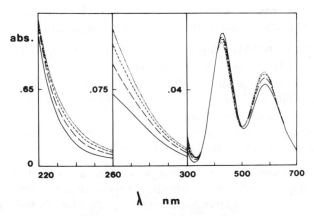

Fig. 4. UV–visible spectra of 3×10^{-3} mol/litre $CrCl_3$ and 6×10^{-3} mol/litre CH_3COONa solution recorded 10 min (——), 130 min (— —), 250 min (- - - - -) and 370 min (.....) after mixing the reagents.

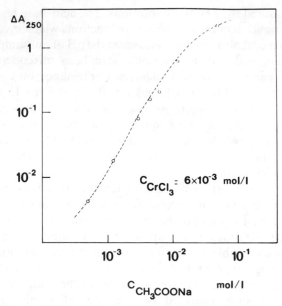

Fig. 5. Experimental ($\bigcirc$) and theoretical (---) variations of ΔA_{250} ($\Delta A = A_{equil} - A_{initial}$): concentration in $CrCl_3$ 6×10^{-3} mol/litre and concentration in CH_3COONa ranging from 5×10^{-4} to 5×10^{-2} mol/litre. The error in the determination of ΔA is the largest when [chromium]/[acetate] $\simeq 1$.

500 is observed. We have compared this increase with the calculated acetate–chromium complex concentration variation, taking into account all the equilibrium reactions and using the following arguments:

—the $CrOOHCCH_3^{3+}$ ion is a very strong acid, $K_A \simeq 4$,[12] and only $CrOOCCH_3^{2+}$ has to be considered;

—the acid–base equilibrium constants of acetate complexes (related to the ionization of the water ligand) have nearly the same values as for the hexaaquochromium ions;[12]

—the modifications of the species concentrations due to the formation of diacetatochromium are negligible.

Figure 5 illustrates this comparison between measurements and computations of the complex concentrations. A very good agreement is observed over the whole range of absorbance variation, showing the validity of both the assumptions and the equilibrium constant values used for this calculation. We have tested different values of K_{A1} and K_{A2}; no significant difference is observed when K_{A2} is changed, but the best fit occurs for K_{A1}

equal to 5.5×10^{-5} instead of 7.9×10^{-5}. It must be noted that using this value in the above calculations does not affect considerably the results.

We have also determined the molar extinction coefficient of the complexes at 250 nm and found it equal to 500. Such a large value corroborates the fact that this wavelength is well appropriated to follow the complexation. Moreover, this coefficient is much larger than the molar extinction coefficients of Cr^{3+} and $CrOH^{2+}$ determined previously, so that it is completely legitimate to neglect the influence of small pH variations during complexation.

4 INTERACTION BETWEEN Cr^{3+} AND HYDROLYSED POLYACRYLAMIDE

The comparison between the complexations observed for a low molecular weight model and for the polymer has been widely discussed by Tsuchida and Nishide.[15] In particular, they have pointed our that the complexation is completely modified when a significant change in the polymer conformation is induced by the formation of complexes. For instance, in the case of dilute solutions of flexible polymers, they report that the apparent constants of complexation are modified in conjunction with the decrease in viscosity, which is associated with the collapse of the chain due to intramolecular crosslinking. This effect is related to the high local ligand concentrations inside the polymer coils in dilute solutions and is further increased by the contraction of the polymer chain.

In the case of our system, we have observed the same kind of behaviour for samples with relatively large [chromium]/[acrylate anion] ratios, i.e. for samples corresponding to the phase diagram region where precipitates are obtained. For instance, Fig. 6 shows the variation of ΔA (defined in Section 3.3) versus the polymer concentration for a constant chromium concentration (1.2×10^{-3} mol/litre). A large discrepancy between the measured and the calculated values is observed, especially for the low polymer concentrations for which the $[\Delta A_{exp}/\Delta A_{calc}]$ ratios are the largest.

The situation is completely different for samples prepared at larger polymer concentrations and low chromium concentrations. Indeed, in the case of more concentrated polymer solutions, the macromolecules are entangled and the concentration can be considered uniform. Moreover, since the [chromium]/[acrylate anion] ratio is small and Cr(III) does not form complexes with acrylamide, the modification of the chain conformation due to complexation is negligible. We have also studied the

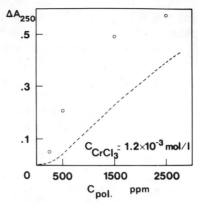

Fig. 6. Experimental (○) and theoretical (---) variations of ΔA_{250} for solutions with 1.2×10^{-3} mol/litre $CrCl_3$ and various polymer concentrations. The large discrepancy observed between the experimental points and the theoretical curve is due to the modification of the chain conformation during complexation.

properties of samples prepared with 2500 ppm hydrolysed polyacrylamide (which corresponds to 7.8×10^{-3} mol/litre carboxylic groups) for various chromium concentrations up to 1.2×10^{-3} mol/litre. The variation of ΔA measured at 250 nm is plotted in Fig. 7. A good agreement is observed with the predicted values, except for the largest chromium concentrations. Beyond 6×10^{-4} mol/litre, which corresponds to a [chromium]/[acrylate

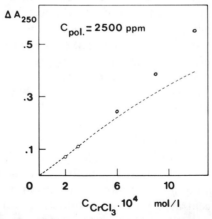

Fig. 7. Experimental (○) and theoretical (---) variations of ΔA_{250} for solutions with 2500 ppm polymer and various $CrCl_3$ concentrations. Below 6×10^{-4} mol/litre the complex concentration is predicted with an accuracy better than 10%.

anion] ratio of 0·08, the relative deviation becomes larger than 10%. For the lower concentrations, the molar extinction coefficient is equal to 1230 and the theoretical curves plotted in Figs 6 and 7 have been computed with this value. The difference of about a factor 2 of the molar extinction coefficient relative to the acetate complexation is easily explained by the changes in the complexes' environment.

We have also measured the variation of the Newtonian viscosity at the end of the complexation reaction as a function of the chromium concentration, using a magnetic sphere levitation rheometer.[16,17] As reported in Fig. 8, a very large increase is observed up to 3×10^{-4} mol/litre $CrCl_3$: above this value a gel phase is obtained. In this range of concentrations the [chromium]/[acrylate anion] ratio is lower than 0·04. Therefore, as previously demonstrated by spectroscopic measurements, the use of sodium acetate as a low molecular weight model for the polymer is

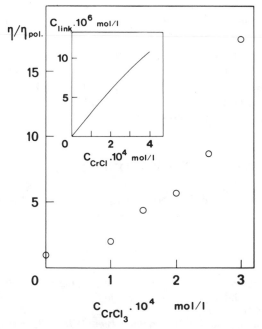

Fig. 8. Variation of the Newtonian viscosity η recorded at the end of the complexation process versus the $CrCl_3$ concentration, for 2500 ppm of polymer. The results are plotted relative to the viscosity η_{pol} of a 2500 ppm polymer solution. Insert: theoretical variation of the crosslink concentration C_{link} versus $CrCl_3$ concentration for 2500 ppm of polymer.

legitimate. So, the number of links, i.e. the number of diacrylate anion–chromium complexes, can be calculated by the same method. The variation of the number of links versus the chromium concentration is monotonous (see insert in Fig. 8). The very large increase in viscosity can therefore be attributed unambiguously to a critical divergence of the viscosity in the vicinity of the sol–gel transition.

5 CONCLUSION

Hydrolysed polyacrylamide–chromium III appears to be a suitable system for the investigation of polymer–ions interactions. Potentiometry and spectroscopy measurements have shown that the concentrations of the different species resulting from complexation between chromium III and sodium acetate agree well with the values calculated from the equilibrium constants found in the literature. For polymer solutions in the dilute regime, an enhancement in the complex formation has been observed as already reported by Tsuchida and Nishide.[15] In the case of relatively high polymer concentrations and low chromium concentrations, we have found that the complex formation is the same as that predicted by the low molecular weight model. This allows the calculation of the number of links formed and consequently the interpretation of rheological measurements in terms of critical phenomena.

ACKNOWLEDGEMENTS

We are grateful to M. Bolte and A. Ricard for very helpful discussions.

REFERENCES

1. Allain, C. and Salomé, L., *Polymer Communications*, 1987, **28**, 109.
2. Truong, N. D., Galin, J. C., François, J. and Pham, Q. T., *Polymer*, 1986, **27**, 467.
3. Taube, H., *Chem. Rev.*, 1952, **50**, 69.
4. Klein, J. and Conrad, K., *Makromol. Chem.*, 1978, **179**, 1635.
5. Dellien, I., Hall, F. M. and Hepler, L. G., *Chem. Rev.*, 1976, **76**, 290.
6. Elving, P. J. and Zemel, B., *J. Am. Chem. Soc.*, 1957, **79**, 1281.
7. Johnson, H. B. and Reynolds, W. L., *Inorg. Chem.*, 1963, **2**, 468.
8. Finholt, J. E., Caulton, K. G. and Libbey, W. J., *J. Am. Chem. Soc.*, 1964, **3**, Notes 1801.

9. Cotton, F. A. and Wilkinson, G., *Advanced Inorganic Chemistry*, 4th edn, John Wiley & Sons, New York, 1980.
10. Sillen, L. G. and Martell, A. E., *Stability Constants of Metal–Ion Complexes*, Chem. Soc., London, 1964.
11. Emerson, K. and Graven, W. M., *J. Inorg. Nucl. Chem.*, 1959, **11**, 309.
12. Deutsch, E. and Taube, H., *Inorg. Chem.*, 1968, **7**, 1532.
13. Hall, H. T. and Eyring, H., *J. Am. Chem. Soc.*, 1950, **72**, 782.
14. Charlot, G., *Les Réactions Chimiques en Solution*, Masson, Paris, 1969.
15. Tsuchida, E. and Nishide, H., *Adv. Polym. Sci.*, 1977, **24**, 1.
16. Gauthier-Manuel, B., Meyer, B. and Pieranski, P., *J. Phys. E: Sci. Instrum.*, 1984, **17**, 1177.
17. Adam, M. & Delsanti, M., *Revue Phys. Appl.*, 1984, **19**, 253.

20

CURED EPOXY RESINS: MEASUREMENTS IN DILUTE AND SEMIDILUTE SOLUTION

E. Wachenfeld-Eisele and W. Burchard

Institute of Macromolecular Chemistry, University of Freiburg, Stefan Meier Str. 31, 7800 Freiburg, FRG

ABSTRACT

The conformations of two types of amine-cured triglycidyl isocyanurate resins in dilute solution were studied and compared with older measurements of different polyhydroxyethers. The influence of branching on the overall dimensions is scarcely detectable. However, branching shows a very pronounced effect on the intrinsic viscosity. The behaviour of the latter results from the inverse proportionality of the intrinsic viscosity to the segment density. The amine-cured epoxy resins showed no swelling up to molecular weights of 10 000; in the higher molecular weight region the swelling ratios increase continuously up to a factor of 9.

Two resins were measured in the semidilute regime. The data obtained lie, as expected, in the region between hard spheres and coils of flexible chains.

1 INTRODUCTION

Epoxy resins have been the subject of intense research for a long time. Mostly, they are prepared by hardening diglycidyl or triglycidyl compounds with amines or anhydrides.

In most studies so far the chemical reactivity, kinetics and clarification of the mechanism have been emphasized. This is understandable in view of the need to control the reaction. On the other hand, the final properties of the resins depend on the conformation and internal structure of the macromolecules, and on the dynamic response of the product under the strain of a time-dependent external force.

Recently we have started a project to study these conformational and

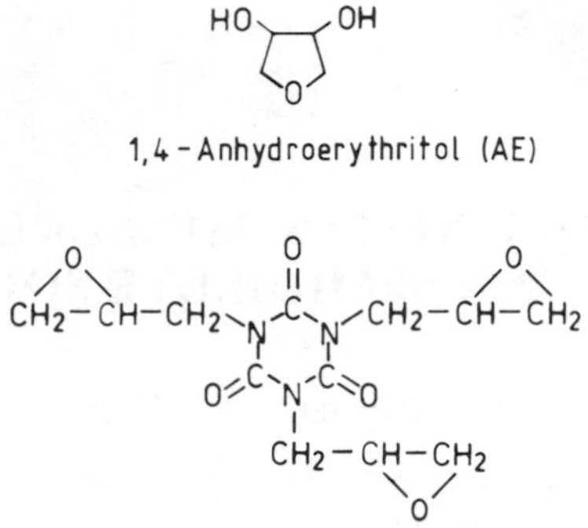

1,4 -Anhydroerythritol (AE)

Triglycidyl isocyanurate (TGI)

Fig. 1.   Chemical formulae of 1,4-anhydroerythritol and triglycidyl isocyanurate.

dynamic properties, of which little is known in the field of epoxy resins. Polyhydroxyethers from bisphenol-A diglycidyl ether (DGEBA) cured with bisphenol-A (BA) or anhydroerythritol (AE) (Fig. 1) had already been prepared and measured in dilute solution. This contribution deals now with two extremely highly branched amine-cured epoxy resins: triglycidyl isocyanurate (TGI) (Fig. 1) has been cured with ethylenediamine (EDA) or with hexamethylenediamine (HMDA). For both resins the dilute solution properties were measured.

In addition, two different TGI/HMDA resins were investigated in semidilute solution; we first report here results and then compare these with theoretical predictions and experimental data from other polymer architectures.

2 DILUTE SOLUTION MEASUREMENTS

Dilute solution properties are characterized by the molecular weight dependence of:

(1)   the radius of gyration $R_g = \langle S^2 \rangle_z^{1/2}$;
(2)   the hydrodynamic radius $R_h = kT/(6\pi\eta_0 D_z)$;
(3)   the intrinsic viscosity $[\eta]$;
(4)   the second virial coefficient A_2; and
(5)   by the shape of the particle scattering factor $P_z(q)$.

The hydrodynamic radius R_h is defined by the reciprocal translational diffusion coefficient D_z through the Stokes–Einstein relationship. The particle scattering factor describes the angular dependence of scattered light or scattered neutrons, where $q = (4\pi/\lambda)\sin\theta/2$ is the magnitude of the scattering vector $\mathbf{q}$ which is related to the scattering angle θ and the wavelength $\lambda (= \lambda_0/n)$ in the medium. n is the solvent refractive index and λ_0 the wavelength in vacuum. Molecular weight, second virial coefficient and diffusion measurements can be carried out by common, i.e. static, and by dynamic light scattering, which with our instrumental set-up[1] could be measured simultaneously. Radii of gyration and, moreover, the shape of particle scattering factors can be determined if $qR_g > 0.5$ or $qR_g > 4$, respectively. The q-values covered by visible light scattering vary between 0·9 and $4 \times 10^{-2}\,\text{nm}^{-1}$. Thus the radius of gyration has to be $R_g > 12\,\text{nm}$ before it can be measured by light scattering, or it has to be $R_g > 100\,\text{nm}$ before the shape of the particle scattering factor can be estimated by light scattering. As can be seen from Fig. 2, the radii of gyration of TGI/EDA and TGI/HMDA resins were large enough to be measured by light scattering, but, in the region of lower molecular weight, only with an appreciable experimental error. For the DGEBA/BA resins the radii of gyration had to be measured by neutron scattering, which was carried out at the Institute Laue-Langevin (ILL) in Grenoble, France, using the D17 goniometer. The intrinsic viscosities were measured, as usual, by capillary viscometry.

Much is known of the mentioned quantities for linear chains. It appears obvious that branching will exert an influence on the dimensions of the

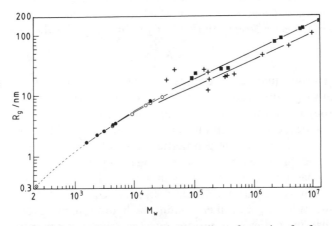

Fig. 2. Molecular weight dependence of the radius of gyration for four epoxies: ○, DGEBA/BA, linear; ●, DGEBA/BA, branched; +, TGI/EDA, ■, TGI/HMDA.

molecule when compared with a linear chain of the same molecular weight. Branching decreases the dimensions of a molecule considerably, and this decrease becomes more pronounced with increasing branching density. Similar behaviour is expected for the intrinsic viscosity, which is related to the radius of gyration by[2]

$$[\eta] = \Phi(R_g^3/M) \tag{1}$$

with a molecular weight independent prefactor Φ which, however, depends on the molecular architecture.

It has become common use to characterize branching by the following three 'shrinking' factors g, g' and h:[3,4]

$$g = R_{g,branch}^2/R_{g,linear}^2 \tag{2}$$

$$g' = [\eta]_{branch}/[\eta]_{linear} \tag{3}$$

$$h = R_{h,branch}/R_{h,linear} = D_{linear}/D_{branch} \tag{4}$$

where all ratios have to be taken at the same molecular weight. The quantities of eqns (2)–(4) are well defined for *regularly branched* structures where no isomers exist and the samples have no molecular weight distribution. Most epoxy resins are formed, however, by a *random* process of reaction, and such randomly branched materials have very broad molecular weight distributions. This fact makes the definition of g, g' and h ambiguous, even if the convention is made that the radii have to be compared at the same weight-average molecular weight.

More instructive than the shrinking factors is the ρ-parameter, defined as the ratio of the geometric to the hydrodynamic radii[5] for the *same sample*

$$\rho = R_g/R_h \tag{5}$$

because here two quantities are compared at the same molecular weight *and* molecular weight distribution. The difference in these two radii results from the hydrodynamic interaction between the segments, which defines the hydrodynamic radius in addition to the topological conditions, where only the latter is determining the geometric radius of gyration. The hydrodynamic interaction increases with increasing segment density; the solvent in the coiled macromolecules becomes immobilized, and the whole macromolecule behaves hydrodynamically like an equivalent homogeneous sphere with radius R_h. For linear chains the hydrodynamic radius is much smaller than the radius of gyration since the solvent can penetrate rather deeply into the coil. Figure 3 tends to explain this behaviour.

The *shape* of a flexible linear chain is that of a random coil. In the

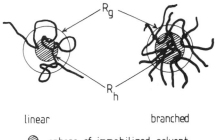

linear   branched

⊘ sphere of immobilized solvent

Fig. 3. Illustration of the radius of gyration and the hydrodynamic radius in a linear and a branched sample.

unperturbed state, i.e. for $A_2 = 0$, this coil has a radial segment density distribution $W(R)$ around the centre of mass which is close to a Gaussian distribution. For highly branched chains the molecule can be expected to have the shape of a swollen sphere with a segment density distribution that remains fairly constant in the central part but shows an exponential decay at the periphery, which results from dangling chains. Figure 4 demonstrates schematically the two density distributions in comparison with the equivalent density in a homogeneous sphere.

2.1 Results
Epoxy resins from TGI, which is trifunctional, and from EDA and HMDA, respectively, which are tetrafunctional, give highly branched structures

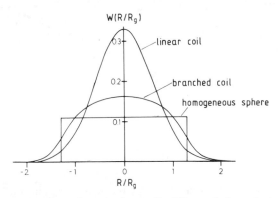

Fig. 4. Density profiles of segments in a coil of linear chains, of a branched chain and of a homogeneous sphere. The density distributions are normalized to $\int W(x)4\pi x^2 \, dx = 1$.

with a huge polydispersity. The DGEBA/BA resins are of widely different branching densities depending on the catalyst[6-10] and the curing temperature. The branching density of the DGEBA/AE resins[11] is rather low.

Figure 5 gives the result of the hydrodynamic radius R_h as a function of the molecular weight M_w. The use of the hydrodynamic radius instead of the directly measured translational diffusion coefficient D_z is preferred here, since the diffusion coefficient depends on the solvent viscosity and the temperature of measurement, whereas the hydrodynamic radius is independent of these variables. Figure 2 shows the same plot for the radius of gyration. For the TGI/EDA and TGI/HMDA resins the radii of gyration were measured by light scattering, R_g of the DGEBA/BA resins with less experimental error by SANS.[12] Again the four curves of the different series lie close together and allow no clear distinction.

The influence of branching becomes detectable when the ρ-parameter is considered. Figure 6 demonstrates the result if the smoothed curves of Fig. 2 and Fig. 5 are taken.

One notices a slight increase in ρ with growing molecular weight M_w for the TGI/EDA resins (a). For the TGI/HMDA resins this increase is much more pronounced (b). These resins behave totally differently from the branched DGEBA/BA resin (d), which seems to approach a plateau value; this is lower than that of the linear resin (c). The increase of ρ at very large molecular weights of TGI/EDA and TGI/HMDA resins seems to indicate a change in the reaction mechanism during the course of the reaction.

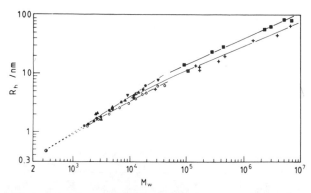

Fig. 5. Molecular weight dependence of the hydrodynamic radius R_h for six different epoxy resins: ▲, DGEBA/AE, cured at 75°C; ▼, DGEBA/AE, cured at 100°C; ○, DGEBA/BA, linear; ●, DGEBA/BA, branched; +, TGI/EDA; ■, TGI/HMDA.

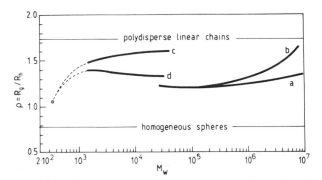

Fig. 6. Molecular weight dependence of the ρ-parameter for four different resins: (a) TGI/EDA; (b) TGI/HMDA; (c) DGEBA/BA, linear; (d) DGEBA/BA, branched.

This conjecture is confirmed by the molecular weight dependence of the intrinsic viscosity, which is shown in Fig. 7. Evidently branching has a very pronounced effect here.

2.2 Discussion

The results raise the question of why branching has so little influence on the dimensions and why the branching effect becomes so clearly detectable in the intrinsic viscosity.

We start the interpretation with $[\eta]$. For polydisperse samples eqn (1) has to be replaced by[13,14]

$$[\eta] = \Phi'(R_g^3)_n/M_n \tag{6}$$

where $\Phi' \neq \Phi$ for branched materials, and the subscript n denotes the number average. It will be recognized that

$$N_A(4\pi/3)R_g^3/M = v_{app} = 1/d_{app} \tag{7}$$

characterizes a volume that is spanned by a molecule of the molecular weight M and thus represents a reciprocal apparent density. We now turn to Einstein's[15] well-known formula for the intrinsic viscosity of homogeneous spheres

$$[\eta]_{sphere} = 2\cdot5/d \tag{8}$$

where d is the physical density of the material in the sphere. Highly branched macromolecules may be considered as being spherical in shape, which allows us to make the following approximation:

$$[\eta] = 2\cdot5/d_{app} \tag{9}$$

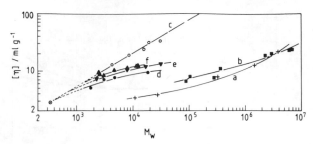

Fig. 7. Molecular weight dependence of the intrinsic viscosity for six epoxy resins: (a) TGI/EDA, (b) TGI/HMDA; (c) DGEBA/BA, linear; (d) DGEBA/BA, branched; (e) DGEBA/AE, cured at 100°C; (f) DGEBA/AE, cured at 75°C.

We then notice the following:

(1) For the TGI/EDA samples one finds up to $M_w = 10^4$ g mol^{-1} only a small decrease in the density to 0·74. This value is near the random closest sphere packing of monomers. The samples are obviously not swollen at low molecular weights. At larger molecular weights, however, the swelling ratio increases considerably to a factor of about 9 (density about 0·1), and this indicates longer sections of flexible sub-chains in the resin or, in other words, the extent of branching decreases with increasing size.

(2) For the TGI/HMDA system we do not have samples of $M_w \simeq 10^4$ g mol^{-1}. At $M_w = 10^5$ g mol^{-1} the segment density is about 0·4 and thus already in the range of weakly swollen spheres. The decrease in density with increasing M_w is not as pronounced as for TGI/EDA resins but also goes down to densities of about 0·1 for $M_w > 10^6$ g mol^{-1}.

(3) The apparent density decreases with the square root of the molecular weight for the linear resins. This behaviour is typical for random coils of linear chain molecules, and is predicted from Kuhn's square root law $R_g \sim \sqrt{M}$.[16]

(4) The density of the branched DGEBA/BA resins is reduced to 0·5 for the smallest and 0·26 for the highest molecular weights. For the less branched DGEBA/AE samples, the segment density is reduced to 0·3 and 0·2, respectively. These values mean that the molecules are weakly swollen spheres with swelling ratios of 2–5.

The insensitivity of the overall dimensions and the translational diffusion coefficient to branching is explained in Fig. 8. The straight lines are the molecular weight dependences of D_z and $\langle S^2 \rangle_z$ for monodisperse linear samples. A monodisperse branched sample of the same M_n should show a shrinking effect. Thus the diffusion would be quicker and the radius of

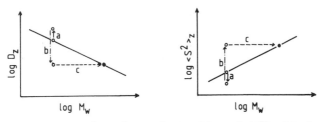

Fig. 8. Influence of polydispersity on the D_z–M_w- and $\langle S^2 \rangle_z$–M_w-dependences.

gyration smaller (step a). In a polydisperse sample of the same M_n, this shrinking effect is overcompensated by the effect that comes from the z-average (step b). At the same time the molecular weight is shifted to higher values because of the weight averaging (step c). Finally the D_z–M_w- and $\langle S^2 \rangle_z$–M_w-dependences of linear monodisperse and branched polydisperse materials are very close together.

We carried out calculations of the z-average diffusion coefficient on the basis of the cascade branching theory for one linear and two branched resins with low and high branching densities. The result is shown in Fig. 9 and is in full agreement with the experimental findings. Kajiwara[17] has shown that the diffusion increases as expected if fractions of randomly branched samples are considered, where in the unfractionated samples almost no change in D_z was observed.

As a last example we compare a particle scattering factor of a branched DGEBA/BA resin that was measured by SANS[12] and of a high molecular weight TGI/HMDA resin ($14 \cdot 0 \times 10^6$ g mol^{-1}) determined by light

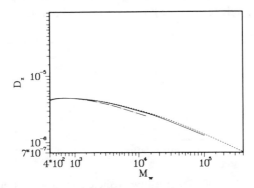

Fig. 9. Calculated molecular weight dependence of the diffusion coefficient for a linear resin (- - - - -), the DGEBA/AE resin cured at 100°C (———) and for a highly branched resin (—·—).

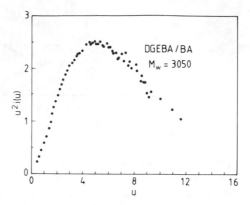

Fig. 10. Kratky plot of the particles scattered from a highly branched DGEBA/BA resin. Measurements were performed by SANS at ILL in Grenoble. $i(u)$ is the non-normalized scattering intensity.

scattering with those calculated by the branching theory (Figs 10 and 11). Long linear chains with polydispersity $M_w/M_n = 2$ and randomly branched polymers[18] approach in the normalized Kratky plot a plateau of $P_z(u)u^2 \rightarrow 3$, where $u = R_g q$.

In the calculated curves the typical plateau is not observed. They pass through a maximum and decay at larger q-values, which results from the finite dimensions of the monomeric units and is more pronounced for the low molecular weights. A similar maximum is observed for the DGEBA/BA resin of $M_w = 3050$ g mol^{-1}, in good agreement with theory. For the TGI/HMDA resin a maximum is not clearly displayed. This results from the long wavelength of the light (488 nm), which is too large compared

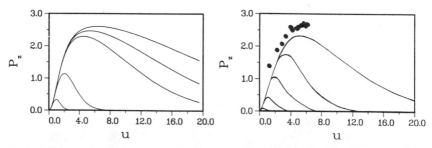

Fig. 11. Kratky plots of theoretically predicted particle scattering factors for DGEBA/AE resins of weight-average molecular weights 330, 1200, 5000, 20 000 and 100 000, respectively: $u = R_g q$. Left: linear resins; right, branched resins and (●) TGI/HMDA resin of 14×10^6 g mol^{-1}.

with the radius of gyration of the sample. In any case, the height of the maximum is higher than the calculated ones, but still lower than 3. This is just what is expected by theory for such a high molecular weight.

3 MEASUREMENTS IN SEMIDILUTE SOLUTION

The study of semidilute solutions has enjoyed increasing interest in recent years. This is partially due to the development of the dynamic or quasi-elastic light scattering (QELS) technique.[19] In QELS, no external force is applied to the system and all dynamic quantities are obtained from concentration fluctuations.

Recently, the behaviour of different polymer architectures in semidilute solution has been investigated in our research group. Here the behaviour of two different TGI/HMDA resins is discussed.

Applying scaling arguments, de Gennes[20] predicted for linear flexible chains in a good solvent and in the semidilute regime an exponent of 1·25 for the osmotic compressibility $(M/RT) \, \partial\pi/\partial c$ when $X = A_2 M_w c = c/c^* \gg 1$. Recently, Ohta and Oono[21] derived an expression from renormalization group theory which describes the whole region from dilute to semidilute solutions. Freed[22] predicted a stronger increase of the compressibility for more compact architectures like star-branched polymers or microgels. Hard spheres should show a diverging compressibility at $X \simeq 2·96$.[23,24]

3.1 Osmotic Compressibility

Semidilute solutions are of rather low concentration, but nevertheless show significant difference in behaviour compared with dilute solutions. The change in properties occurs around the coil-overlap concentration $c^* = 1/(A_2 M_w)$. This definition of c^* is only one of many which differ by constant factors.

As already mentioned, the scaling theory by de Gennes predicts for the concentration dependence of the normalized zero-angle scattering intensity a limiting exponent of 1·25 for flexible linear chains. The theory by Oono *et al.* reaches this asymptote for $X \simeq 20$, but experiments with high molecular weight polystyrene show at this point a still increasing function with maximum exponent 1·46 (Fig. 12, curve c).[25,26] Curves a and b (Fig. 12) correspond to the TGI/HMDA samples with a molecular weight of $2·3 \times 10^6$ and $7·0 \times 10^6 \, g \, mol^{-1}$, respectively. Up to a value of $X \simeq 1·5$, the curves a and b coincide pretty well with the theoretical curve d of Oono.[21] From that point on, a and b show a much steeper increase. We expected this result,

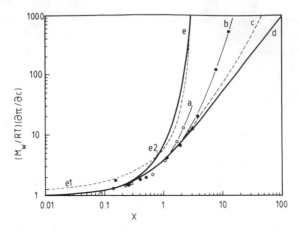

Fig. 12. Reduced osmotic compressibility $(M_w/RT)(\partial\pi/\partial c)$ as a function of $X = A_2 M_w c$ for: (a) TGI/HMDA, $M_w = 2\cdot3 \times 10^6$ g mol^{-1}; (b) TGI/HMDA, $M_w = 7\cdot0 \times 10^6$ g mol^{-1} and (c) polystyrene of different molecular weight;[25,26] (d) Ohta and Oono theory;[21] (e2) virial expansion for hard spheres quoted by Ferry;[23] (e1) free volume theory[34,24] for hard spheres; (e) curve for hard spheres averaged from (e1) and (e2).

because the TGI/HMDA resins are extremely highly branched and thus should resemble impenetrable spheres. The fact that the resin of higher molecular weight shows the weaker increase is consistent with the viscosity measurements, where a decreasing segment density with increasing molecular weight was found. Thus the increase in the osmotic compressibility should become steeper with decreasing molecular weight. Similar behaviour was found for polyfunctional stars with a highly crosslinked centre.[27]

3.2 Radius of Gyration

Theoretically, the radius of gyration at a finite concentration c and at concentration zero is expected to be a weakly decreasing function,[28,29] i.e. $\langle S^2 \rangle_c / \langle S^2 \rangle_0 \simeq 1$. From the initial slope of the angular-dependent curve in the Zimm plot, a value $\langle S^2 \rangle_{app}$ can be determined. $\langle S^2 \rangle_{app}$ is related to the thermodynamic correlation length ξ_T[30] by

$$\langle S^2 \rangle_{app} = 3\xi_T^2 \tag{10}$$

and to $\langle S^2 \rangle_c$ as follows:[31,32]

$$\langle S^2 \rangle_c / \langle S^2 \rangle_0 = (\langle S^2 \rangle_{app} / \langle S^2 \rangle_0)(M_w/RT)(\partial\pi/\partial c) \tag{11}$$

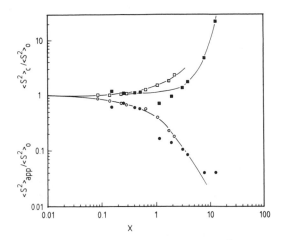

Fig. 13. Apparent mean-square radius of gyration $\langle S^2 \rangle_{app}$ (lower curves) and $\langle S^2 \rangle_c$ (upper curves) normalized by the zero concentration mean square radius of gyration $\langle S^2 \rangle_0$ as a function of X. $\langle S^2 \rangle_{app}$: $\bigcirc$, TGI/HMDA of $M_w = 2\cdot3 \times 10^6$ g mol^{-1}; $\bullet$, TGI/HMDA of $M_w = 7\cdot0 \times 10^6$ g mol^{-1}. $\langle S^2 \rangle_c$: $\square$, $M_w = 2\cdot3 \times 10^6$ g mol^{-1}; $\blacksquare$, $M_w = 7\cdot0 \times 10^6$ g mol^{-1}.

Figure 13 shows the X dependence of $\langle S^2 \rangle_{app}/\langle S^2 \rangle_0$. As expected $\langle S^2 \rangle_{app}$ decreases with increasing concentration, but the expected behaviour for $\langle S^2 \rangle_c$ to be constant was not found. Instead we find an increase of $\langle S^2 \rangle_c$ with increasing concentration. So far we have no explanation for this fact.

3.3 Diffusion Coefficient

Recently Oono et al.[33] derived an expression from renormalization group theory for the X-dependence of the diffusion coefficient of a flexible linear chain in a good solvent. Our experimental data are shown in Fig. 14. The theoretical curve increases much more steeply, but still reaches the limiting

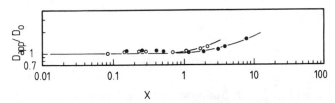

Fig. 14. Apparent diffusion coefficient D_{app} (at zero scattering angle) normalized by D_0 (at zero concentration) as a function of X for ($\bigcirc$) TGI/HMDA $M_w = 2\cdot3 \times 10^6$ g mol^{-1} and for ($\bullet$) TGI/HMDA $M_w = 7\cdot0 \times 10^6$ g mol^{-1}.

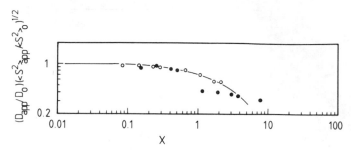

Fig. 15. Ratio of the thermodynamic shielding length to the hydrodynamic correlation length as a function of X for the same samples as in Fig. 14.

exponent of 0·75 predicted by the scaling theory[20] at even higher values of X than for the osmotic compressibility. Similar deviation from theory is found for completely different polymer architectures.

3.4 Correlation Lengths

In principle, three different correlation lengths have to be distinguished,[32] i.e.

(a) ξ, the mean distance between two points of the entanglement;
(b) ξ_T, a thermodynamic correlation length, defined by eqn (10);
(c) ξ_h, a hydrodynamic shielding length.

In earlier theories these three quantities were assumed to be proportional to each other, though not identical in length. One can now define the normalized correlation lengths $\xi_T^* = \xi_T/R_g$ and $\xi_h^* = \xi_h/R_h$. The value

$$\xi_T^*/\xi_h^* = (D_{app}/D_0)(\langle S^2 \rangle_{app}/3\langle S^2 \rangle_c)^{1/2} \qquad (12)$$

is expected to be constant. Experiments show, however, that this is not observed for the present samples. It can be seen from Fig. 15 that the thermodynamic value is influenced differently by the concentration than the hydrodynamic value.

ACKNOWLEDGEMENTS

We are grateful to Dr M. Budnowski (Henkel KGaA, Düsseldorf, FRG) and Professor F. Lohse (Ciba-Geigy, Basel, Switzerland) for supplying us with the triglycidyl isocyanurate. The work was kindly supported by the Deutsche Forschungsgemeinschaft within the scheme SFB 60.

REFERENCES

1. Bantle, S., Schmidt, M. and Burchard, W., *Macromol.*, 1982, **15**, 1604.
2. Fox, T. G. and Flory, P. J., *J. Chem. Phys.*, 1949, **53**, 197.
3. Zimm, B. H. and Stockmayer, W. H., *J. Chem. Phys.*, 1949, **17**, 1301.
4. Stockmayer, W. H. and Fixman, M., *Ann. NY Acad. Sci.*, 1953, **57**, 334.
5. Burchard, W., Schmidt, M. and Stockmayer, W. H., *Macromol.*, 1980, **13**, 580, 1265.
6. Burchard, W., Bantle, S. and Zahur, S. A., *Makromol. Chem.*, 1981, **182**, 145.
7. Bantle, S. and Zahur, S. A., *J. Am. Chem. Soc. ACS Sympos. Series*, 1983, **221**, 245.
8. Bantle, S. and Burchard, W., *Polymer*, 1987, **27**, 728.
9. Burchard, W., Bantle, S. and Wachenfeld-Eisele, E., *Makromol. Chem. Macromol. Symp.*, 1987, **7**, 55.
10. Bantle, S. and Burchard, W., *Polymer*, in preparation.
11. Wachenfeld-Eisele, E. and Burchard, W., *Polymer*, 1987, **28**, 817.
12. Bantle, S., Hässlin, H. W., ter Meer, H.-U., Schmidt, M. and Burchard, W., *Polymer*, 1982, **23**, 1889.
13. Newman, S., Krigbaum, W. R., Laugier, C. and Flory, P. J., *J. Polymer Sci.*, 1954, **14**, 451.
14. Marriman, J. and Hermans, J. J., *J. Phys. Chem.*, 1961, **65**, 385.
15. Einstein, A., *Ann. Phys.*, 1910, **33**, 1270.
16. Kuhn, W., *Kolloid.-Z.*, 1934, **68**, 2.
17. Kajiwara, K., *Polymer*, 1971, **12**, 57.
18. Kajiwara, K., Burchard, W. and Gordon, M., *Br. Polymer J.*, 1970, **2**, 110.
19. Burchard, W., *Chimia*, 1985, **39**, 10.
20. de Gennes, P. G., *Scaling Concepts in Polymer Physics*, Cornell University Press, Ithaca, New York, 1979.
21. Ohta, T. and Oono, Y., *Physics Letters*, 1982, **89A**, 460.
22. Cherayil, B. J., Bawendi, M. G., Miyake, A. and Freed, K. F., *Macromol.*, 1986, **19**, 2770.
23. Ferry, J. D., *Adv. in Protein Chem.*, 1948, **4**, 1.
24. Kirkwood, J. G., Maun, E. K. and Alder, B. J., *J. Chem. Phys.*, 1950, **18**, 1040.
25. Burchard, W. and co-workers, unpublished data.
26. Wiltzius, P., Haller, H. R., Cannell, D. S. and Schaefer, D. W., *Phys. Rev. Lett.*, 1983, **51**, 1183.
27. Lang, P. and Burchard, W., unpublished data.
28. Daoud, M., Cotton, J. P., Farnoux, B., Jannink, G., Sarma, G., Benoit, H., Duplessix, R., Picot, C. and de Gennes, P. G., *Macromol.*, 1975, **8**, 804.
29. Albrecht, A. C., *J. Chem. Phys.*, 1957, **27**, 1014.
30. Edwards, S. F., *Proc. Phys. Soc. (London)*, 1985, **83**, 5293.
31. Wenzel, M., Burchard, W., and Schätzel, K., *Polymer*, 1986, **27**, 195.
32. Coviello, T., Burchard, W., Dentini, M. and Crescenzi, V., *Macromol.*, 1987, **20**, 1102.
33. Oono, Y.m Baldwin, P. R. and Ohta, T., *Phys. Rev. Lett.*, 1984, **53**, 2149.
34. Eyring, H. and Hirschfelder, J. O., *J. Phys. Chem.*, 1937, **41**, 249.

21

NETWORKS AS THE BASIS OF PRE-THICKENING SMC

S. F. Bush, J. M. Methven and D. R. Blackburn

Polymer Engineering Division, UMIST, PO Box 88, Manchester M60 1QD, UK

ABSTRACT

The formation of a temporary network by a saturated polyester added to an unsaturated resin appears to provide both an attractive route to the pre-thickening of such resins when used in Sheet Moulding Compounds, and a means of controlling their shrinkage during moulding. It is believed that this is the first report of a successful method of achieving both goals with a single material. The network is based upon the formation of precrystalline clusters of polymer chain segments. The density and character of the network are determined by the compatibility of the saturated polyester and the unsaturated resin.

1 INTRODUCTION

This work was initiated by the commercial availability of a novel class of unsaturated resins known as oligourethane acrylates. These resins may be considered roughly as the urethane analogues of vinyl ester resins and, in common with vinyl esters, offer the prospect of improved chemical resistance over conventional unsaturated polyesters (alkyds), used in glass-reinforced plastics constructions. The structures of these three resin types are shown in Fig. 1, where it can be seen that the improved chemical resistance of both the urethane acrylates and vinyl esters is due to their having considerably fewer hydrolysable groups (ester links) in the main chain.

The particular concern of this paper is with the class of glass-reinforced plastics known as Sheet Moulding Compounds (SMC). These materials

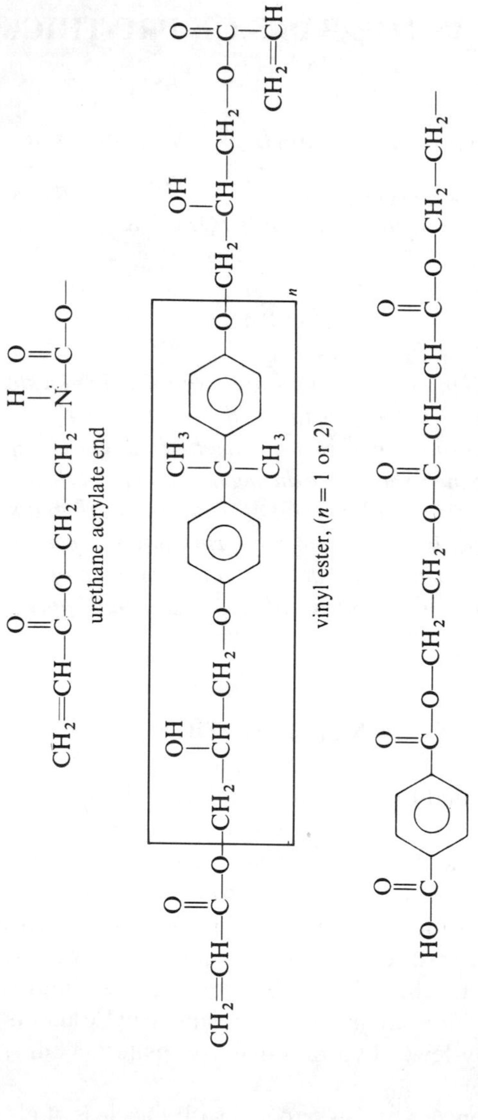

Fig. 1. Anticipated structures of terminal segments of each material.

TABLE 1
General Purpose SMC Formulation

Ingredients	Composition (% by weight)
Unsaturated polyester dissolved in styrene monomer[a]	25
Shrinkage control additive[a]	5
Catalyst[a]	0·3–0·5
Fillers[a] (e.g. chalk)	40–50
Release agent[a]	0·4–0·5
Thickening agent	0·2–0·4
Glass (25 mm)	25

[a]Such ingredients are first mixed together in a high shear mixer.

consist of a combination of unsaturated resin, particulate fillers, chopped glass fibres as reinforcement, and a variety of other constituents, the nature and purpose of which are described in due course. A typical compound formulation is shown in Table 1.

An electron micrograph of a section through a moulded sheet is shown in Fig. 2.

Fig. 2. Electron micrograph of SMC showing the presence of glass fibres (magnification × 540).

The fundamental requirements of any SMC are:

(a) It must be handleable (i.e. relatively tack-free) at room temperature so that it can be cut and draped to fit a particular mould.

(b) Under prescribed (compression) moulding conditions, the sheet must flow uniformly to fill the mould with no segregation of the components shown in Table 1.

(c) On filling the mould, the unsaturated resin component must crosslink to give it a permanent form. It should be noted that in the formulation of Table 1 the unsaturated polyesters crosslink through the styrene present.

A typical unsaturated polyester based compound such as that shown in Table 1 has an initial viscosity (measured at ambient temperature without the glass reinforcement) of around 200 P (20 Pa s), while in order to achieve (a) above, industry practice suggests a requirement of around 10 000 P (1 kPa s measured under the same conditions). Thus, in order to achieve both (a) and (b), two distinct steps are required:

(i)   The unsaturated resin must be thickened at room temperature to obtain the desired viscosity for handling.

(ii) The viscosity must decrease sharply to facilitate flow during moulding.

The first step is known as 'pre-thickening' of an SMC and, as described in the next section, is based traditionally on the reaction of residual carboxylic acid groups in the unsaturated polyester resin (see Fig. 1), with oxides and hydroxides of Group II metals.

By contrast, the approach to pre-thickening described in this chapter is to achieve step (i) by means of an organic polymer additive which at room temperature forms a temporary 'informal' network. At temperatures appropriate to step (ii), this informal network will be disbanded, causing the viscosity to decrease sufficiently for the flow process to occur.

Experiments have been conducted with all three resins mentioned above, with particular emphasis on the urethane acrylate (henceforth uracrylate) system. The use of such resins may be expected to provide improvements in chemical resistance, end-use temperature, fire performance and moulding cycle times over those commonly found with compounds based on unsaturated polyester resins. In addition, the lower viscosity of uracrylates compared with unsaturated polyesters may be expected to result in more effective wetting contact with the reinforcing glass fibres in the compounds, and hence give improved mechanical properties. Finally, it must be realised that since uracrylates possess neither terminal nor pendant carboxylic acid residues, they cannot be pre-thickened by the conventional metal oxide route and are currently excluded from SMC manufacture.

Following some background information on the manufacture of SMC (Section 2), the results of the novel network approach are described and discussed in Section 3. The conclusions and future prospects for the application of the method and its implications are discussed in Section 4.

2 SHEET MOULDING COMPOUNDS

2.1 Manufacture and Moulding
The manufacture of an SMC based on the composition shown in Table 1 is shown schematically in Fig. 3 and consists of four basic steps:

(a) High shear mixing of the particulate fillers with the resin(s).
(b) Metering of this paste onto the glass fibres which are chopped *in situ* from rovings.
(c) Consolidation of and removal of adventitious air from the resulting sheet.
(d) Pre-thickening of the sheet prior to moulding.

The resulting sheet is cut to fit a particular mould so that it covers around 90% of the mould (cavity) area, and the appropriate amount of material is obtained by stacking multiple plies. The sheet is moulded at pressures of between 4 and 7 MPa, at temperatures between 130 and 170°C.

2.2 Pre-thickening
Pre-thickening of the SMC paste is based upon a chemical reaction between the residual (terminal) carboxylic acid residues in the unsaturated polyester

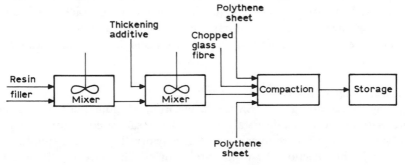

Fig. 3.   Schematic of SMC manufacture.

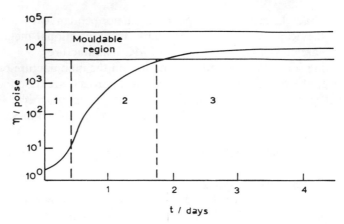

Fig. 4.   Typical thickening curve for SMC (viscosity vs. time). Section 1, induction period; section 2, rapid viscosity rise; section 3, stable viscosity region.

and a Group II metal oxide such as magnesium oxide (MgO). This is added to and blended with the SMC paste immediately before the paste is transferred to the SMC machine. In the final stage of manufacture the sheet is stored for some days to allow the pre-thickening reaction to take place. A typical thickening curve is shown in Fig. 4.

The rate of increase in viscosity is chosen to establish the final sheet viscosity as quickly as possible, while still facilitating sufficient wetting of the glass fibres by the paste. The chemical and physical processes involved are indicated in Fig. 5.

The effect of the chemical reaction (Fig. 5(a)) is to create a labile network by crosslinking the polyester chains via complex metal salts (shown schematically in Fig. 5(b)). The extent of this reaction is dependent upon the level of carboxylic acid groups in the resin, and this must be carefully monitored for consistent pre-thickening behaviour. In practice it is also found that the rate and extent (final viscosity) of the reaction are influenced by both the particle size of the pre-thickening agent and the level of water in the resin.

2.3 Shrinkage of SMC
During the moulding of an SMC, the unsaturated monomer reacts in the presence of the peroxide catalyst with the unsaturated residues of the polymer to form a permanent, covalent network (Fig. 6) (e.g. Ref. 1).

During this reaction the resin shrinks in volume by up to 10%, and unchecked this would not only reduce the fidelity with which the moulding

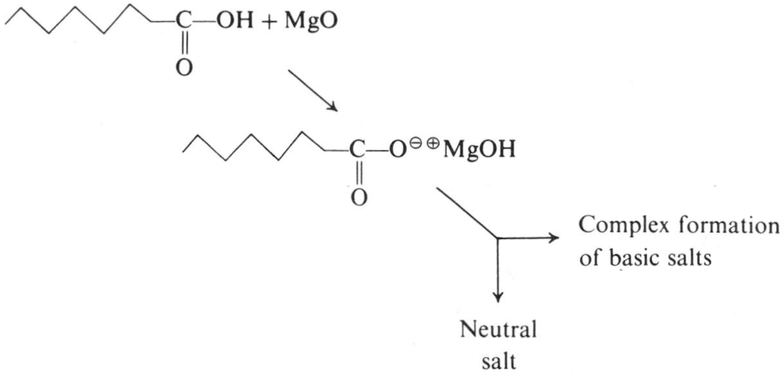

(a)

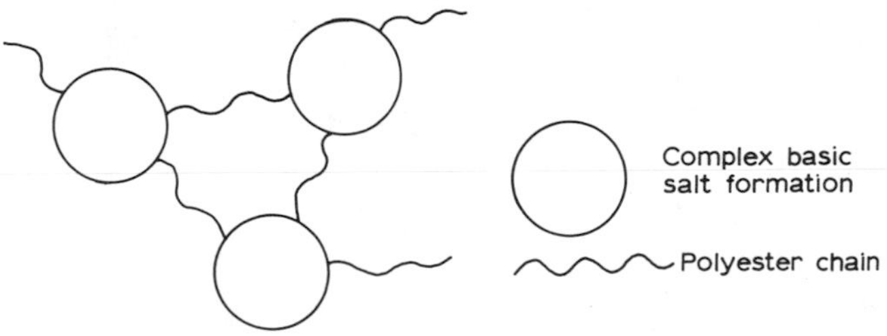

(b)

Fig. 5. (a) Pre-thickening reaction of conventional SMC. (b) Representation of possible MgO-thickened polyester network.

compound reproduced the mould dimensions, but would also render the surface of the moulding compound somewhat unattractive by highlighting the presence of the reinforcing fibres.

Hitherto, control of moulding shrinkage has been effected by adding a solution of a thermoplastic in styrene to the SMC formulation. The solution commonly contains around 30% by weight of the thermoplastic. Appropriate thermoplastics include polystyrene, polyvinyl acetate, poly-caprolactone, polymethyl methacrylate, and more recently, certain polybutadienes. Typically, the ratio of unsaturated polyester resin to the solution of thermoplastic is between 90:10 and 60:40 by weight.

Various mechanisms have been proposed to account for this control of shrinkage (e.g. Ref. 2). The view taken here is that, as the styrene and unsaturated polyester resin crosslink (Fig. 6), its solubility parameter falls from that of the unreacted polyester (Fig. 6(a)) to one much closer to that of the added thermoplastic, which consequently swells the network.[3,4] This is

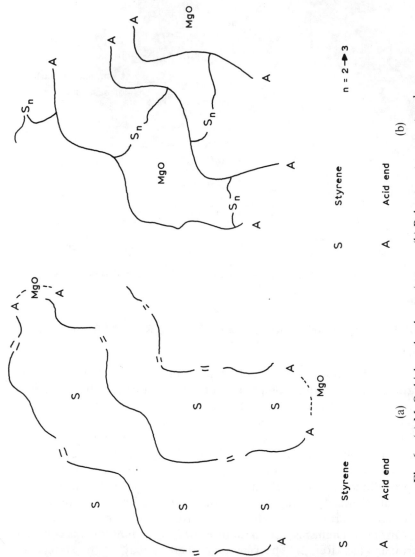

Fig. 6. (a) MgO-thickened polyester/styrene. (b) Polyester–styrene network.

reasonable because the solubility parameter corresponding to the structure of Fig. 6(b) is estimated to be around 20 $(MPa)^{1/2}$, while that of a typical thermoplastic additive (e.g. polystyrene) is 19 $(MPa)^{1/2}$, which is a value close enough for swelling to take place.* The swelling pressure within the network thus offsets the shrinkage pressure arising from the network formation and thermal contraction.

When the solubility parameters (δ) of the network-forming resin and the thermoplastic polymer are sufficiently different, this rationale would predict that phase separation occurs as the network is completed. This is indeed reported to be the case[6] in experiments on epoxides, toughened by inclusion of polybutadiene domains. As the epoxide network ($\delta = 22$ $(MPa)^{1/2}$) is formed, the polybutadiene ($\delta = 17$ $(MPa)^{1/2}$) separates into rubbery domains.

3 URACRYLATE PRE-THICKENING

Two approaches have been taken in the work reported here to effect pre-thickening of the uracrylate resin (i.e. a mixture of oligourethane acrylates and unreacted methyl methacrylate):

(i) The addition of polyurethane precursors (polyol, chain extender, diisocyante, catalyst) to a low viscosity paste of uracrylate and filler, to form a polyurethane network of low crosslink density, independent of the uracrylate. The result of this reaction was expected to be a sparse elastomeric polyurethane network swollen with the uracrylate paste and hence a material possibly of a rubbery or leathery consistency analogous to a pre-thickened polyester based SMC. On cure of the uracrylate resin, the resulting compound would thus yield two interpenetrating networks (IPNs).

(ii) The addition of a saturated polyester which, while a crystalline solid at room temperature, would melt well below the cure temperature, and in the molten state be compatible with the base resin. The expectation was that, on recooling to room temperature, highly dispersed crystalline domains would be formed by the saturated polyester, thereby creating a network analogous to that formed by the MgO centres in Fig. 5(b). It was also assumed that the heat of fusion of the candidate material would have to be sufficiently low to ensure that in the presence of the uracrylate only a

---

* The solubility parameters of the various oligomers used in the work have been estimated from the Group Methods of Ref. 5.

small proportion of the polymer chain units would form these crystalline domains. The material selected to meet these criteria was polyethylene adipate (PEA), the structure of which is shown below:

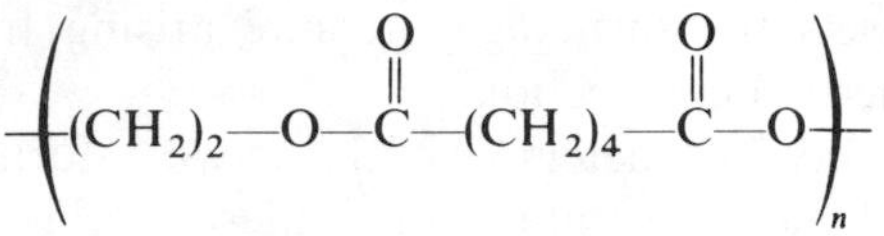

$$\left(\!\!(CH_2)_2\!-\!O\!-\!\overset{\displaystyle O}{\overset{\displaystyle \|}{C}}\!-\!(CH_2)_4\!-\!\overset{\displaystyle O}{\overset{\displaystyle \|}{C}}\!-\!O\!\right)_{\!\!n}$$

polyethylene adipate

3.1 IPN Formation

Table 2 shows the composition and properties of the polyurethanes employed. In the case of the uracrylate, little reaction of the urethane components was observed at temperatures of between 20 and 50°C, this upper limit determined by loss of monomer. While some evidence of a viscosity increase was observed in an unsaturated polyester resin with the polyurethane precursors, this may have been due, in part at least, to the reaction of the isocyanate with residual hydroxyl groups in the polyester. Such reactions will lead to chain extension as opposed to network formation, and are hence unlikely to lend themselves to effective pre-thickening.

No reason for this lack of significant reaction of the polyurethane precursors in the resins has been deduced so far, even though the formulation derived from the materials in Table 2 produces a satisfactory result in the absence of uracrylate.

TABLE 2
Materials used for PU IPN Systems

Material	Molecular weight or equivalent M_n (g mol^{-1})	Functionality f_n
Isocyanate, RMA 400 (Upjohn), MDI (methylene diisocyanate) based	159	2
Polyol, M111 (Lankro), polyethylene oxide tipped, polypropylene oxide triol	2036	2·76
Chain extender, ethylene glycol (HOCH$_2$CH$_2$OH)	62	4

TABLE 3
Polyethylene-adipate-based SMC Formulations

Material	% by weight		
Uracrylate resin dissolved in methyl methacrylate monomer	29·4	25·7	22·0
Filler (hydrocarb)	36·4	36·4	36·4
Trigonox (catalyst)	0·8	0·8	0·8
Zinc stearate (mould release agent)	1·1	1·1	1·1
Polyethylene adipate	7·3	11·0	14·7
Glass mat	25·0	25·0	25·0

3.2 Addition of a Saturated Polyester: PEA

The saturated polyester used in this work was a commercial grade of polyethylene adipate (PEA) of number-average molecular weight 2000. Table 3 shows the SMC formulations based on this material.

Since the additive is a solid at ambient temperature with a melting point of around 50°C, it was first melted and blended with the resin/filler combination, and the resulting mixture spread onto the appropriate quantity of chopped strand glass mat kept at this temperature by means of a hot table. The SMC so prepared was then allowed to cool to ambient temperature between sheets of polythene and cellophane.

PEA was chosen, as its solubility parameter of 20 $(MPa)^{1/2}$ is close enough to that of the uracrylate $(20·5 (MPa)^{1/2})$ to give full fluid miscibility with the latter. In addition, its heat of fusion was thought sufficiently low to restrict its tendency to crystallize in the presence of the uracrylate.

The results of adding PEA to the uracrylate resin in proportions of between 20% and 40% by weight were indeed found to transform a resin with a viscosity of around 1 P (0·1 Pa s) to a coherent but malleable sheet of perhaps 10 000 P (10 kPa s). As expected, over the range of additive proportions applied, the greater the proportion of additive to resin, the stiffer the sheet. In all cases a satisfactory pre-thickening was obtained. The mechanical properties of the fully cured sheet (Table 4) compared

TABLE 4
Mechanical Properties of Moulded Uracrylate Compounds Compared with Conventional SMC

	Uracrylate resin	Conventional SMC
Tensile strength, σ_{FRAC} (MPa)	55	50–60
Tensile modulus, E (GPa)	15	10–15

Fig. 7.   Electron micrograph of uracrylate SMC (original magnification × 950).

favourably with those of conventional unsaturated polyester based SMC materials.

A scanning electron micrograph of an unpolished section through the sheet is shown in Fig. 7.

To further explore the basic concept, PEA was added in controlled proportions to a standard SMC unsaturated polyester typically used in SMC manufacture, and to a vinyl ester (Fig. 1). Both materials have significantly different solubility parameters from PEA, and these and the appearances of the derived SMC sheets are given in Table 5.

The unsaturated polyester and the vinyl ester are clearly less compatible with the PEA than the uracrylate is, but in different directions. In the case of

TABLE 5
Solubility Parameters of Materials used

Material	Solubility parameter δ $(MPa^{1/2})$	Effect of PEA addition to pre-thickened sheet at room temperature
Polyethylene adipate (PEA)	20·0	—
Vinyl ester	18·8	Floppy
Uracrylate	20·6	Malleable
Unsaturated polyester	22·0	Stiff

the unsaturated polyester, the difference in solubility parameter appears to provoke extensive segregation of the PEA in crystalline domains, leading to the more brittle character of the sheet. In the case of the vinyl ester, however, it is thought that the PEA cannot crystallize, and instead participates in the hydrogen bonding, i.e. between the hydroxyl groups on the vinyl ester chain and the ester groups elsewhere, which is responsible for the relatively high viscosity (100 Pa s) of the vinyl ester on its own. In fact the PEA acts as a plasticizer.

3.3 Shrinkage

It may be expected that, on the basis of the view given in Section 2.4, the addition of PEA to the uracrylate resin would resist shrinkage during the formation of the resin network. This arises because the similarity of the solubility parameters for the PEA and uracrylate will ensure that the molten PEA will swell the network at reaction temperatures (of about 140°C). On cooling to room temperature, the network will interfere with any PEA crystallization, thereby maintaining the swelling pressure, which in turn offsets the shrinkage pressure. Such is found to be the case. In fact a small net expansion was found on cooling (Table 6).

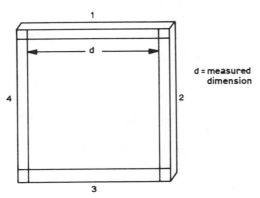

TABLE 6
Plane Dimension Changes for Panels Moulded from Uracrylate Resins

Side	Size of tool (mm)	Size of panel (mm)	Dimension change (%)[a]
1	139·90	139·97	+0·050
2	139·89	139·96	+0·050
3	139·88	139·90	+0·014
4	139·89	139·94	+0·036

[a] Average dimension change = +0·038%.

4 CONCLUSIONS

The usefulness of a saturated polyester additive for both pre-thickening and shrinkage control of SMC has been demonstrated. The mechanism of the interactions involved in the thickening has been qualitatively interpreted in terms of the formation of temporary 'informal' networks, the nature and density of which are dependent upon the compatability of the unsaturated resin and the polymeric additive.

The shrinkage resistance of the cured PEA–uracrylate system has been ascribed to the swelling of the permanent (covalent) network formed during the cure. As the swelling behaviour and the formation of the informal networks both depend upon the solubility parameters of the resin and the additive being similar, all the three requirements defined in Section 1 are achieved with one additive.

One of the attractions of this method of pre-thickening is that it is virtually instantaneous when compared with other methods based on Group II metal oxides (Section 2), which can require some days of storage to reach the required viscosity.

The results of adding PEA to a standard unsaturated polyester resin, and to a vinyl ester resin, show the effects of increasing the solubility parameter difference of resin to additive. These effects are consistent with the reasoning that led to the choice of the PEA to be used with the uracrylate. This indicates that thickening additives based on the same principles may be chosen for other resin systems.

REFERENCES

1. Brydson, J. A., *Plastics Materials*, Butterworth, London, 4th edn, 1982.
2. Burns, R., *Polyester Molding Compounds*, Marcel Dekker Inc., New York, 1982.
3. Flory, P. J., *Principles of Polymer Chemistry*, Cornell University Press, Ithaca, New York, 1953.
4. Hildebrand, J. H. and Scott, R. L., *Solubility of Non-electrolytes*, 3rd edn, Rheinhold, New York, 1950.
5. Van Krevelin, D. W., *Properties of Polymers*, 2nd edn, Elsevier, Amsterdam, 1977, Chapter 8.
6. Williams, R. J. J., Barrajo, J., Adabbo, H. E. and Rojas, A. J., *ACS Sym. Ser.*, 1984, **208**, 195–213.

22

SPECIAL FEATURES OF NETWORK BUILD-UP IN CURING OF POLYEPOXIDES BASED ON N,N-DIGLYCIDYLANILINE DERIVATIVES

KAREL DUŠEK and LIBOR MATĚJKA

Institute of Macromolecular Chemistry, Czechoslovak Academy of Sciences, 162 06 Prague, Czechoslovakia

ABSTRACT

The curing of N-N-*diglycidylaniline (DGA) and* N,N,N′,N′-*tetraglycidyldiaminodiphenylmethane with polyamines differs from that of diglycidyl ether of bisphenol-A by a high tendency to cyclization and by interdependence of reactivities of the glycidyl groups bound to the same nitrogen atom. Cyclization is characterized by formation of small cycles by intramolecular epoxy–amine addition or etherification. The reaction of the monoadduct of DGA and a secondary amine bearing another epoxy group with another secondary amine molecule is considerably faster than that of DGA with the same secondary amine. This feature is explained by intramolecular catalysis of the addition by the OH group in the monoadduct. The interplay of the substitution effects in the amine group and in DGA as reflected in gelation is complex, however. The development of a statistical branching theory taking these features into account is outlined.*

1 INTRODUCTION

Polyepoxides based on diglycidylaniline and its derivatives represent an important class of epoxy resins used in high-performance composites (mainly $N,N,N′,N′$-tetraglycidyl-4,4′-diaminodiphenylmethane, TGDDM) or as reactive diluents for epoxy resins (N,N-diglycidylaniline, DGA):

335

Diglycidylaniline serves also as a model for curing TGDDM and possibly other N,N-diglycidylamine derivatives. Diglycidylamines differ from the bisphenol-A type diepoxides by the steric closeness of glycidyl groups, which may have an effect on network formation. In this contribution, features of the reaction of DGA with amines are summarized and the conclusions are used in an analysis of gelation and network build-up.

2 FEATURES OF THE REACTION OF N,N-DIGLYCIDYLANILINE WITH AMINES

2.1 Cyclization

Cyclization has been found to be very weak in curing of bisphenol-A type epoxy resins.[1] The steric closeness of the epoxy groups in N,N-diglycidylamines makes possible the formation of small rings. An increased tendency to cyclization in TGDDM–diamine systems was already mentioned by Morgan.[2] Model reactions of DGA with aniline showed that in addition to linear oligomers (both monomers are bifunctional) cyclic products are also formed, and among these one of which the concentration is relatively large. This product was identified as an eight-membered ring compound (**III**);[3]

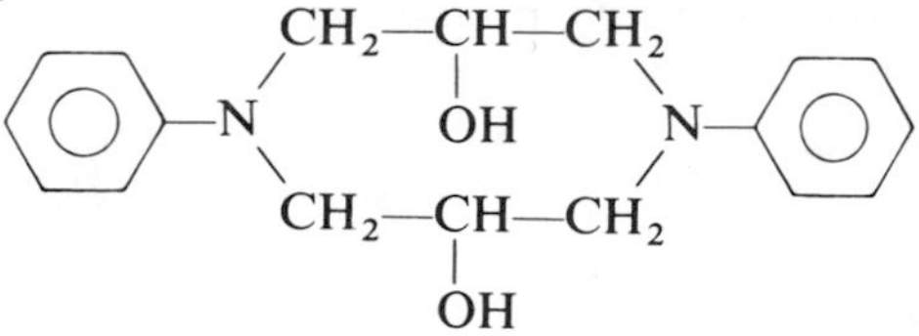

III

This compound cannot enter any further addition reaction except etherification under certain conditions (excess of epoxy groups).

At an excess of epoxy groups, cyclic structures can be formed by internal etherification. Diglycidylaniline and N-methylaniline were used as model compounds.[3] It has been found that a seven-membered ring of structure **IV** is formed in a relatively high yield:

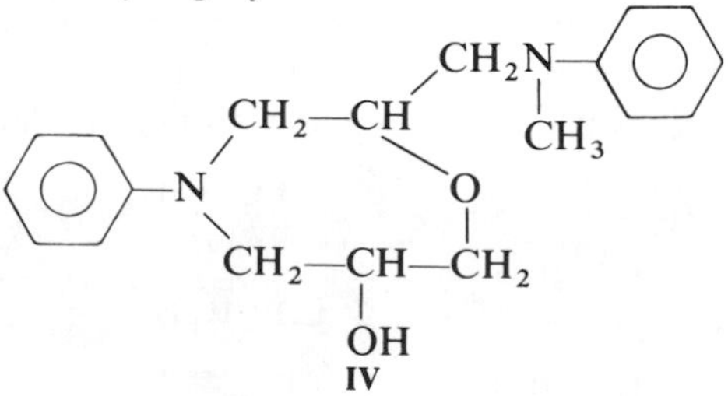

IV

Thus, cyclization is a typical feature of the reaction of N,N-diglycidyl-amines with primary or secondary amines. It lowers the effective functionality of the monomers and retards gelation.

2.2 Dependence of Reactivity of Epoxy Groups

The dependence of reactivities of epoxy groups in DGA was determined by studying the kinetics of the reaction of DGA with N-methylaniline (NMA) and aniline. Under conditions where etherification does not interfere, two addition products are formed: DGA.NMA (monoadduct) and DGA. 2NMA (diadduct). Rate constants were determined using HPLC.[4] For the reaction

$$\text{DGA} + \text{NMA} \xrightarrow{\ k_2\ } \text{DGA.NMA} \qquad (1)$$

$$\text{DGA.NMA} + \text{NMA} \xrightarrow{\ k_4\ } \text{DGA.2NMA} \qquad (2)$$

the ratio of rate constants k_4/k_2 for the uncatalyzed and catalyzed (by OH groups) reactions, respectively, varies between 3 and 16. The kinetics are complicated by the existence of two steroisomers of DGA and its reaction products[5] which have different reactivities. Nevertheless, it is clear that $k_4/k_2 > 1/2$, i.e. the reactivity of the epoxy group in the monoadduct is much higher than that of the epoxy group in DGA. This finding can be explained by an intramolecular catalysis of the addition reaction by the OH group of the monoadduct.

When bifunctional DGA reacts with bifunctional aniline, linear oligomers as well as cyclic products mentioned above are formed. Since there is a substitution effect in DGA as well as in the amino group of aniline,[6,7] these substitution effects could be independent or dependent. In the former case, the chain growth is determined by two ratios of the rate constants and in the latter case by at least by four constants or three ratios (cf. the following reactions):

$$\text{E}_0 + \text{A}_0 \xrightarrow{\ k_1\ } \text{E}_1 + \text{A}_1 \qquad (3)$$

$$\text{E}_0 + \text{A}_1 \xrightarrow{\ k_2\ } \text{E}_1 + \text{A}_2 \qquad (4)$$

$$\text{E}_1 + \text{A}_0 \xrightarrow{\ k_3\ } \text{E}_2 + \text{A}_1 \qquad (5)$$

$$\text{E}_1 + \text{A}_1 \xrightarrow{\ k_4\ } \text{E}_2 + \text{A}_2 \qquad (6)$$

A_i and E_i are, respectively, amino group and diepoxide unit, with i reacted functionalities. If the substitution effect within the diepoxide and amine were independent, we would have $k_2/k_1 = k_4/k_3 = \rho$ and $k_3/k_1 = k_4/k_2 = \kappa$, so that $k_4/k_1 = \kappa\rho$.

However, the cyclization and stereoisomer problems are not considered in this scheme.

3 NETWORK FORMATION

3.1 Development of Theory

In this section, gelation and network formation from the bifunctional DGA or tetrafunctional TGDDM and a monoamine or diamine will be considered. It is assumed that the reactivities of amino groups in the diamine and the reactivities of diglycidylamine groups in TGDDM are independent. Therefore, the reaction between bifunctional DGA and aniline is considered first, taking into account cyclization and the reactivity problem.

Let us briefly recall the simple case of network build-up from a diamine and a diepoxide in which also the reactivities of epoxy groups are independent and cyclization is negligible. Such a situation arises in the curing of the diglycidyl ether of bisphenol-A (DGEBA).[1] Then, in reactions (3)–(6), $k_1 = k_3$ and $k_2 = k_4$, so that a single ratio ρ determines the distribution of A_0, A_1 and A_2, i.e. the fractions of primary, secondary and tertiary amino groups.[6] This distribution is the single piece of information necessary for the calculation of various network parameters, as explained elsewhere.[1]

For the present case, where cyclization and complex reactivity interdependence exist, a theoretical approach to network formation will only be outlined, because more experimental data on model systems are still needed. Since the system is composed of two monomers with dependent reactivities of groups, and long-range correlations due to cyclization are to be considered, no exact branching theory is available (cf. Ref. 8). The following simplifications are possible:

(a) The formation of cycles is limited to the smallest cycle, i.e. the population of larger cycles is insignificant. Then, one can solve the problem using the kinetic method. The kinetic equations (infinite in number) for all species can be formulated and possibly transformed into a single equation for the degree-of-polymerization generating function (cf., for example, a similar but simpler problem treated by Dušek and Šomvársky).[9]

(b) A statistical build-up of structures from monomer units or larger fragments is employed. This simplification is possible, if the long-range effect of reactivity correlation is not too serious (cf., for example, the case of step polyaddition of a trifunctional monomer by Mikeš and Dušek[10]).

Alternatively, one can examine the effect of increasing the size of fragments from which the structures are built up. By increasing the fragment size, the results converge to the correct solution of the kinetic scheme. The build-up from units or fragments has many advantages, for instance it can describe the structure of the gel. Therefore, we will briefly analyze the case (b) of the statistical build-up using the theory of branching processes employing the probability-generating functions (pgf) as a tool (cf. Ref. 1). Several cases of increasing complexity will be outlined:

(1) Build-up from monomer units, including the smallest cycle. The building units are monomer units with zero, one and two reacted functional groups and a fraction of units that are a part of the smallest cycle. The pgfs for the number and type of bonds issuing from a unit in the root $(g = 0)$ can be formulated as

$$F_{0A}(z) = a_0 + a_1 z_E + a_2 z_E^2 + a_{2c} z_{Ec} \qquad (7)$$

$$F_{0E}(z) = e_0 + e_1 z_A + e_2 z_A^2 + e_{2c} z_{Ac} \qquad (8)$$

where the subscripts A and E refer to the monoamine and diepoxide component, respectively; z is the variable of the pgf and the subscript denotes the direction of the bond; a_0, a_1, and a_2 are fractions of A_0, A_1 and A_2, respectively, and the same notation holds for the fractions of diepoxide units. The fractions a_{2c} and e_{2c} refer to the diepoxide and amine units consumed in the DGA–amine ring. We have $n_A a_{2c} = n_E e_{2c}$ (n_A and n_E are molar fractions of components A and E), because the cycle is a 1:1 reaction product.

From eqns (7) and (8), one obtains by differentiation the pgfs for the number of bonds issuing from units on generation $g > 0$:

$$F_A(z) = (\partial F_{0A}(z)/\partial z_E)N = (a_1 + 2a_2 z_E)/(a_1 + 2a_2) \qquad (9)$$

$$F_E(z) = (\partial F_{0E}(z)/\partial z_A)N = (e_1 + 2e_2 z_E)/(e_1 + 2e_2) \qquad (10)$$

$$F_{Ac}(z) = (\partial F_{0A}(z)/\partial z_{Ec})N = 1 \qquad (11)$$

$$F_{Ec}(z) = (\partial F_{0E}(z)/\partial z_{Ac})N = 1 \qquad (12)$$

where N is a normalizer, so that $F(z = 1) = 1$. The distributions a_i and e_i are obtained from kinetic equations.

(2) Build-up from monomer units, where cycles larger than the smallest one are taken into account. For the treatment of cyclization of longer range, the spanning-tree approximation is used. This means that the smallest ring is still calculated exactly but the effect of larger rings is summed up in units that have reacted twice: one bond may be still effective

in branching and the other may be ring forming. Then, the pgfs F_{0A} and F_{0E} are modified to yield

$$F_{0A}(z) = a_{002} + a_{101}z_E + a_{110}z_E + a_{200}z_E^2 + a_{2c}z_{Ec} \tag{13}$$

$$F_{0E}(z) = e_{002} + e_{101}z_A + e_{110}z_A + e_{200}z_A^2 + e_{2c}z_{Ac} \tag{14}$$

In eqns (13) and (14), a_{ijk} and e_{ijk} are fractions of units having $i+j$ groups reacted, of which i are engaged in inter- and j in intramolecular bonds, and k groups have not reacted. These fractions are to be obtained by solving differential equations, similarly to the case of polyurethane networks.[11] Beyond the gel point, only elastically inactive rings are to be counted. The pgfs for units in generation $g > 0$ are obtained by differentation and renormalization similarly to case (1).

(3) Build-up of the structures from larger fragments, e.g. from dyads. In terms of reactions (3)–(6), the following molecules and fragments (dyads) are used for the build-up of the network:

$$A_0, \ E_0, \ A_1E_1, \ E_2A_2, \ A_1E_2\text{-}, \ \text{-}A_2E_1, \ \text{-}A_2E_2\text{-}, \ E_1A_2\text{-}, \ \text{-}E_2A_1$$

with

$$[A_1E_2\text{-}] = [\text{-}E_2A_1] = [E_2A_1] = e_2(a_1) = a_1(e_2)$$
$$[\text{-}A_2E_1] = [E_1A_2\text{-}] = [A_2E_1] = a_2(e_1) = e_1(a_2)$$

and E_2A_2 is the smallest ring (the fractions of units in the ring are a_{2c} and e_{2c} as before).

The pgfs for units in the root can now be formulated as follows:

$$\begin{aligned} F_{0A}(z) = &\ a_0 + a_1(e_1)z_{(A1)E1} + a_1(e_2)z_{(A1)E2} \\ &+ a_2(e_1)[\phi_1 z_{(A2)E1} + \phi_2 z_{(A2)E2}]z_{(A2)E1} \\ &+ a_2(e_2)[\phi_1 z_{(A2)E1} + \phi_2 z_{(A2)E2}]z_{(A2)E2} + a_{2c}z_{Ec} \end{aligned} \tag{15}$$

$$\begin{aligned} F_{0E}(z) = &\ e_0 + e_1(a_1)z_{(E1)A1} + e_1(a_2)z_{(E1)A2} \\ &+ e_2(a_1)[\psi_1 z_{(E2)A1} + \psi_2 z_{(E2)A2}]z_{(E2)A1} \\ &+ e_2(a_2)[\psi_1 z_{(E2)A1} + \psi_2 z_{(E2)A2}]z_{(E2)A2} + e_{2c}z_{Ec} \end{aligned} \tag{16}$$

$$\phi_1 = [A_2E_1]/([A_2E_1] + [A_2E_2])$$
$$\phi_2 = 1 - \phi_1$$
$$\psi_1 = [E_2A_1]/([E_2A_1] + [E_2A_2])$$

where $x(y)$ means the fraction of X units whose neighbours are Y units: this fraction is equal to the fraction of $[XY] = [YX]$ dyads. These fractions are obtained from the corresponding kinetic and balance equations and are determined by the four rate constants of reactions (3)–(6) and a rate constant for the formation of the smallest cycle. The variable $z_{(X)Y}$ is assigned to a bond extending from unit X on generation g to unit Y on generation $g + 1$.

The pgfs for the number of bonds issuing from the units in generation $g > 0$ are obtained by differentiation and renormalization. Thus, for example

$$F_{(A1)E1}(z) = (\partial F_{0E}(z)/\partial z_{(E1)A1})N = 1$$
$$F_{(A2)E1}(z) = (\partial F_{0E}(z)/\partial z_{(E1)A2})N = 1$$
$$F_{(A1)E2}(z) = (\partial F_{0E}(z)/\partial z_{(E2)A1})N$$
$$= \frac{e_2(a_1)[2\psi_1 z_{(E2)A1} + \psi_2 z_{(E2)A2}] + e_2(a_2)\psi_1 z_{(E2)A2}}{e_2(a_1)(1 + \psi_1) + e_2(a_2)\psi_1},$$

etc. The following modification is to be made with regard to polyfunctional systems in passing (diamine and tetraepoxide): since it is assumed that the reactivity of amino groups in diamine and diepoxide groups in tetraepoxide are independent, the pgfs for units in the root are obtained by squaring the pgfs (9) and (10), or (15) and (16). The pgfs for units in generation $g > 0$ are obtained by differentiating the new functions obtained by squaring. This procedure does not apply to case (2) because the intensity of cyclization involving cycles larger than the smallest one depends on the functionality of the monomer.

3.2 Gelation Experiments

Gelation data[12] on a number of systems have been analyzed in order to reveal the difference between the systems based on DGEBA and DGA or TGDDM. In Fig. 1, the critical molar ratio necessary for gelation is plotted as a function of dilution by an inert diluent. One can notice a striking difference between the dilution dependence of r_c, which characterizes the extent of cyclization. DGEBA systems are dilution independent, which proves that cyclization is very weak. The DGA and TGDDM systems undergo a considerable decrease of the critical ratio $r_c = 2[NH_2]/[epoxy]$ with increasing dilution, which is caused by increasing cyclization. The value of r_c extrapolated to 'infinite' concentration of functional groups should not be affected by cyclization. It is only a function of the functionality of the monomers and reactivities of the groups. At present, we can only test whether or not the gelation experiments are in agreement with the assumption of independence of the substitution effect in the amino group and DGA.

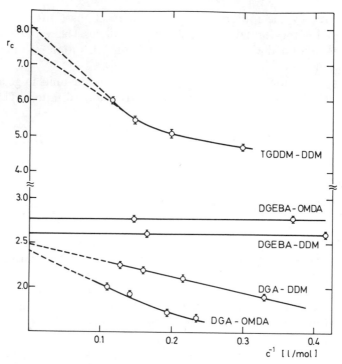

Fig. 1. The critical molar ratio of amine to epoxy functionalities, $r_c = 2[NH_2]/[epoxy]$, plotted against the reciprocal concentration of functional groups. For abbreviations see Table 1.

The critical conversion is determined by the relation $F'_A F'_E = 1$, where F'_A and F'_E are values of derivatives of the pgfs F_A and F_E for $z = 1$.[1] For an excess of amine and full conversion of epoxide groups, from eqn (10) $F'_E = 1$ because $e_2 = 1$ and $e_1 = 0$. Consequently, the extrapolated value of r_c should be independent of the substitution effect in DGA and it should be given by eqn (61) of Ref. 1.

Therefore, for the same diamine, the r_c values should be the same as in the case of DGEBA; it follows from Table 1 that they differ in fact. The assumption of the independence of the substitution effect does not seem to be fulfilled and one should examine method (3). There is still a possibility that the extrapolation is not correct, because the concentration of the smallest cycles may not follow the assumed dependence on dilution. This possibility will be checked in the future, by measuring the concentration of cycle **III** as a function of dilution.

TABLE 1

Critical Molar Ratios of Amino to Epoxy Functionalities, $r_c = 2[NH_2]/[epoxy]$, Necessary for Gelation, Extrapolated to $c^{-1} = 0$ (cf. Fig. 1)

Polyepoxide–diamine[a]	r_c (extrapolated)
DGEBA–DDM	2·61
DGEBA–OMDA	2·77
DGA–DDM	2·48
DGA–DDS	2·16
DGA–OMDA	2·41
TGDDM–DDM	7·4–8·0

[a] DGEBA, diglycidyl ether of Bisphenol-A; DDM, 4,4'-diamino-diphenylmethane; OMDA, octamethylene-diamine; DGA, N,N-diglycidylaniline; DDS, 4,4'-diaminodiphenylsulphone; TGDDM, N,N,N',N'-tetraglycidyl-4,4'-diaminodiphenylmethane.

4 CONCLUSIONS

The curing of N,N-diglycidylamine epoxy resins differs from that of DGEBA resins by a high tendency toward the formation of small cycles and by an interdependence of the reactivities of neighbouring vicinal epoxy groups. Gelation is even more complicated, because the substitution effects in amino groups and N,N-diglycidylamine groups are not independent.

REFERENCES

1. Dušek, K., Network formation in curing of epoxy resins, *Adv. Polym. Sci.*, 1986, **78**, 1.
2. Morgan, R. J., Structure–property relations of epoxies used as composite materials, *Adv. Polym. Sci.*, 1985, **72**, 1.
3. Matějka, L., Tkaczyk, M., Pokorný, S. and Dušek, K., Cyclization in the reaction between diglycidylaniline and amine, *Polym. Bull.*, 1986, **15**, 389.
4. Matějka, L., Pokorný, S. and Dušek, K., Mechanism and kinetics of the reaction between diglycidylaniline and aromatic amines, *Crosslinked Epoxies*, Sedláček, B. and Kahovec, J. (Eds), W. deGruyter, Hamburg, 1987, p. 241.
5. Doskočilová, D., Matějka, L., Pokorný, S., Březina, M., Štokr, J., Dobáš, I. and Dušek, K., Curing of epoxy resins. Conformational structure and reactivity of stereoisomers in the model reaction of diglycidylaniline with N-methylaniline, *Polym. Bull.*, 1885, **14**, 123.

6. Dušek, K., Bleha, M. and Luňák, S., Curing of epoxide resins. Model reactions of curing with amines, *J. Polym. Sci., Polym. Chem. Ed.*, 1977, **15**, 2393.
7. Rozenberg, B. A., Kinetics, thermodynamics and mechanism of reactions of epoxy oligomers with amines, *Adv. Polym. Sci.*, 1985, **75**, 113.
8. Dušek, K., Formation–structure relationships in polymer networks, *Brit. Polym. J.*, 1985, **17**, 185.
9. Dušek, K. and Šomvársky, J., Build-up of polymer networks by initiated polyreactions. 1. Comparison of kinetic and statistical approaches to living polymerization type build-up, *Polym. Bull.*, 1985, **13**, 313.
10. Mikeš, J. and Dušek, K., Simulation of the polymer network formation by the Monte-Carlo method, *Macromolecules*, 1982, **15**, 93.
11. Dušek, K. and Vojta, V., Concentration of elastically active network chains and cyclization in networks obtained by alternating stepwise polyaddition, *Brit. Polym. J.*, 1977, **9**, 164.
12. Matějka, L., Dušek, K., and Dobáš, I., Curing of epoxy resins with amines. Gelation of polyepoxides derived from diglycidylaniline, *Polym. Bull.*, 1985, **14**, 309.

23

THE INFLUENCE OF VITRIFICATION ON THE FORMATION OF DENSELY CROSSLINKED NETWORKS USING PHOTOPOLYMERIZATION

J. G. Kloosterboer and G. F. C. M. Lijten

Philips Research Laboratories, PO Box 80000, 5600 JA Eindhoven, The Netherlands

ABSTRACT

Photopolymerization of diacrylates near ambient temperature may result in the formation of densely crosslinked glassy polymers. Isothermal vitrification restricts the ultimate conversion of reactive groups. This will affect the mechanical properties of the product. The rate of monomer conversion during vitrification cannot be described with ordinary chemical kinetics but it can be fitted with a Williams–Watts equation, similar to volume relaxation. With diacrylates even the chemistry will change during vitrification, since direct termination is gradually replaced by termination through hydrogen transfer.

1 INTRODUCTION

Photopolymerization of di(meth)acrylates can be used for the manufacturing of optical discs[1,2] as well as for the replication of special aspherical lenses for the laser read-out systems used with these discs.[3,4] In these and other applications a long-term stability of the replicated surface is required. With certain lenses, for example, the desired surface contour has to be approximated to within $0.1\ \mu$m.[5] This can be achieved with densely crosslinked glassy polymers. The main advantages of photopolymerization are its high speed at moderate temperatures and the easy external control of the process. During the filling of a mold such as that used for lens replication, no reaction will occur, but after switching on the UV light the process starts immediately and at a high rate.

Crosslinked glasses are conventionally made by polymerizing at a high temperature followed by cooling, optionally annealing and further cooling to ambient temperature. With photopolymerization, both the chemical reaction and the vitrification are carried out isothermally, at or near ambient temperature. Isothermal vitrification, i.e. vitrification through structural modification only, introduces some special problems and features. It affects the physical properties, the chemical kinetics and even the chemistry itself. These three topics will be discussed in this contribution. We have used several model monomers, all of which can be obtained in a pure state (Scheme 1). These monomers were photopolymerized in the bulk,

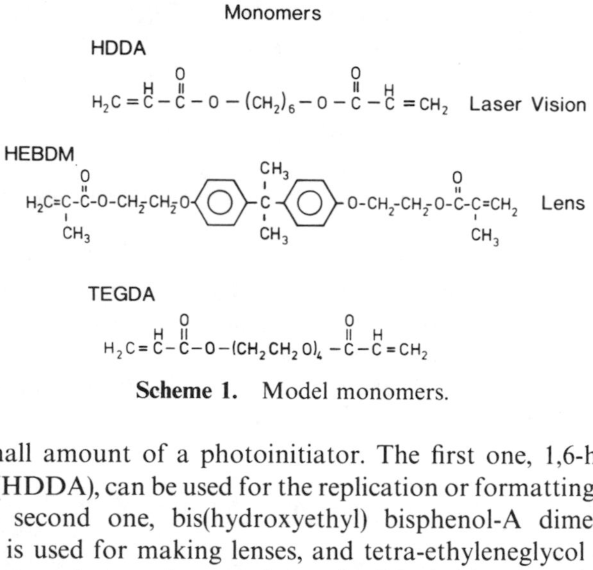

Monomers

HDDA

$$H_2C = \overset{H}{\underset{\displaystyle}{C}} - \overset{O}{\underset{\displaystyle}{C}} - O - (CH_2)_6 - O - \overset{O}{\underset{\displaystyle}{C}} - \overset{H}{\underset{\displaystyle}{C}} = CH_2 \quad \text{Laser Vision}$$

HEBDM

$$H_2C = \overset{O}{\underset{CH_3}{C}} - \overset{O}{\underset{\displaystyle}{C}} - O - CH_2CH_2O - \bigcirc - \overset{CH_3}{\underset{CH_3}{C}} - \bigcirc - O - CH_2 - CH_2 - O - \overset{O}{\underset{CH_3}{C}} - C = CH_2 \quad \text{Lens}$$

TEGDA

$$H_2C = \overset{H}{\underset{\displaystyle}{C}} - \overset{O}{\underset{\displaystyle}{C}} - O - (CH_2CH_2O)_4 - \overset{O}{\underset{\displaystyle}{C}} - \overset{H}{\underset{\displaystyle}{C}} = CH_2$$

Scheme 1. Model monomers.

using a small amount of a photoinitiator. The first one, 1,6-hexanediol diacrylate (HDDA), can be used for the replication or formatting of optical discs. The second one, bis(hydroxyethyl) bisphenol-A dimethacrylate (HEBDM) is used for making lenses, and tetra-ethyleneglycol diacrylate (TEGDA) was chosen since its long, flexible bridge introduces more mobility into the network. α,α-Dimethoxy-α-phenylacetophenone (DMPA) and 2-hydroxy-2-methyl-1-phenylpropane-1-one (HMPP) were used as photoinitiators. Experimental procedures have been published elsewhere.[6-9]

2 RESULTS AND DISCUSSION

2.1 Vitrification and Conversion
In the bulk polymerization of the model compounds, physical changes (vitrification) limit the chemical conversion (Table 1) and thereby affect the

TABLE 1

Maximum Extent of Reaction (DSC, $0.2\,mW$ cm^{-2}, 4 w% DMPA)

	Temperature (°C)	
	20	80
HDDA	0·80	0·89
HEBDM[a]	0·48	0·72
TEGDA	0·89	0·97

[a] In Ref. 6, somewhat higher values were quoted since the monomer used there still contained a significant amount of monoester.

end properties. The maximum extents of reaction were determined by isothermal calorimetry and we consider the reaction as virtually finished when the rate has dropped below our detection limit, i.e. when it has decreased by 2–3 decades.[6]

During the polymerization the molecular mobility decreases due to interconnection of the molecules as well as due to the shrinkage, caused by the interconnection. Since shrinkage cannot keep up with conversion, a temporary excess of free volume results.[6] Therefore, by increasing the light intensity the portion of the reaction that will occur in this unrelaxed high-volume state will also increase, and due to the correspondingly enhanced mobility a higher maximum extent of reaction will be observed with DSC (Table 2).

The higher extent of reaction, obtained at the higher intensity, is reflected in the physical properties as well. For HDDA, for example, the temperature

TABLE 2

Maximum Extent of Reaction (DSC, 20°C, 4 w% DMPA)

	Intensity ($mW\ cm^{-2}$)		
	0·002	0·02	0·2
HDDA	0·65	0·72	0·80
HEBDM[a]	0·3[b]	0·4[b]	0·48
TEGDA	0·84	0·87	0·89

[a] See footnote to Table 1.
[b] Inaccurate value due to very low rate of reaction.

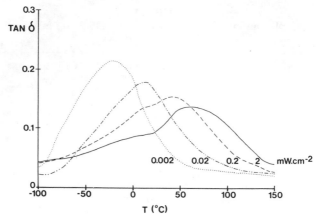

Fig. 1. Mechanical loss at 1 Hz as a function of temperature for samples of HDDA, photopolymerized at 20°C using various light intensities. Photoinitiator: 0·25 w% DMPA. Required exposure times were read from DSC curves. They decrease with increasing intensity: 25, 15, 12 and 7·5 min.

of maximum mechanical loss at 1 Hz, $T(\tan \delta_{max})$, varies by 85°C over an intensity range of 3 decades (Fig. 1). In this experiment the times of irradiation have been obtained from the DSC experiment: irradiation was stopped when no reaction could be detected anymore. This means that the sample which was exposed at $2 \, \mathrm{mW \, cm^{-2}}$ has received the highest dose. Results obtained with equal doses will be shown below.

The physical process of shrinkage is thus coupled to some extent to the chemical process of polymerization, but not so strongly that shrinkage is a measure for conversion, as is normally observed with liquid systems. The coupling of the two processes is shown schematically in Fig. 2.

The mutual dependence of shrinkage, conversion and their respective rates may be compared with physical aging (Fig. 2(a)). Physical aging is a self-decelerating process due to the closed-loop dependence of free volume V_f, segment mobility μ and rate of shrinkage dV_f/dt.[10] The mobility of the polymer segments is the important microscopic parameter. Physical aging is usually studied in polymers which have already been formed. However, a closed-loop dependence is expected to exist during isothermal bulk polymerization as well (Fig. 2(b)).[11] During reaction there exists an additional and similar dependence of conversion x and rate of conversion dx/dt on the mobility μ of unreacted groups, either present as pendant double bonds or as free monomer. Here, too, the mobility is the important parameter. Since chemical reaction immediately generates free volume,

a. Physical aging

b. (Photo)polymerization

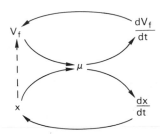

Fig. 2. Schematic relation between segmental mobility μ, free volume V_f and double bond conversion x during physical aging and photopolymerization.

there is also a direct relation between x and V_f. Chemical reaction drives the shrinkage process but a high extent of reaction reduces the rate of shrinkage. A high extent of volume shrinkage in turn reduces the rate of chemical reaction.

In Fig. 3 the results obtained with TEGDA are summarized.[7] Double bond conversion increases with light intensity and with temperature. When the samples which were polymerized at 20°C are heated in the dark and under nitrogen, additional heat is liberated. This effect is indicative of the presence of trapped radicals, which are remobilized by heating and then continue to react.[6,11]

Another feature is the endothermic effect, observed upon heating after decay of the trapped radicals by introducing air. This effect is connected with the free volume state of the glass. The free volume state of a glass is determined by the rate of cooling from the melt and the time of annealing, i.e. by its thermal history. A slowly cooled or well-annealed glass shows an endothermic effect when heated rapidly; at first the mobility cannot adapt itself quickly enough to the higher temperature, and this causes first a delayed growth and then an overshoot in heat capacity c_p. This and related phenomena have been extensively discussed by Petrie and other workers.[12,13] Unannealed, rapidly quenched glasses do not show this behavior. Isothermally produced glasses have no thermal history but an 'exposure time history'. Their time history is determined by the actual amount of shrinkage divided by the maximum amount of shrinkage that corresponds to the extent of conversion. Our glasses were made by isothermal polymerization at a rate which decreased continually beyond the maximum, and then they were heated at a constant rate of 10 K min^{-1}. They behave like well-annealed glasses and show a considerable endothermic effect. Under nitrogen the endothermic effect of relaxation is opposed by the exothermic effect of reaction, so the additional conversions depicted in Fig. 3(b) are probably a lower limit. The effects are not necessarily additive since chemical changes may have accompanied the oxidation of the radicals.

2.2 Vitrification and Rate of Polymerization

The self-retarding character of the process causes the reaction to continue at a rate below the limit of detection of the DSC, but changes still show up in the mechanical properties, notably in $T(\tan \delta_{max})$, the temperature of maximum mechanical loss. This can be seen when the experiments shown in

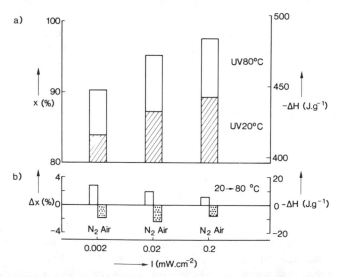

Fig. 3. (a) Maximum extent of reaction x_{max} measured with DSC at 20 and 80°C for various light intensities. TEGDA with 3·4 w% DMPA. Intensities are shown at the bottom. (b) Heat effects observed upon heating from 20 to 80°C. Left: in N_2, just after polymerization. Right: the same but after one night storage in air.

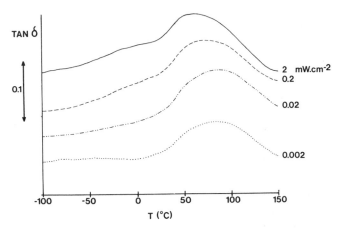

Fig. 4. Same experiments as in Fig. 1, but now for equal doses. Exposure times: 7500, 750, 75 and 7·5 min.

Fig. 1 are repeated under slightly different conditions. In Fig. 4 the irradiations at the lower intensities have been continued until equal doses were applied in all cases. Obviously, the reaction has continued and now the lowest intensity even seems to be the most effective, presumably since a low intensity favors a long kinetic chain length and thereby a high extent of reaction.

The chemical reaction not only continues during irradiation but also in the dark. This is due to the presence of long-living trapped radicals. In HDDA these radicals persist for several months after irradiation if the sample is kept at room temperature and under vacuum,[11,14] even in the presence of a large amount of unreacted monomer. This is ascribed to inhomogeneity.[8,9,15-17] The formation of inhomogeneous networks has been modeled by Boots.[9,17]

If we allow the samples to stand before extraction and determination of free monomer, we observe virtually no decrease of the radical concentration, but the concentration of free monomer [M] decays in the dark via a self-decelerating process (Fig. 5). If ordinary kinetics would apply, log [M] vs. t would be linear. Here it is the other way round: [M] vs. log t is almost linear over more than 4 decades. The first half-time is about 20 h and the next one 300 h. Physically this could mean that the monomer molecules are distributed among traps and that the molecules from the shallowest traps escape and diffuse to a reactive radical site more quickly than the other ones. Reaction will deepen the traps of the remaining molecules, so the relaxation time of the system will increase continuously.

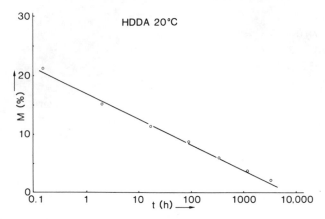

Fig. 5. Decay of extractable monomer M as a function of time. HDDA, photopolymerized for 90 s at 20°C and stored under vacuum at room temperature. Photoinitiator: 0·25 w% HMPP, 0·2 mW cm^{-2}.

Many relaxation phenomena in polymeric glasses can be described in an empirical way using the Williams–Watts relaxation function $\phi(t)$,[18] in which a parameter β accounts for the width of the relaxation spectrum:

$$\phi(t) = \exp -(t/\tau)^{\beta}$$

β is an adjustable parameter, $0 < \beta \leq 1$. τ is also an adjustable parameter which has the form of a characteristic time. Stress relaxation,[19] creep recovery,[10] dielectric relaxation[18] and volume recovery[20] data have all been fitted with such a function. This purely empirical equation has also been connected with several kinds of models, used for the description of polymeric glasses. Models used include diffusion,[21] molecular kinetics[22] and stochastic ones.[21] However, in view of the limited data and restricted knowledge of the combined, but certainly not co-operative, processes of physical aging and chemical conversion, we have so far only attempted to fit our data with the empirical equation by choosing $[M]_t/[M]_0$ as the relaxation function:

$$[M]_t/[M]_0 = \exp -(t/\tau)^{\beta} \qquad \ln \ln [M]_0/[M]_t = \beta \ln t - \beta \ln \tau$$

Figure 6 shows that the chemical reaction data do fit with the Williams–Watts equation. The slope is found to be 0·25 and the intercept has a value of 125 h. This fit is not surprising since so many phenomena have been fitted with this empirical equation.[23] However, zero-, first- or second-order plots did not yield straight lines. From this plot it therefore

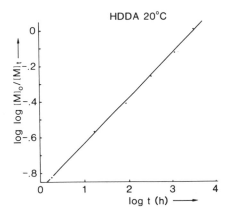

Fig. 6. Data of Fig. 5 plotted according to the Williams–Watts equation: $\beta = 0.25$, $\tau = 125\,h$.

appears that the chemical kinetics of monomer consumption in a vitrifying system are largely determined by the physical process of decay of free volume. A complicating factor is, however, that free volume is not only destroyed by overall shrinkage but is also generated by the chemical reaction itself (Fig. 2(b)). For comparison, it may be noted that Struik reported that β has a value of about 1/3 for creep recovery in many different materials,[10] and that similar values have been observed for many kinds of relaxation in a large variety of vitreous materials, including metals and low-molecular weight glasses.[23] According to Hodge and Berens, typical values for β are in the range of 0·4–0·6.[20] Our rather low value of 0·25 points to a very broad distribution of relaxation times.

2.3 Change of Chemistry During Vitrification

During radical chain photopolymerization, radicals are generated by UV light and destroyed by termination. With acrylates, termination proceeds almost exclusively by direct combination of two radicals.[24] In addition, chain transfer may occur, for example to the polymer. Since usually an active radical is regenerated in this step, it has little influence on the kinetics,[25] unless a vitrifying network is being made. When most of the radicals are hooked up to a denser and denser macroscopic network, direct combination becomes more and more difficult. Now transfer through hydrogen abstraction may maintain a certain mobility of radical sites and thereby provide an additional termination path. When termination by direct combination becomes more and more severely suppressed, the

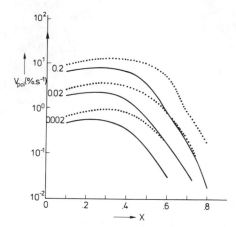

Fig. 7. Rate vs. extent of reaction of HDDA, recorded at 0·2, 0·02 and 0·002 mW cm^{-2} and 20°C. Photoinitiator, 4 w% HMPP;——, 1,6-hexanediol diacrylate;·····, α,α'-dideutero-1,6-hexanediol diacrylate.

hydrogen-transfer assisted termination gradually takes over. This possibility has been checked by deuteration of HDDA at its most sensitive α-positions.[26] This suppresses transfer to the polymer since deuterium abstraction requires a higher activation energy than hydrogen abstraction. If we now compare the rates of the deuterated and the normal compound, plotted in Fig. 7 as a function of double bond conversion, we see that the deuterated compound reacts considerably faster than the normal one, at each of the light intensities used. Most important is that the isotope effect, and with it the relative importance of hydrogen transfer, increases with conversion. Initially it is about 1·4 and it increases up to a factor of 5 upon overall vitrification. So here we see that physical changes, caused by the chemical reaction, may in turn change the chemistry itself.

REFERENCES

1. Van den Broek, A. M., Haverkorn van Rijsewijk, H. C., Legierse, P. E. J., Lippits, G. J. M. and Thomas, G. E., *J. Rad. Cur.*, 1984, **11**(1), 1.
2. Kloosterboer, J. G. and Lippits, G. J. M., *J. Rad. Cur.*, 1984, **11**(1), 10.
3. Zwiers, R. J. M. and Dortant, G. C. M., in *Integration of Fundamental Polymer Science and Technology*, Kleintjens, L. J. and Lemstra, P. J. (Eds), Elsevier Applied Science Publishers, London, 1986, p. 673.
4. Zwiers, R. J. M. and Dortant, G. C. M., *Appl. Opt.*, 1985, **24**, 4483.

5. Visser, D., Gijsbers, T. G. and Jorna, R. A. M., *Appl. Opt.*, 1985, **24**, 1848.
6. Kloosterboer, J. G., Van de Hei, G. M. M., Gossink, R. G. and Dortant, G. C. M., *Polym. Commun.*, 1984, **25**, 322.
7. Kloosterboer, J. G. and Lijten, G. F. C. M., *Polymer*, 1987, **28**, 1149.
8. Kloosterboer, J. G., Van de Hei, G. M. M. and Boots, H. M. J., *Polym. Commun.*, 1984, **25**, 354.
9. Boots, H. M. J., Kloosterboer, J. G., Van de Hei, G. M. M. and Pandey, R. B., *Brit. Polym. J.*, 1985, **17**, 219.
10. Struik, L. C. E., *Physical Aging in Amorphous Polymers and Other Materials*, Elsevier, New York, 1978.
11. Kloosterboer, J. G., Van de Hei, G. M. M. and Lijten, G. F. C. M., in *Integration of Fundamental Polymer Science and Technology*, Kleintjens, L. J. and Lemstra, P. J. (Eds), Elsevier Applied Science Publishers, London, 1986, p. 198.
12. Petrie, S. E. B., *J. Macromol. Sci.-Phys.*, 1976, **B12**, 225.
13. Wunderlich, B., in *Thermal Characterization of Polymeric Materials*, Turi, E. A. (Ed.), Academic Press, New York, 1981, p. 171.
14. Kloosterboer, J. G., Lijten, G. F. C. M. and Greidanus, F. J. A. M., *Polym. Commun.*, 1986, **27**, 268.
15. Dušek, K., in *Developments in Polymerisation—3*, Haward, R. N. (Ed.), Elsevier Applied Science Publishers, London, 1982, p. 143.
16. Funke, W., *Plast. Rubb. Process. Applic.*, 1983, **3**, 243.
17. Boots, H. M. J., in *Integration of Fundamental Polymer Science and Technology*, Kleintjens, L. J. and Lemstra, P. J. (Eds), Elsevier Applied Science Publishers, London, 1986, p. 204.
18. Williams, G. and Watts, D. C., *Trans. Far. Soc.*, 1970, **66**, 80.
19. Knott, W. F., Hopkins, I. L. and Tobolsky, A. V., *Macromolecules*, 1971, **4**, 750.
20. Hodge, I. M. and Berens, A. R., *Macromolecules*, 1982, **15**, 762.
21. Simha, R., Curro, J. G. and Robertson, R. E., *Polym. Eng. Sci.*, 1984, **24**, 1071.
22. Chow, T. S., *Polym. Eng. Sci.*, 1984, **24**, 1079.
23. Struik, L. C. E., *Europhys. Conf. Abs.*, 1980, **A4**, 135.
24. Bamford, C. H., Dyson, R. W. and Eastmond, G. C., *Polymer*, 1969, **10**, 885.
25. Odian, G., *Principles of Polymerization*, 2nd edn, Wiley, New York, 1981, p. 273.
26. Kloosterboer, J. G. and Lijten, G. F. C. M., *Polymer Commun.*, 1987, **28**, 2.

CHARACTERISATION OF POLYMER NETWORKS

24

ORIENTATIONAL BEHAVIOUR OF FREE POLYMER CHAINS DISSOLVED IN A STRAINED NETWORK: A DEUTERIUM MAGNETIC RESONANCE INVESTIGATION

B. Deloche, P. Sotta

*Laboratoire de Physique des Solides, CNRS–LA2,
Université de Paris-Sud, 91405 Orsay, France*

J. Herz and A. Lapp

*Institut Charles Sadron (CRM-EAHP), CNRS-ULP,
6 rue Boussingault, 67083 Strasbourg, France*

ABSTRACT

The ^2H-NMR technique is used to monitor the orientational behaviour of short labelled polymer chains dissolved in an unlabelled network under uniaxial stress. The methodology exploits the observation of residual nuclear interactions which are related to the degree of induced order. When the host matrix of polydimethylsiloxane (PDMS) is uniaxially deformed, an orientational anisotropy is induced at the segmental level of the PDMS probe chain. This orientation of the uncrosslinked chains demonstrates that orientational correlations take place in the deformed network. From a comparison with results established on crosslinked chains, it appears that these interchain effects are a dominant contribution to the orientation process of the effectively elastic chains.

1 INTRODUCTION

The main purpose of this work is to develop an experiment giving better understanding of the chain ordering in strained rubbers. For that reason, let us consider a linear polymer chain dissolved into a network made of

Fig. 1. Schematic diagram of a probe chain dissolved in a rubber matrix. The molecular weight of the dissolved chain is supposed to be lower than the average molecular weight between entanglements M_c in the melt of the material which composes the matrix.

chemically identical chains; the dissolved chains are short enough to be 'free', i.e. devoid of any entanglements with the network structure. The basic question concerning such a system, shown schematically in Fig. 1, is the following: does the equilibrium configuration of the free chain remain random as the rubber matrix is uniaxially deformed? In other words, is there an orientational anisotropy induced at the segment level of the probe chain when the system is elongated? Among the various microscopic techniques sensitive to the chain anisotropy, deuterium NMR (^{2}H-NMR) has emerged as a powerful tool[1] and can be used here to monitor the orientational behaviour of the probe chain.

In this paper we extend our previous ^{2}H-NMR experiments performed on elastic chains[2] to mobile chains dispersed inside a rubber matrix. Specifically, we report herein direct measurements of segment ordering carried out on linear free chains of labelled polydimethylsiloxane

(PDMS(D)) incorporated in an elongated network of PDMS. The ^{2}H-NMR spectra reveal some induced orientational anisotropy in the probe chain. Then the degree of anisotropy is contrasted with that of the crosslinked chains of the network.

2 EXPERIMENTAL

2.1 Samples

Experiments were performed on end-linked polydimethylsiloxane (PDMS) networks swollen with linear perdeuterated chains $(Si(CD_3)_2\text{---}O\text{)}_n$, (PDMS) (D)). The precursor chains of the network and the linear chains, used as NMR probes, have the same molecular weight ($M_n = 10\,500$); in both cases, the molecular weight distribution is about 1·7. Let us note also that the molecular weight of the linear chains is lower than the average molecular weight between entanglements M_c in the melt of such materials ($M_c \sim 27\,000$ for PDMS). The synthesis of the network and the deuteration follow procedures outlined in detail in Ref. 3. We just recall that, for the present investigation, a tetrafunctional network was used, prepared by an end-linking reaction in the bulk.

2.2 ^{2}H-NMR

The general features of ^{2}H-NMR in anisotropic fluids have been developed in Ref. 4. The sensitivity of the technique to investigate a rubbery medium has been already shown by studies performed on labelled networks.[2,5] We simply recall here that, when rapid molecular uniaxial reorientations take place, the observed quadrupolar interaction is not averaged to zero. The effect of this residual interaction is to split the liquid-like NMR line into a doublet, whose spacing Δv may be written in frequency units as

$$\Delta v = \tfrac{3}{2}qP_2(\cos\Omega)\langle P_2(\cos\theta(t))\rangle \tag{1}$$

where q designates the static quadrupolar coupling constant, i.e. the quadrupolar coupling before any averaging process: $q \simeq 175\,\text{kHz}$ for the methyl deuterons in PDMS(D). θ is the instantaneous angle that the C–D bond makes with the symmetry axis of the sample (i.e. the direction of the applied constraint), and Ω that between the magnetic field and the symmetry axis. So, a measurement of the splittings Δv gives access directly to the so-called orientational order parameter $S = \langle P_2(\cos\theta(t))\rangle$ of the C–D bond relative to the symmetry axis. The current resolution available for observing Δv (typically 1 Hz) in conjunction with the inherently large value of the C–D quadrupolar constants ($q \sim 200\,\text{kHz}$) enables one to appreciate orientational order differing from the isotropic disorder by as little as a few parts in 10^5.

2.3 Experimental Conditions

Initially, a preweighed sample ($20\,\text{mm} \times 6\,\text{mm} \times 1\,\text{mm}$) of dry PDMS rubber is covered with a drop of deuterated linear chains in the melt ($T_g \simeq -120°\text{C}$): the corresponding weight fraction is 9%. The labelled chains are short enough to be perfectly compatible with the host matrix [6] and free to diffuse through the network structure [7] with a diffusion coefficient D of about $10^{-12}\,\text{m}^2\,\text{s}^{-1}$:[8] after some days they have totally penetrated into the matrix and the surface of the film appears dry. Finally, a neutron scattering study carried out on the relaxed sample shows that no phenomena of demixing or appreciable effects of inhomogeneities occurred in such a homopolymer solution.[9]

Network deformation was performed as described earlier.[2] ^{2}H-NMR spectra are obtained after signal averaging from FT-NMR equipment operating with a conventional electromagnet. The magnetic field is perpendicular to the principal strain direction in the case of a uniaxially elongated sample.

Finally, it should be pointed out that the deuterated probe chains may be removed from the matrix: no ^{2}H-NMR signal was detected after immersion of the sample in a good solvent for 48 h. This indicates that no permanent linking occurred between the probe chain and the matrix.

3 RESULTS

Figure 2 shows the transformation in the ^{2}H-NMR spectrum of the PDMS(D) probe chain when the sample is uniaxially elongated (extension ratio $\lambda > 1$) or compressed ($\lambda > 1$). Just as in the case of crosslinked network chains previously studied,[2] the spectrum changes from a single narrow line in the relaxed state to well-resolved doublets on deforming the matrix. Moreover, no change in the magnitude of the doublet spacing is observable over a time span of about 14 h. Additional experiments performed at a fixed λ and for various angles Ω between the direction of the applied constraint and the magnetic field show that the dependence $\Delta\nu(\Omega)$, reported in Fig. 3, reproduces exactly the $|P_2(\cos\Omega)|$ curve as in eqn (1). Typical spectra for $\Omega \simeq 55°$ and $\Omega = 90°$ are shown in Ref. 10.

Free chains (φ = 0.92)

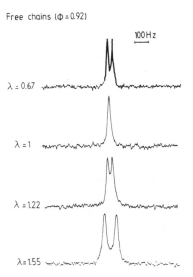

Fig. 2. 13 MHz ²H-NMR spectra of perdeuterated free chains of polydimethyl-siloxane (PDMS) dispersed in a PDMS network as the sample is uniaxially compressed ($\lambda < 1$) or elongated ($\lambda > 1$) along a direction perpendicular to the steady magnetic field ($\Omega = 90°$). Both probe chains and crosslinked chains have the same size ($M_n = 10\,500$). ϕ is the volume fraction of the polymeric network.

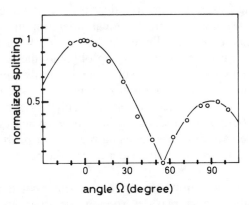

Fig. 3. Variation of the normalized splitting as a function of the angle Ω between the applied uniaxial constraint and the steady magnetic field. The continuous line represents the function $|(3\cos^2\Omega - 1)/2|$.

4 DISCUSSION

Due to the rapid Brownian motion of probe chain segments in the relaxed sample, the ^{2}H-NMR spectrum for $\lambda = 1$ is a single narrow line (motionally narrowed) characteristic of an isotropic liquid (^{2}H-NMR relaxation measurements show that the resonance linewidth in the relaxed state is homogeneous and due to the transverse relaxation T_2 only).[11] On the other hand, the appearance of well-defined splittings for $\lambda \neq 1$, in conjunction with the angular dependence $\Delta v(\Omega)$, shows that the quadrupolar interactions are no longer averaged by the rapid molecular motions (chain isomerizations and larger-scale chain reorientations) in the constrained state. These observations mean that the probe chain acquires an anisotropic orientational conformation as it diffuses through the deformed network: the chain segments undergo uniaxial reorientational diffusion around the direction of the stress axis. Another feature of the observed averaging process in the constrained state is that the spectra are composed of a *single* doublet. This implies that, within the linewidth limit of this structure, the various methyl groups are equivalent on the time scale of the ^{2}H-NMR and that the degree of segmental orientation is quasi-uniform along the contour length of the probe chains. The *permanent* orientational order parameter $\langle P_2 \rangle$ induced at the chain segment level can be easily deduced from eqn (1) exactly as it has been done in Ref. 12; for instance, at $\lambda = 1.22$ the order parameter of the methyl axis, i.e. the Si–CD$_3$ bond, is $\sim 0.9 \times 10^{-3}$ and that of the chain segment connecting two adjacent oxygens in the PDMS is $\sim 1.8 \times 10^{-3}$.

The surprising feature of our investigation is that the chain segments acquire a *permanent* non-zero orientational order although the extremities of the probe chains are not linked to the network structure. Similar observations have been made recently on oligobutadienes dispersed in a polybutadiene network.[13] These results unequivocally show that orientational correlations between chains take place in the deformed network and contribute to the segment orientation phenomenon. Interactions between neighbouring chain segments have been hypothesized before to account for observations in studies using the solvent-probe method,[1] but until now there has been no direct evidence that the polymer itself exhibits such interactions.

From our result, it is quite reasonable to infer that the orientational order quoted above also plays a role in the orientation process of crosslinked chains. Then a comparison of the orientational order of crosslinked and of uncrosslinked chains would enable us to characterize the strength of the intermolecular effects and to appreciate their importance on the segmental orientation in the elastic chains. In order to establish a relevant comparison we have used two identical networks, i.e. networks characterized by the same equilibrium swelling $\Phi_e (= 0.165)$.[14] (From the work of Dubault et al.[14] it is clear that the swelling equilibrium volume fraction of the polymer, Φ_e, is the variable well adapted to characterizing the real topological structure of the network with regard to the induced chain ordering.) One network contains a known fraction ($\sim 20\%$) of perdeuterated elastic chains while the other is unlabelled. These two samples were swollen with labelled and unlabelled PDMS free chains, respectively; the contour lengths of the two kinds of probe chains are identical and the volume fraction of network chains is obviously the same in both cases. Figure 4 shows the quadrupolar splittings of the crosslinked and uncrosslinked chains (Δv_c and Δv_u, respectively) versus $\lambda^2 - \lambda^{-1}$ for both samples, in the small deformation limit. Both Δv_c and Δv_u are linear in this strain function. Moreover, at the same λ, the observed ratio $\Delta v_c / \Delta v_u$ is close to 1, as anticipated in Ref. 12. The fact that the host matrix and the guest probe exhibit the same strain dependence and reflect a similar degree of order is striking. This is indicative of very strong orientational couplings

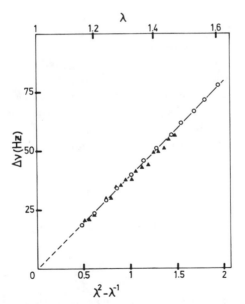

Fig. 4. Probe chain (○) and elastic chain (▲) quadrupolar splittings Δv versus $\lambda^2 - \lambda^{-1}$.

between chains, uncrosslinked or crosslinked. In particular, this means that most of the anisotropy of the crosslinked chains, as manifested by the splitting in the NMR lineshape, is related to intermolecular effects rather than to the conventional end-to-end stretching of the isolated chains.

Finally, our results may be summarized by writing the following relationship between the order parameters of the uncrosslinked chains (S_u) and the crosslinked ones (S_c):

$$S_u = KS_c \tag{2}$$

where K is a coupling factor close to unity, independent of λ in the low deformation regime. More systematic studies are being pursued in order to specify the dependence of the factor K on the molecular weight and the concentration of probe chains.

5 CONCLUDING REMARKS

This preliminary NMR study on polymer chains dissolved in a rubber matrix shows that they are oriented when the system is uniaxially deformed. The sensitivity of the observed effects suggests well that a probe chain, that is structurally identical to the network chains, is an ideal non-perturbative NMR probe for testing the orientational properties of strained elastomers. From this piece of work it appears clear also that the stress-induced orientation in a rubber is not a single-chain process, but involves strong orientational couplings between chain segments. The exact nature of these couplings remains to be elucidated. Finally, an extrapolation of our findings to the case of dangling chains of the network enables us to consider that, due to the presence of orientational correlations, they exhibit the same behaviour as the probe chains.

ACKNOWLEDGEMENTS

We are very grateful to A. Dubault for stimulating discussions.

REFERENCES

1. Deloche, B. and Samulski, E. T., Short range nematic like orientational order in strained elastomers: a deuterium magnetic resonance study, *Macromolecules*, 1981, **14**, 575–81.

2. Deloche, B., Beltzung, M. and Herz, J., Segmental order in uniaxially constrained polydimethylsiloxane network: a deuterium magnetic resonance study, *J. Phys. Lett.*, 1982, **43**, 763–9.

3. Beltzung, M., Picot, C., Rempp, P. and Herz, J., Investigation of the conformation of elastic chains in polydimethylsiloxane networks by small angle neutron scattering, *Macromolecules*, 1982, **15**, 1594, 1600.

4. Samulski, E. T., Investigations of polymer chains in oriented fluid phases with deuterium nuclear magnetic resonance, *Polymer*, 1985, **26**, 177–89.

5. Gronski, W., Stadler, R. and Jacobi, M. M., Evidence of non affine and inhomogeneous deformation of network chains in strained rubber elastic networks by deuterium magnetic resonance, *Macromolecules*, 1984, **17**, 741–8.

6. Lapp, A., Picot, C. and Benoit, H., Determination of the Flory's interaction parameters in miscible polymer blends by measurements of the apparent radius of gyration, *Macromolecules*, 1985, **18**, 2437–41.

7. Brochard, F., Polymer networks swollen by a homopolymer solution, *J. Phys.*, 1981, **12**, 505–11.

8. Garrido, L., Mark, J. E., Clarson, S. J. and Semlyen, J. A., Studies of cyclic and linear polydimethylsiloxanes: diffusion coefficients from network sorption measurements, *Polymer*, 1984, **25**, 218–20.

9. Boué, F., Farnoux, B., Bastide, J., Lapp, A., Herz, J. and Picot, C., Free polymer chains dissolved in a strained elastomer: neutron scattering test of the anisotropy, *Europhysics Lett.*, 1986, **1**, 637–45.

10. Deloche, B., Dubault, A., Herz, J. and Lapp, A., Orientation of free polymer chains dissolved in a strained elastomer: a deuterium magnetic resonance study, *Europhysics Lett.*, 1986, **1**, 629–35.

11. Sotta, P., Deloche, B., Herz, J., Lapp, A., Durand, D. and Rabadeux, J. C., Evidence for short-range orientational couplings between chain-segments in strained rubbers: a deuterium magnetic resonance investigation, *Macromolecules*, 1987, **20**, 2769–74.

12. Toriumi, H., Deloche, B., Herz, J. and Samulski, E. T., Solvent versus segment orientation in strained swollen elastomeric network, *Macromolecules*, 1985, **18**, 304–5.

13. Jacobi, M. M., Stadler, R. and Gronski, W., *Networks '86—8th Polymer Networks Group Meeting*, abstracts.

14. See, for example Dubault, A., Deloche, B. and Herz, J., Effect of cross-linking density of the orientational order generated in strained networks: a deuterium magnetic resonance study, *Polymer*, 1984, **25**, 1405–10; Effects of trapped entanglements on the chain ordering in strained rubbers: a deuterium magnetic resonance investigation, *Macromolecules*, 1987, **20**, 2096–99.

25

POLYMER COIL RELAXATION IN UNIAXIALLY ELONGATED POLY(ETHYLETHYLENE) OBSERVED BY SMALL-ANGLE NEUTRON SCATTERING

KELL MORTENSEN, WALTHER BATSBERG

Risø National Laboratory, DK-4000 Roskilde, Denmark

OLE KRAMER

University of Copenhagen, DK-2100 Copenhagen, Denmark

and

LEWIS J. FETTERS

Exxon Research and Engineering Co., Annandale, New Jersey 08801, USA

ABSTRACT

We report small-angle neutron scattering (SANS) studies on the relaxation of the static formfactor of the chains in an amorphous 2 megadalton poly(ethylethylene-co-ethylene) sample exposed to a step strain. The extension ratio was $\lambda = 3$. Just after the uniaxial elongation the chains deform affinely. Shortly thereafter, however, the SANS data indicates chain contraction as assumed by Doi and Edwards. Some randomization is observed long before the reptation mechanism allows complete disentanglement.

1 INTRODUCTION

The viscoelastic properties of melts and concentrated solutions of polymers exhibit dramatic differences in behaviour, depending on the time scales involved. At short times, the polymer melt behaves as a rubbery solid, whereas it flows like a viscous liquid at long times. These peculiar

369

properties, which are ascribed to the temporary network of the highly entangled chains, presently attract a great deal of interest.

Doi and Edwards[1] published in 1978, on the basis of the so-called tube model,[2] an extended theory, in which they predict novel behaviour of the chain conformation during stress relaxation. The present study is an attempt to verify one of the assumptions, chain retraction, by using the small-angle neutron scattering (SANS) technique on an amorphous polymer exposed to a step strain in simple extension.

2 MODEL FOR POLYMER RELAXATION

In a polymer melt, the polymer chains are highly restricted in their motions due to the completely entangled nature of polymer coils of high molecular weight. Since the chains cannot intersect one another, lateral diffusion of the chain backbone meets high resistance while diffusion along the overall contour of the chain backbone meets little resistance. The chains behave in some respects as if they were enclosed in a tube along the overall contour of the chain.[2] The diameter is related to the entanglement spacing and is expected to be of the order of 50 Å. According to the theory of de Gennes,[3,4] the individual chain wriggles rapidly along itself inside its tube (reptation). In order to change its shape completely, it must, e.g. by the process of reptation, disengage itself from the original tube and thereby rearrange its conformation. Assuming Brownian motion along the tube, the time for complete disengagement is given by[5]

$$\tau_{dis} = (L^2/\pi^2) \times N_0 \zeta_0 / k_B T \tag{1}$$

where $L = Na$ is the arc length of the tube, N being the number of primitive steps of length a; N_0 is the number of monomer units in the chain; ζ_0 is the monomer friction coefficient; k_B is Boltzmann's constant and T is absolute temperature. Since both L and N_0 are proportional to the molecular weight, M, it follows that τ_{dis} is proportional to the cube of M.

Doi and Edwards extended the reptation idea by including two additional relaxation mechanisms. Upon a step strain, the linear chain density along the chain contour becomes non-uniform. Some parts of the chain become extended while others become compressed. The first process in the Doi and Edwards model is a very fast equilibration of chain segments between 'entanglement points'. The second process is the so-called tube contraction or chain retraction process. It was proposed partly in an attempt to explain observed non-linear viscoelastic behavior of polymer melts at large strains.[1]

It can easily be shown that the average linear chain density along the chain contour is smaller in the deformed state than in the undeformed isotropic state. Doi and Edwards proposed that the chain returns to a uniform linear chain density equal to that of the isotropic state before reptation begins. The chain retraction process causes a shrinkage of the chain inside its 'tube'.

The return to equilibrium by the chain retraction process is governed by Rouse diffusion of the entire chain, leading to the following expression for the relaxation time:[1,5]

$$\tau_{eq} = (\langle R^2 \rangle / 6\pi^2) \times N_0 \zeta_0 / k_B T \tag{2}$$

where $\langle R^2 \rangle = Na^2$ is the average square end-to-end distance. It appears that τ_{eq} is proportional to the square of the molecular weight. Combination of eqns (1) and (2) leads to the ratio

$$\tau_{dis} / \tau_{eq} = 6N \tag{3}$$

Thus, for high molecular weight flexible polymers, the two characteristic times, and thereby the two relaxation phenomena, are well separated.

Figure 1 shows schematically how the chain should change its conformation under a macroscopic deformation. Immediately after a step strain, the chain is affinely deformed. Before reptation begins the chain should shrink and become smaller, parallel as well as perpendicular to the direction of elongation. After that, the chain returns to the equilibrium isotropic conformation by reptation.

Doi and Edwards suggested already in their original paper that a powerful check of the theory would be to measure the radius of gyration tensor versus time after a step strain. For a polymer melt, that can be done only by use of the neutron scattering technique.

It should be pointed out that one should expect two additional relaxation mechanisms not discussed by Doi and Edwards. As mentioned above, deformation of the coils causes the linear chain density to become non-uniform at a lower average value than in the isotropic state. The first additional relaxation process should be a rearrangement of the chain along its overall contour in order to assume uniform linear chain density with fixed positions of the chain ends. Unlike the first process of the Doi and Edwards model, the proposed additional process might involve the diffusion of chain segments from one network strand to its nearest neighbours. The process should be molecular-weight independent, since it involves relatively short sections of the chain only. Such local rearrangements seem to take place at the entrance to the rubber plateau.[6]

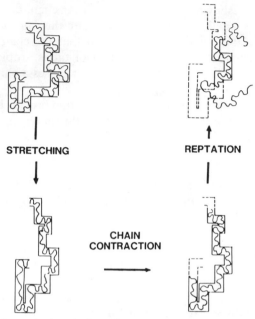

Fig. 1.   Schematic description of the Doi–Edwards relaxation scheme.

The relaxation of the chain ends has not been discussed by Doi and Edwards. If one realizes that a polymer chain can be considered a two-arm star, there is an additional potential relaxation mechanism, namely the chain-end retracing mechanism which was also proposed by de Gennes.[7] Relaxation by the retracing mechanism depends exponentially on arm molecular weight. This should give rise to a very gradual relaxation, similar to that observed in the rubber plateau. The proposed process would also be of importance for the small-angle neutron scattering experiment. It would cause some randomization of the deformed polymer coil ellipsoids before chain retraction and reptation come into play.

3 NEUTRON SCATTERING TECHNIQUE

Small-angle neutron scattering provides a useful technique for determining the conformation of polymer chains. Contrast between the molecules is obtained by using deuterium (D) substitution for hydrogen (H). Chemically, deuteration of molecules is a nearly ideal labelling technique, with no significant changes in properties. Physically, one should, however, be aware of possible changes, e.g. in the glass transition temperature.

Moreover, it has recently been shown that high molecular weight polymers tend to phase separate with an upper critical temperature.[8]

The intensity of scattered neutrons versus scattering vector (q) is governed by the differential scattering cross-section per unit volume, $d\Sigma/d\Omega$. In a thermodynamically ideal mixture of monodisperse pairs of protonated and deuterated chains, this may be expressed as

$$d\Sigma/d\Omega = KN_0 P(q) \tag{4}$$

where $N_0 = N_H = N_D$ is the degree of polymerization and K is the contrast factor:

$$K = [m(b_D - b_H)]^2 \times n \times \phi(1 - \phi) \tag{5}$$

where ϕ is the volume fraction of the deuterium component; n is the number of monomer units per unit volume; m is the number of hydrogens which are substituted with deuterium per monomer; and b_D and b_H are the coherent scattering lengths of deuterium and hydrogen, respectively. $P(q)$ is the Fourier transformation of the spatial distribution function of the molecule. For a Gaussian chain, $P(q)$ is given by the Debye formula[9]

$$P(q) = 2u^{-2}[\exp(-u) - 1 + u] \tag{6}$$

where $u = q^2 R_G^2$, R_G being the radius of gyration. For $u < 1$, eqn (6) reduces to the Guinier formula, $P(q) = \exp(-q^2 R_G^2/3)$.

The scattering function of affinely deformed Gaussian chains cannot be solved analytically. Several authors have proposed various approaches for extracting the radius of gyration tensor on the basis of scattering data. These are, for example, linear fits to a Guinier plot ($\log I$ vs. q^2), a Zimm plot (I^{-1} vs. q^2), a square-root plot ($I^{-1/2}$ vs. q^2) or a Kratky plot ($q^2 I$ vs. q).

The small-angle neutron scattering technique has already been used to study the effects predicted by Doi and Edwards.[10,11] Boué and co-workers made an extensive study on amorphous polystyrene containing deuterated chains. However, even with an extension ratio of 3, they did not succeed in observing the chain retraction phenomenon. Boué and co-workers suggested that a possible reason for not observing the chain retraction process could be related to the relatively large tube diameter of polystyrene.

In order to observe the chain retraction process, the polymer must fulfil the following criteria:

(i) It must be amorphous.
(ii) It must be of high molecular weight with a narrow molecular weight distribution.
(iii) The hydrogenated and deuterated chains must form an ideal solution.

The first criterion is obvious as one cannot expect the model to be valid for partly crystalline material. The second criterion ensures separation of the two time constants involved ($\tau_{dis}/\tau_{eq} \propto M$). As discussed above, the neutron scattering technique depends on the contrast between deuterated and hydrogenated chains. However, blends of high molecular weight deuterated and hydrogenated polymers exhibit a phase diagram with an upper critical temperature.[8] In order to ensure ideal mixing, one should therefore have low concentrations. It also seems beneficial if only some of the hydrogens in the polymer chain have been substituted with deuterium. On the other hand, sufficient numbers of deuterium atoms in the scattering volume are required to obtain adequate scattering intensity.

4 EXPERIMENTAL

4.1 Polymer

The polymer used for the small-angle neutron scattering experiments was poly(ethylethylene-co-ethylene). This polymer was made by hydrogenation and deuteration, respectively, of the corresponding high-vinyl poly-butadiene. Only two out of every eight hydrogens were substituted by deuterium in the deuterated chains (Fig. 2). Laser light scattering on the high-vinyl parent polybutadiene gave a molecular weight of $2 \cdot 0 \times 10^6$. The glass transition temperature of the poly(ethylethylene-co-ethylene) is approximately 240 K. The rubber plateau modulus is about 0·27 MPa.[12] With a molecular weight of 2×10^6, the number of primitive path steps, N, is approximately 250. It means that $\tau_{dis}/\tau_{eq} = 1500$.

The high-vinyl polybutadiene was made by anionic polymerization in cyclohexane. The initiator was n-butyllithium while tetramethylethylene-diamine was used to control the polybutadiene microstructure. Proton NMR indicated a microstructure of 88% vinyl units and 12% 1,4 units. Size exclusion chromatography indicated a M_w/M_n ratio of less than 1·1.

$$
\begin{array}{cc}
\begin{array}{c}
\text{H}\quad\text{H} \\
|\quad\;\;| \\
-\text{C}-\text{C}- \\
|\quad\;\;| \\
\text{H}\;\;\text{H-C-H} \\
|\;\;\;\;\; \\
\text{H-C-H} \\
| \\
\text{H}
\end{array}
&
\begin{array}{c}
\text{H}\quad\text{H} \\
|\quad\;\;| \\
-\text{C}-\text{C}- \\
|\quad\;\;| \\
\text{H}\;\;\text{H-C-D} \\
|\;\;\;\;\; \\
\text{H-C-D} \\
| \\
\text{H}
\end{array} \\
\text{(a)} & \text{(b)}
\end{array}
$$

Fig. 2. Hydrogenated (a) and deuterated (b) ethylethylene monomeric units.

A 1-mm-thick film of poly(ethylethylene-co-ethylene) was cast from a 2–3% solution in heptane. The concentration of partially deuterated chains was approximately 2%. Stress-relaxation measurements were made on thin strips at different temperatures above the glass transition temperature.

The sample for small-angle neutron scattering was made by stretching a large strip of film (22 mm × 125 mm) to an extension ratio of 3 in simple extension at 248 K. After stretching, plates were glued onto both sides of the stretched film with ethylcyanoacrylate glue (Fig. 3). Then the film was cut at the upper and lower edges of the sample holder before the assembly was cooled to below the glass transition temperature. The latter was done

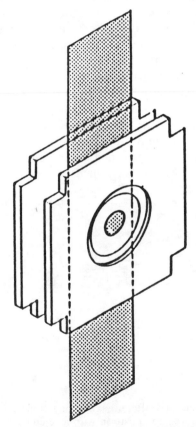

Fig. 3. Sample holder for the recording of SANS spectra of stretched samples. The two halves are glued to the stretched film with ethylcyanoacrylate glue at 248 K.

to prevent further relaxation during the neutron scattering experiment. The stretching and glueing operations, combined, took about 300 s. The sample holder was made from polymethylmethacrylate and had 12 mm holes in the middle for the neutron beam. On the front side there was, in addition, a cadmium disc with an 11 mm aperture.

4.2 Small-Angle Neutron Scattering

The scattering experiments were carried out at the SANS instrument at the cold source of Risø National Laboratory. The diameter of the neutron beam aperture in front of the sample was 7·3 mm, the entrance aperture after the mechanical velocity selector was 16 mm in diameter, and the beam-stop on the detector was 40 mm in diameter. The distance between entrance aperture and sample was 2·25 m and between sample and detector 3·0 m. The scattered neutrons were detected by a 40 cm × 40 cm area-sensitive ^{3}He-proportional counter. The measurements were carried out with neutrons of wavelength 15 Å and a wavelength distribution $\Delta\lambda/\lambda = 0·18$. The obtained momentum transfer was in the range $q = 0·03$–$0·4\,nm^{-1}$. The two-dimensional scattering data were corrected for inhomogeneities in the detector efficiency by using the incoherent scattering from water. Incoherent scattering from the sample was subtracted from the spectra using scattering results from a sample without deuterated chains.

SANS experiments were performed on an undeformed as well as on a stretched sample. The SANS measurements on the stretched polymer were

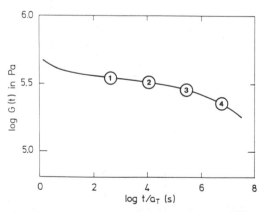

Fig. 4.   Stress-relaxation curve after uniaxial elongation by a factor of 1·5 obtained by using time–temperature superposition with a reference temperature of 253 K. The numbers indicate the four stages of relaxation at which SANS experiments were performed.

all made at a temperature of about 225 K, i.e. well below T_g. It means that no relaxation has taken place during the recording of the SANS spectra which took about 8 h for each spectrum. Relaxation in the strained state was performed in the SANS equipment at various temperatures above T_g. The time–temperature superposition principle was used to get the corresponding points on the stress-relaxation curve shown in Fig. 4. SANS spectra of the undeformed sample were also made at different temperatures in order to investigate whether the blend of 2% deuterated and 98% hydrogenated chains formed an ideal mixture.

5 RESULTS AND DISCUSSION

The radially averaged scattering intensity versus the scattering vector is shown in Fig. 5 for the undeformed sample. Curve fitting to the Debye function (eqn (6)), shown as the solid curve in the figure, gives a radius of gyration $R_G = 380$ Å. This is in perfect agreement with the coil size in a θ-solvent calculated on the basis of molecular weight.[13] This result indicates that the deuterated chains are independent random coils.

Below T_g, the scattering intensity was observed to be independent of temperature. The scattering intensity was clearly dependent on temperature above T_g. A 7% increase in forward scattering intensity was observed when the temperature was decreased from $T_g + 60$ K to T_g. This

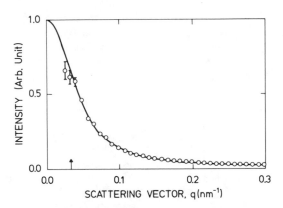

Fig. 5. Scattering intensity in arbitrary units versus scattering vector of the isotropic polymer. The solid line shows the Debye function with $R_g = 380$ Å. The arrow indicates the maximum q-value of the beam-stop.

dependence on temperature may reflect spinodal fluctuations, in agreement with an upper critical solution temperature as observed by Bates *et al.*[8]

In the present study on the relaxation of deformed coils, all spectra were recorded well below T_g. In principle, composition fluctuations should therefore contribute the same to all the spectra. However, the fact that increased relaxation of the stretched sample was obtained by allowing it to relax at higher and higher temperatures for each new relaxation, each time followed by quenching to below T_g, could possibly make a difference. But it could not change the conclusions made below.

Figure 6 shows the scattering profiles parallel and perpendicular to the stretching direction for the spectrum with the least amount of relaxation, i.e. 300 s at 253 K. These profiles are obtained by using only three rows of detector channels along each of the two directions. It means that the statistics of the data are much worse than for the undeformed sample, which exhibits a simple circular symmetry. The value of the radius of gyration in the direction perpendicular to the stretching direction is found to be approximately 200 Å, indicating affine deformation of the coils. The radius of gyration in the direction parallel to the stretching direction is expected to be of the order of 1000 Å, which is too large to be determined from the present data.

Better statistics could in principle be obtained by increasing the deuterium substitution in the deuterated molecules or by using higher concentrations of the deuterated molecules. The maximum contrast would be obtained by a 50% solution of fully deuterated chains in fully

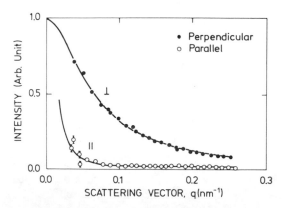

Fig. 6.   Scattering intensity profile in arbitrary units, parallel (∥) and perpendicular (⊥) to the stretching direction versus scattering vector. The solid lines show the Debye function with R_g equal to 1000 Å and 200 Å, respectively.

hydrogenated chains. However, this does not seem to be a possible route because of the danger of demixing of the deuterated and hydrogenated chains.[8]

Another approach for improving the statistics in the present case is to use more of the data of the area detector by fitting the contour levels to ellipses. This work is still in progress, but Fig. 7 shows preliminary results. The eccentricity is defined here as the ratio of the short axis to the long axis of the ellipse. The insert in Fig. 7 shows a small change in eccentricity as the relaxation progresses. The observed small but distinct increase in the value of the eccentricity indicates that some randomization is occurring long before the sample breaks. The chain retraction process of the Doi and Edwards model should reduce the size of the ellipsoid without changing the eccentricity. Complete randomization by reptation should make the eccentricity become equal to one. This could not be observed in the present study since the samples always broke at the entrance to the terminal zone.

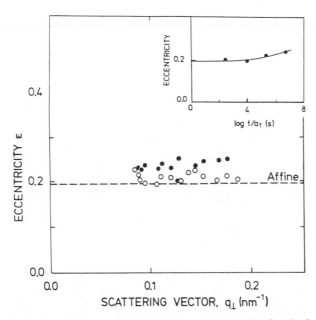

Fig. 7. Eccentricity of ellipses of constant scattering intensity for the first (○) and the last (●) spectra obtained on the stretched polymer, plotted versus scattering vector in the direction perpendicular to the stretching direction. The insert shows the q-averaged eccentricity plotted against stress relaxation time for the four stages of relaxation.

The background contributes nearly 50% to the measured scattering intensities in the present study. Any error in the determination of the background could therefore change the shapes of the spectra substantially. Another serious problem is the lack of good statistics of the data shown in Fig. 6 and in similar plots for subsequent stages of relaxation. Taken together, the two problems prevent the calculation of R_g in the perpendicular direction with any kind of confidence. It is necessary to use a more sensitive method of comparison, namely subtraction of subsequent spectra from each other.

Subtraction of the first spectrum, corresponding to point No. 1 in Fig. 4,

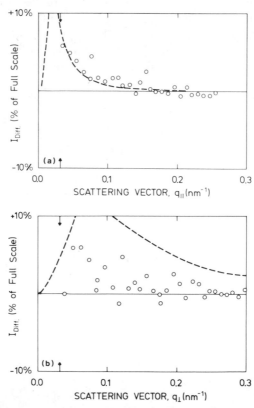

Fig. 8. Difference spectra obtained by subtraction of the first from the last spectrum plotted versus the scattering vector, (a) parallel and (b) perpendicular to the stretching direction. The dashed curves show the calculated Doi–Edwards contraction, using the Debye formula (eqn (6)) as an approximation for the scattering functions. The arrows indicate the maximum q-value of the beam-stop.

from the last spectrum, corresponding to point No. 4, shows a very interesting result, namely that the difference spectrum shows positive values all over the detector plane. This result is in agreement with the expected effect of the proposed chain retraction process. It is in direct contradiction to a randomization process, which would convert the stretched ellipsoid back to the undeformed spherical coil. The latter process would give rise to a negative difference in the perpendicular direction and a positive difference along the parallel direction, only. It has not been possible to find an explanation for the observed difference spectra other than the chain retraction process. We therefore consider these results to be the first direct evidence of the chain retraction process proposed in the Doi and Edwards model. However, it should be emphasized that the results are preliminary since they are based on one series of spectra only.

Figure 8 shows the difference spectrum profiles in the parallel and perpendicular directions. The statistics for the difference profiles are of course even worse than for the profiles shown in Fig. 6, but it is clear that the difference profiles are positive in both directions. Figure 8 also shows a first attempt to compare the observed profiles with the prediction of the Doi and Edwards theory. The dashed lines show the calculated difference profiles, using the Debye function (eqn (6)). The experimental results are qualitatively in good agreement with the prediction of theory. However, the observed effect is smaller than the prediction. Several factors may contribute to the lack of quantitative agreement: (i) the Debye function should not be expected to be completely correct for deformed coils; (ii) it is unlikely that the point of maximum contraction has been reached in one of the two last spectra; (iii) some randomization may have taken place either by reptation or by some other process. Relative to the predicted values, some randomization would increase the values of the difference in the parallel direction and decrease the values in the perpendicular direction, as observed. Concentration fluctuations may also influence the result as discussed above.

6 CONCLUSION

For the first time, direct evidence of the chain retraction process proposed in the Doi and Edwards model has been observed. Small-angle neutron scattering experiments were performed at different stages of relaxation on a stretched sample of poly(ethylethylene-co-ethylene) with a molecular weight of 2×10^6. The observed effects are only semi-quantitatively in agreement with theoretical predictions.

ACKNOWLEDGEMENT

This work was supported by a grant from the Danish Council for Science and Industrial Research under the FTU program. The SANS facility at Risø was established with support from the Danish and the Swedish Natural Science Research Councils.

REFERENCES

1. Doi, M. and Edwards, S. F., *J. Chem. Soc. Faraday II*, 1978, **74**, 1789, 1802, 1818; 1978, **75**, 38.
2. Edwards, S. F., *Proc. Phys. Soc. London*, 1967, **92**, 9.
3. de Gennes, P. G., *J. Chem. Phys.*, 1971, **55**, 572.
4. de Gennes, P. G., *Scaling Concepts in Polymer Science*, Cornell University Press, Ithaca, New York, 1979.
5. Graessley, W. W., *Adv. Polym. Sci.*, 1982, **47**, 67.
6. Kramer, O., *Brit. Polym. J.*, 1985, **17**, 129.
7. de Gennes, P. G., *J. Phys. (Paris)*, 1975, **36**, 1199.
8. Bates, F. S., Wignall, G. D. and Koehler, W. C., *Phys. Rev. Lett.*, 1985, **55**, 2425.
9. Debye, P. J., *Phys. Colloid. Chem.*, 1947, **51**, 18.
10. Boué, F., Nierlich, M., Jannink, G. and Ball, R., *J. Phys. (Paris)*, 1982, **43**, 137.
11. Boué, F., Osaki, K. and Ball, R. C., *J. Polym. Sci., Polym. Phys. Ed.*, 1985, **23**, 833, and references therein.
12. Carella, J. M., Graessley, W. W. and Fetters, L. J., *Macromolecules*, 1984, **17**, 2775.
13. Mays, J., Hadjichristidis, N. and Fetters, L. J., *Macromolecules*, 1984, **17**, 2723.

26

THE SCATTERING OF LIGHT BY SWOLLEN NETWORKS

R. S. Stein, V. K. Soni,* H. Yang†

*Polymer Research Institute, University of Massachusetts,
Amherst, Massachusetts 01003, USA*

and

Burak Erman

*School of Engineering, Bagazici University,
Bebek 80815, Istanbul, Turkey*

ABSTRACT

Swollen networks scatter more light than solutions of linear molecules of the same concentration. This occurs because of the concentration fluctuations associated with the variation of the local degree of swelling related to variations in the local crosslink density. Measurements are reported on model end-linked bimodal networks of controlled heterogeneity, and compared with theoretical estimates of the effect. The observed discrepancy is indicative of the need for an improvement in concepts of microdeformation of networks during swelling.

SCATTERING FROM SOLIDS

It is recognized that homogeneous media do not scatter radiation.[1] This may be demonstrated from the fact that the total scattered amplitude may be described by

$$E = \sum \rho_j \exp\left[-i(\vec{q} \times \vec{r}_j)\right] \tag{1}$$

* Present address: Shell Development Company, Westhollow Research Center, PO Box 1380, Houston, Texas 77001, USA.
† Present address: Research Laboratories, Eastman Kodak Research, Rochester, New York 14650, USA.

where ρ_j is the scattered amplitude arising from the jth volume element, $\vec{q}$ is the scattering vector defined by

$$\vec{q} = (2\pi/\lambda)(\vec{s}_1 - \vec{s}_0) \tag{2}$$

having the magnitude

$$q = |\vec{q}| = (4\pi/\lambda)\sin(\theta/2) \tag{3}$$

where λ is the wavelength of the radiation within the medium, $\vec{s}_1$ and $\vec{s}_0$ are unit vectors along the incident and scattered rays, respectively, $\vec{r}_j$ is a vector from the origin (arbitrary) to the jth scattering element, and θ is the scattering angle (between $\vec{s}_0$ and $\vec{s}_1$).

For a uniform medium, $\rho_j = \rho_0$ and is a constant throughout the medium. The sum of the exponential over an infinite medium will then be zero. Physically, this implies that for any scattered ray arising from a particular volume element, there will be a corresponding ray of equal amplitude arising from another volume element differing in phase by 180°, so as to destructively interfere with the first. Hence, in a medium with dimensions much larger than λ, volume elements can be paired in this manner, demonstrating that there should be no residual scattering.

In a real medium, ρ_j will not be constant so that the amplitude of the scattered rays from different volume elements will differ. Hence, cancellation will not be complete but will depend upon how ρ_j depends upon $\vec{r}_j$. This is defined by the structure of the system.

The fluctuation in ρ_j can be represented by

$$\eta_j = \rho_j - \langle \rho \rangle \tag{4}$$

where $\langle \rho \rangle$ is the mean value of ρ_j. For a homogeneous medium, η_j is zero. It is positive or negative with equal probability for a real medium.

Debye and Bueche[2] analyzed the scattering from a heterogeneous medium and showed that

$$\mathcal{R}(\vec{q}) = K\langle \eta^2 \rangle \int \gamma(\vec{r}) \exp[-i(\vec{q} \times \vec{r})] \, d^3\vec{r} \tag{5}$$

Here, $\mathcal{R}(\vec{q})$ is the 'Rayleigh factor' defined by

$$\mathcal{R}(\vec{q}) = I_s(\vec{q})p^2/I_0 V_s \tag{6}$$

which describes the scattering power of the system. Here, $I_s(\vec{q})$ is the scattered intensity at $\vec{q}$, I_0 is the incident intensity, p is the sample-to-detector distance, and V_s is the scattering volume. $\mathcal{R}(\vec{q})$ is a property of the system and should be independent of the experimental procedure for its measurement. The vector $\vec{r}$ is $(\vec{r}_j - \vec{r}_k)$ and is the vector distance between the

jth and kth elements, and is independent of the location of the origin. The correlation function $\gamma(\vec{r})$ is defined by

$$\gamma(\vec{r}) = \langle \eta_j \eta_k \rangle_{\vec{r}} / \langle \eta^2 \rangle \tag{7}$$

where $\langle \ \rangle_{\vec{r}}$ designates an average over all pairs of volume elements separated by $\vec{r}$. It is evident that when $r = 0$, $\gamma(0) = 1$, while $\gamma(\infty) = 0$, since in a medium with no long-range order there is no correlation between η_j and η_k when the volume elements are infinitely separated.

For media with no macroscopic orientation, $\gamma(\vec{r})$ depends only upon the magnitude of $\vec{r}$, so that one may integrate over its angular coordinates, giving

$$\mathscr{R}(q) = 4\pi K \langle \eta^2 \rangle \int_0^\infty \gamma(r) \frac{\sin(qr)}{(qr)} r^2 \, \mathrm{d}r \tag{8}$$

Here, the scattering depends upon two material variables; $\langle \eta^2 \rangle$ describing the amplitude of the fluctuations and $\gamma(r)$ describing their geometry. For a homogeneous medium, $\langle \eta^2 \rangle = 0$ and the scattering vanishes.

The function $\gamma(r)$ can be obtained from Fourier inversion of $\mathscr{R}(q)$, and for many systems is found to be approximated by an empirical exponential function

$$\gamma(r) = \exp[-r/a_C] \tag{9}$$

The quantity a_C is called a correlation length and serves as a measure of the spatial extent of the fluctuation. For this, eqn (8) becomes

$$\mathscr{R}(q) = K' a_C^3 \langle \eta^2 \rangle / [1 + q^2 a_C^2]^2 \tag{10}$$

so that a plot of $\mathscr{R}(q)^{-1/2}$ vs. q^2 should be linear, permitting a ready means of determining $\langle \eta^2 \rangle$ and a_C. This is called a 'Debye–Bueche plot'. It is evident that $\mathscr{R}(q)$ depends upon both of these parameters so that, for example, at $q = 0$,

$$\mathscr{R}(0) = K' a_C^3 \langle \eta^2 \rangle \tag{11}$$

It should be emphasized that these equations are general and independent of the kind of radiation being used. The definitions of K (or K') and ρ_j (or $\langle \eta^2 \rangle$) are radiation specific. The choice of the kind of radiation to use is made so as to maximize $\langle \eta^2 \rangle$ (contrast) and so that λ leads to a variation of $I(\theta)$ with θ over a reasonable range of angle. This implies that λ should be of the order of the dimensions of the heterogeneity.

Besides being dependent upon the type of radiation, ρ_j depends upon the material property which is fluctuating. It is evident that it will depend upon the gravimetric density changes which affect the electron density (important for X-rays), the polarizability or refractive index (for light scattering) and atomic nucleus concentration (for neutron scattering). Furthermore, for a multiphase system, these also depend upon fluctuations in composition. The scattering of light is also affected by fluctuations in anisotropy and optic axis orientation for anisotropic systems. These latter are dependent upon the polarization of the light.

SCATTERING FROM NETWORKS

Networks are characterized by the presence of crosslinks which link the molecules together in a three-dimensional structure. The concentration of crosslinks is usually relatively low and often not more than a per cent or two. Consequently, unless the scattering power of the crosslinks is very high, there is usually not a significant difference between the scattering of a network and that of the corresponding uncrosslinked polymer. Thus, scattering experiments on such systems do not yield significant information about the network structure. There have been a few cases where the crosslinks have been labelled so as to increase their scattering power. For example, neutron scattering studies have been conducted on polystyrene networks which have been crosslinked with deutero-divinylbenzene.[3]

An approach for learning about network structure where labeling is not required is to observe the scattering increase occurring upon swelling.[4] As has been shown by Flory and Rehner,[5] the degree of swelling of a network depends upon its degree of crosslinking, with the more highly crosslinked polymer swelling less. Thus, it follows that an inhomogeneously crosslinked network will swell inhomogenously. If the refractive index of the solvent differs from that of the polymer, this gives rise to refractive index fluctuations leading to enhanced light scattering. Thus, the measurement of such scattering serves to characterize the amount and size of such concentration fluctuations, which are related to the fluctuations of crosslink density.

Light scattering characterizes fluctuations having dimensions of the order of the wavelength of light. Far smaller fluctuations may be studied with neutron scattering where the wavelength is in the range of 0·1–2 nm. In this case, contrast may be achieved by swelling with a deuterous solvent.

SWELLING OF HOMOGENEOUS AND HETEROGENEOUS NETWORKS

The equilibrium swelling of a network is characterized by the equality of chemical potentials of the solvent within, $(\mu_1)_n$, and outside, $(\mu_1)_o$, the swollen network:[5]

$$(\mu_1)_n = (\mu_1)_o \tag{12}$$

where it is normally assumed that the free energy of the network, G_n, arising from the free energy of mixing, $(\Delta G)_M$, and the free energy of elastic deformation, $(\Delta G)_E$, are additive, leading to

$$(\mu_1)_n = \left(\frac{\partial(\Delta G)_n}{\partial n_1}\right)_{T,P} = \left(\frac{\partial(\Delta G)_M}{\partial n_1}\right)_{T,P} + \left(\frac{\partial(\Delta G)_E}{\partial n_1}\right)_{T,P} \tag{13}$$

There has been some concern about this assumption of additivity.[6]

The free energy of mixing is usually described in terms of a mean-field theory such as that of Flory[7] and Huggins[8] (FH), in which

$$(\Delta G)_M = RT[n_1 \ln \phi_1 + n_2 \ln \phi_2 + \chi \phi_1 \phi_2] \tag{14}$$

where n_1 and n_2 are the numbers of moles of solvent and solute, respectively, and ϕ_1 and ϕ_2 are their volume fractions. The interaction parameter between solute and solvent is designated χ. Since, for a network, the polymer molecular weight is essentially infinite, the number of moles of polymer, n_2, is negligible as compared with n_1, the number of moles of solvent.

In the 'pure' FH theory, χ describes only pairwise enthalpic interactions and is a constant, independent of concentration. In fact, χ is often found to be concentration dependent, reflecting the inadequacy of the FH postulates and an entropic contribution to χ.

The elastic contribution to $(\mu_1)_n$ is dependent upon the postulates of elasticity theory. In its original formulation by Kuhn,[9] the assumptions are that:

(1) The force is of entropic origin.
(2) The displacement of crosslinks during the deformation is affine.
(3) The entropy change with deformation can be calculated from intermolecular considerations alone. Average intermolecular interactions do not change with deformation.
(4) The effect of chain entanglement on the entropy of deformation can be neglected.

The theory was subsequently modified by James and Guth (JG),[10] who relaxed the assumption of affine deformation of crosslinking points and demonstrated that only the average position of the crosslinks deformed affinely. The crosslinks were allowed to fluctuate about this average position, subjected only to the constraint, transmitted through the chains, imposed by the displacement of the sample boundaries. This gave rise to a lower entropy reduction upon stretching with a modulus of half of that predicted by the Kuhn approach for a tetrahedral network.

Various workers attempted to include the effects of chain entanglements.[11]. Trapped entanglements, occurring between two crosslinking points, were distinguished from others which could disentangle with time. The earlier theories postulated that the trapped entanglement served as a 'physical crosslink', and that their number should be added to that of the chemical crosslinks in calculating the modulus. It was pointed out,[12] however, that the entropy reduction resulting from an entanglement was less than that resulting from a crosslink, in that slippage between entangled chains may occur during deformation.

The effect of interchain interaction was treated differently in the theory of Flory and Erman (FE).[13] Rather than considering the entanglements as additional crosslinks, they included their effects as modifying the mobility of the chemical crosslinks in a manner described by a parameter, κ. This permitted behavior intermediate between that of the Kuhn and the JG theories in a manner dependent upon deformation. They also introduced a second parameter, ξ, to allow for non-affine deformation of the average position of the crosslinks. The theory has been remarkably successful in phenomenologically accounting for the mechanical behavior of rubbers, but there is need for further defining the molecular nature of the parameters.

In the Flory–Rehner (FR) theory of swelling, the FH equation for the free energy of mixing was combined with the Kuhn equation for the elastic free energy to give for the volume fraction of rubber, ϕ_2, in the swollen network

$$-[\rho V_1 \phi_2^{1/3}/M_C] = \ln(1 - \phi_2) + \phi_2 + \chi \phi_2^2 \qquad (15)$$

where ρ is the density of the polymer, M_C its molecular weight between crosslinks and V_1 the molar volume of the solvent.

This equation has provided a convenient means for measurement of M_C. For high degrees of swelling, the equation can be expressed as the approximation

$$\phi_2^{5/3} = \frac{\rho V_1}{M_C(1/2 - \chi)} \qquad (16)$$

This predicts that the degree of swelling increases with an increasing M_C and an increasing interaction between the solvent and the polymer (more negative χ). The theory may be reformulated in terms of the JG or the FE theories, the latter including the parameters κ and ξ.

In these equations, the effect of a distribution in M_C was not considered. It is evident that if there are local regions of varying M_C, the local ϕ_2 will also vary. The effect was (somewhat naively) described by Stein[4] who, by differentiating eqn (16), showed that

$$\Delta\phi_2 = [(3/5)\phi_2/\langle M_C\rangle]\,\Delta M_C \tag{17}$$

The fluctuation in ϕ_2 was related to the fluctuation in refractive index, ΔN, assuming linear additivity of the refractive indices of the solvent, N_1, and solute, N_2, according to

$$N = \phi_1 N_1 + \phi_2 N_2 \tag{18}$$

so that

$$
\begin{aligned}
\langle(\Delta N)^2\rangle &= (N_2 - N_1)^2\langle\Delta\phi_2\rangle^2 \\
&= [(9/25)(N_2 - N_1)^2][\langle(\Delta M_C)^2\rangle/\langle M_C\rangle^2]\phi_2^2
\end{aligned} \tag{19}
$$

By identifying $\langle\eta^2\rangle$ of the DB theory with $\langle(\Delta N)^2\rangle$, one obtains

$$\mathscr{R}(0) = [(9/25)(N_2 - N_1)^2]K'a_C^3[\langle(\Delta M_C)^2\rangle/M_C^2]\phi_2^2 \tag{20}$$

In addition to the $\langle\eta^2\rangle$ arising from the network heterogeneity, refractive index fluctuations will also occur because of thermal concentration fluctuations, just as in a polymer solution. Hence, scattering arises from this cause, even if the network is unimodal, and its contribution should be subtracted from the observed scattering (assuming additivity). Contributions also arise from density fluctuations, associated with the compressibility of the system. Since at low degrees of crosslinking the compressibility will not be much affected by crosslinking, the above subtraction also suffices to remove the density fluctuation contribution.

At higher degrees of crosslinking, one must be concerned with the scattering by the crosslink point itself. This may be appreciable if the crosslink is large in volume (as scattering depends on V^2) or if its refractive index is appreciably different from that of the polymer matrix. This may be the case for divinyl benzene crosslinked polystyrene or triisocyanate crosslinked hydroxyl terminated polybutadiene. In the case of the poly(dimethylsiloxane) (PDMS)–silane crosslinked networks described in this work, the crosslinker is small and of the same chemical nature as the matrix, so a problem is not likely to arise. Evidence is that there is not appreciable difference in scattering between the crosslinked and uncross-linked unswollen polymer.

To a good approximation, the concentration fluctuation contribution to the scattering from a swollen network will be close to that of a solution of the same concentration of the swollen network. Concentration fluctuations may be described, following Einstein,[14] by

$$\langle (\Delta c)^2 \rangle = \frac{RTc}{-(\partial \mu_1 / \partial c)_{T,P}} \tag{21}$$

The derivative $(\partial \mu_1 / \partial c)_{T,P}$ may be theoretically evaluated using FH theory, or else a measurement may be made on a (uncrosslinked) solution of the same concentration. Actually, this procedure is not quite correct since thermal fluctuations in the network will be retarded by the total change in free energy with concentration, including the elastic as well as the mixing contributions. The measurement on the solution only includes the mixing contribution and not the elastic, so that its scattering will be somewhat greater than the thermal concentration contribution to the scattering from the corresponding swollen network. One can theoretically account for this difference and apply a correction. Although the results depend upon the elastic model chosen, the correction is usually small except at high degrees of crosslinking.

After correcting the measured $\mathscr{R}(0)$ in consideration of the above, one may obtain $\mathscr{R}(0)$ and a_C by carrying out a DB plot using the corrected value in eqn (10). The parameters thus obtained serve to characterize $\langle (\Delta M_C)^2 \rangle / \langle M_C \rangle^2$. The application of this approach and its limitations will be discussed subsequently.

DISTRIBUTION OF M_C AND BIMODAL NETWORKS

A unimodal network is one in which all chains have the same M_C. A bimodal one contains chains with two different values of M_C. If these are M_L (for the molecular weight of the long chains) and M_S (for the short ones), and the number fractions of the two species are x_L and x_S, respectively, then the number average value of M_C is

$$\langle M_C \rangle = x_L M_L + x_S M_S \tag{22}$$

For such a network, it follows that

$$\langle (\Delta M_C)^2 \rangle = x_L x_S (M_L - M_S)^2 \tag{23}$$

In practice, one usually produces such bimodal networks by 'end-linking' through a crosslinker the chains of end-reactive linear polymers of molecular weights M_L and M_S.[15] While, ideally, these prepolymers are monodisperse, real polymers have a finite width molecular weight distribution. If these are described as Schultz distributions[16] with width parameters b_L and b_S, then eqn (23) may be generalized to give

$$\langle(\Delta M_C)^2\rangle = x_L x_S (M_L - M_S)^2 + x_L M_L^2/b_L + x_S M_S^2/b_S \qquad (24)$$

Thus, by this means, it should be possible to prepare networks of controlled $\langle(\Delta M_C)^2\rangle$ and test eqn (20). Such a test will be described and will indicate the need for improved theory.

A parameter in eqn (20) is the correlation length, a_C. This characterizes the size of the heterogenous regions and depends upon the disposition of chains of different length. It may be determined experimentally from a DB plot using eqn (10). Its molecular origin depends upon the mechanism of the crosslinking reaction.

EXPERIMENTAL

Previously reported experimental studies[17] involved use of radiation or peroxide crosslinked polybutadienes or triisocyanate end-linked hydroxyl terminated butadienes. These early experiments did not lead to well-defined networks and could not be used to verify the theory presented here.

The studies reported in this work were made on bimodal networks prepared from narrow distribution PDMS terminated on both ends with vinyl groups. They were end-linked using the tetrafunctional silane crosslinker $Si—[—O—Si(CH_3)_2H]_4$ using a Pt catalyst. Experimental details are published elsewhere.[18] All reactants were obtained from Petrarch Systems, Bristol, Pennsylvania, USA. Characteristics of the PDMS prepolymers are given in Table 1. Advantages of employing this system are:

(1) Ready control of $\langle M_C \rangle$.
(2) Simple and convenient chemistry.
(3) Low viscosity of prepolymers permitting cleaning by filtration.
(4) Excellent optical clarity of polymers.
(5) Chemical similarity of crosslinker and matrix.

Disadvantages are:

(1) A finite polydispersity of the PDMS ($M_w/M_n \sim 1·8$).
(2) Deuterous monomer of PDMS is expensive, so that small-angle neutron scattering with labelled chains is not convenient.
(3) Prepolymers contain $\sim 3\%$ non-reactive cyclics.

TABLE 1
Characteristics of the PDMS Prepolymers

Sample	η (centistokes)	M_n	M_w	$[M_w/M_n]$
PDMS 4–6	5	770	1 450	1·88
PDMS 1 000	1 000	22 500	40 600	1·80

While it is often assumed that the crosslinker reacts stoichiometrically and that the achieved $\langle M_C \rangle$ agrees with that theoretically calculated, complications arising from side reactions have been demonstrated.[19] Nevertheless, this system appears to be one of the best for approaching the ideal of a model network.

Equilibrium swelling and modulus measurements and the networks are reported elsewhere.[18,20] They were found to agree acceptably with theory, exhibiting swelling and moduli intermediate between those predicted by the Kuhn and JG theories and fitable with a suitable choice of the FE parameters.

Light scattering measurements were performed in the laboratory of K. Langley of the University of Massachusetts Physics Department, USA,

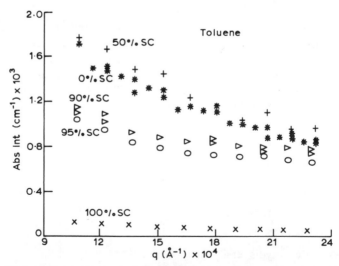

Fig. 1. The variation of the corrected $\mathcal{R}(q)$ with q for various percentages of short chains (% SC) for a bimodal network of $M_S = 770$, $M_L = 22\,500$ swollen to equilibrium with toluene.

utilizing an Ar ion laser (514·5 nm) with a photon counter, a Langley–Ford digital correlator run off of a Tektonix computer.[18] The sample cell was held in a thermostatted vat containing an excess of solvent. Absolute values of $\mathscr{R}(q)$ were obtained by calibrating against toluene. Both solution and swollen network measurements were made with this apparatus. Solutions were held in 10-mm-diameter test tubes. Tilted rectangular swollen network samples were employed, depending upon the q-range being studied. As angular resolution was not critical, measurements were made with large aperture detection so as to minimize speckle pattern effects.

Flat samples were of the order of 5 mm thick and held in a specially designed cell. Corrections were applied for reflection and refraction following the procedure described by Stein and Keane.[21]

Measurements were made on dry uncrosslinked and crosslinked samples as well as on solutions of the same concentration as the swollen networks, so as to evaluate the various contributions to the scattering as discussed above. Typical plots of the corrected $\mathscr{R}(q)$ vs. q for bimodal networks of various x_S (% SC), prepared with PDMS of number-average molecular weights $M_L = 22\,500$ and $M_S = 770$, swollen to equilibrium with toluene, are presented in Fig. 1. It is evident that:

(1) The scattering in all cases decreases uniformly with increasing q.
(2) The scattering from the 100% SC network is least and that from the 50% most. It is noted that the 100% SC network is swollen least.
(3) The 0% SC network scatters much more than the 100% SC but less than the 50% SC.

A typical DB plot for the 50% SC in toluene is given in Fig. 2. The plot is seen to be linear, indicating the appropriateness of the empirical exponential correlation function. Values of a network heterogeneity parameter, H, defined by

$$H = [\langle(\Delta M_C)^2\rangle^{1/2}]/\langle M_C\rangle \tag{25}$$

and a_C were obtained using eqn (10) and are presented in Table 2, where they are compared with theoretical values calculated from the composition of the networks using eqns (20) and (24). It is noted that similar experimental results are obtained for samples swollen in both toluene and benzene.

It is evident that the sizes of the regions that give rise to the scattering are appreciable, and of the order of 30–40 nm. The size is several times that of molecular dimensions, suggesting that the regions consist of many

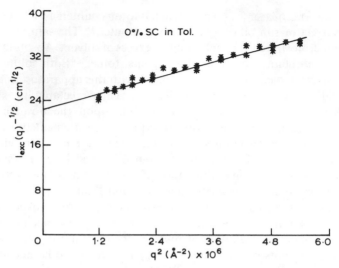

Fig. 2.   A typical Debye–Bueche plot for the above 50% SC network swollen to equilibrium in toluene.

crosslinks. A striking observation is the large discrepancy between the experimental and theoretical values of H. One sees that:

(1) $H(\text{exp.}) \ll H(\text{theory})$. This means that the experimentally observed scattering is much less than theoretically predicted. Most sources of experimental error would lead to excess scattering and errors in the reverse direction.

(2)   Experimentally, $H(0\% \text{ SC}) \neq H(100\% \text{ SC})$, while the theory predicts equality.

TABLE 2
Network Characteristics Determined by Light Scattering

Sample (% S)	a_C (nm)		$H(exp.)$		$H(calc.)$
	In toluene	In benzene	In toluene	In benzene	
0	32	38	0·049	0·042	0·89
50	37	43	0·040	0·043	1·54
90	36	35	0·026	0·032	3·10
95	35	36	0·027	0·030	3·54
100	38	37	0·007	0·008	0·94

It is evident that there are serious differences between the experimental observations and the theoretical predictions. We believe that these arise primarily as a result of the oversimplified theoretical model employed, as will be discussed in the next section.

DISCUSSION

It is believed that a primary reason for the discrepancy is the use of the FR equation to relate the fluctuation of ϕ_2 to that of M_C. The FR theory is applicable to a homogeneous network in which a chain having a given M_C is in the environment of chains of the same type. Thus, it can account for the difference in ϕ_2 between two unimodal networks, one of which consists only of M_L chains and the other of M_S chains. However, it would not apply to a network in which M_L and M_S chains were mixed together unless they were segregated so as to reside among chains of the same type. This is not usually the case.

The problem may be better understood if one considers the possible configurations of chains about a tetrafunctional crosslink as described in Fig. 3. Depending upon the number of long and short chains joined together at the crosslink, there are five possible combinations, which we may designate as S_4, S_3L, S_2L_2, SL_3 and L_4. The FR theory would approximately describe the difference between S_4 and L_4 type regions but would not correctly describe the intermediate configurations. The use of FR theory implies that one can consider a network consisting of a mixture of long and short chains to be composed of an equal number of S_4 and L_4

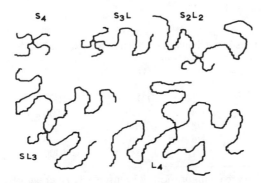

Fig. 3. Possible configurations of long and short chains in a cluster about a tetrafunctional crosslink in a bimodal network.

configurations, neglecting the intermediate ones. Actually, the intermediate configurations in which there are a mixture of long and short chains connected to a crosslink are more probable, since the statistical weight, g, of a configuration consisting of N_S short chains and N_L long ones is

$$g = \frac{(N_S + N_L)!}{N_S! \, N_L!} \qquad (26)$$

assuming equal reactivity of the S and L chains so that a bimodal distribution is followed. Thus an $S_2 L_2$ configuration is six times as probable as an S_4 or an L_4. The values of g subsequently given are normalized such that the sum of the g-values for a given configuration adds up to unity.

It is evident that the swelling in the vicinity of such an intermediate configuration will be intermediate between that of an S_4 and an L_4. Thus the fluctuation in the degree of swelling and hence in $\langle \eta^2 \rangle$ will be less than that calculated. As a first approximation, one might assume that the degree of swelling is related to the average molecular weight of the chains associated with a given crosslinking point. Based upon this hypothesis, the values of g and of $\langle (\Delta M_C)^2 \rangle_j$ associated with each of the configurations are listed in Table 3. The weighted average $\langle (\Delta M_c)^2 \rangle$ is then

$$\langle (\Delta M_C)^2 \rangle = \sum g_j \langle (\Delta M_C)^2 \rangle_j / \sum g_j \qquad (27)$$

This implies that the local degree of swelling in the vicinity of a particular configuration is related to the $\langle (\Delta M_C)^2 \rangle_j$ value for that configuration, which may be readily calculated from the values of N_S and N_L for that configuration. This was calculated from the data in Table 3, from which it is seen that the value of $\langle (\Delta M_C)^2 \rangle$ obtained by averaging over all five

TABLE 3
Characteristics of Network Clusters for $M_S = 1000$, $M_M = 19\,000^a$

Configuration	g_j	$\langle M_j \rangle$	$[\langle M_j \rangle - \langle M_C \rangle]$	$\langle (\Delta M_C)^2 \rangle$
S_4	1/16	1 000	−9 000	81×10^6
$S_3 L$	1/4	5 500	−4 500	20.3×10^6
$S_2 L_2$	3/8	10 000	0	0
SL_3	1/4	14 500	4 500	20.3×10^6
L_4	1/16	19 000	9 000	81×10^6

[a] Average over configurations of cluster: (1) Average of S_4 and L_4: $\langle (\Delta M_C)^2 \rangle = 81 \times 10^6$, $H = 0.90$; (2) Average over all configurations: $\langle (\Delta M_C)^2 \rangle = 20.3 \times 10^6$, $H = 0.45$.

configurations is appreciably less than if only the S_4 and L_4 configurations were considered, as would be equivalent to the original FR calculation. This is in the right direction for reducing the predicted scattering so as to correspond more closely with the experimentally measured values.

The approximate nature of this calculation is evident. It has recognized that the local degree of swelling is not a property of the M_C of an isolated chain, but depends upon the neighboring chains. In the above consideration, the calculation has been extended to a cluster of four chains connected to a crosslinking point, and a lower fluctuation was found. One may logically ask, why stop there? This cluster will exhibit a degree of swelling dependent upon its environment. Thus, an L_4 cluster will swell more if it exists in an environment of other L_4 clusters than if, to consider an extreme case, it existed in a matrix of S_4 clusters which would 'encapsulate' it in a low-swelling environment. It appears that interactions over greater distances must be considered. The effect of doing this will evidently further reduce the calculated extent of swelling fluctuations and predicted scattering. Computer modelling is in progress in an effort to account for this.

A similar problem exists in considering chain deformation. Kuhn theory says that the segmental orientation of a chain relates to its M_C. This is true in a uniform network in which M_C for all chains is the same and all are subjected to an affine deformation of their average crosslink position. In a non-uniform network, this is not true, and an L chain within a L_4 cluster will exhibit different segmental orientation than one within a S_3L cluster. The local deformation of the L chain is obviously restricted by its being associated with three S chains. Thus it appears that single chain segmental orientation is not a property of the chain alone but depends upon its environment. Experimental evidence for this comes from infrared dichroism measurements on bimodal networks, as reported separately.[22] This problem also requires the development of a theory for the 'micromechanics' of non-uniform networks.

While the above considerations point to the source of the discrepancy between the measured and observed scattered intensity, they do not account for the second disagreement, that the measured H for the 'unimodal' S and L networks differ. The inherent molecular weight distributions of these cannot account for the differences. It was noted that measurements were made at equilibrium swelling so that the degree of swelling of the S network was much less than that of the L. The environment of a crosslink point in the two networks is considerably different, in that the L crosslinks will be on chains which are more interpenetrated, so they will be among a larger fraction of crosslinks associated with distantly connected chains (spatial neighbors), as contrasted with the crosslinks in the S network which will reside in an environment richer in crosslinks on connected chains (topological neighbors).

Differences in the environment of crosslinks in different networks may be described by the κ parameter of the FE theory. According to this theory, κ is taken to be proportional to the number of crosslinks in the spatial domain pervaded by the network chains meeting at the junction.[13] Thus, κ is proportional to the ratio of total neighbors to topological neighbors of a junction. This ratio was also introduced recently by Bastide and co-workers[23] in order to explain neutron scattering results on PDMS networks. On a molecular basis, κ reflects the average 'strength' of entanglements operating on junctions. As a further refinement, one might associate different κ_j parameters with clusters of junctions having different mobilities, leading to differences in local swelling and deformation. In as much as κ increases with M_C, larger values of the latter are expected to lead to larger excess scattering. The second parameter, ξ, of the FE theory is directly related to the network unfolding concepts introduced recently.[23] Larger values of ξ lead to higher degrees of rearrangements of junctions, thereby resulting in lower values of $\langle \eta^2 \rangle$.

It appears that the local 'non-affineness' of non-uniform networks could also be described through the formalism of the FE theory in that the κ_j and ξ_j parameters could be considered as a local property. Thus a value of κ_j and ξ_j might be associated with each local configuration. Preliminary calculations of $\langle \eta^2 \rangle$ resulting from local fluctuations in κ and ξ are in agreement with the results of scattering experiments reported in this work. Calculations are performed by varying the parameters κ and ξ about their mean values and evaluating the corresponding fluctuations in ϕ_2. Values of ϕ_2 are determined by solving the FR equation, modified to include the κ and ξ parameters in the elastic potential. Quantitative agreement with observed excess scattering values may be obtained by varying the κ and ξ parameters within limits suggested by previous mechanical and swelling studies on similar networks. While experimental data may be fitted by this formalism, computer modeling may be necessary to associate molecular meaning to these parameters. Experimental measurements of small-angle neutron scattering and infrared dichroism arising from bimodal networks containing labeled chains should provide data to test such theories. Experiments on networks swollen to degrees other than equilibrium are also desirable.

ACKNOWLEDGEMENTS

The work reported in this paper was primarily supported by the Center for University of Massachusetts-Industry Research on Polymers [CUMIRP]. For a portion of the work, CUMIRP was partly supported by a grant from the National Science Foundation.

We are indebted to Professor Kenneth Langley of the University of Massachusetts Physics Department for allowing us to use his light scattering equipment and for much advice concerning measurements. We appreciate the advice of Professor James Mark on the chemistry of preparing PDMS model networks. The authors value several useful discussions with Professor Paul J. Flory concerning the interpretation of the results.

REFERENCES

1. Bhagavantam, S., *The Scattering of Light and the Raman Effect*, Chemical Publishing Co., New York, 1942.
2. Debye, P. and Bueche, A. M., *J. Appl. Phys.*, 1949, **20**, 518.
3. Benoit, H. et al., *J. Polym. Sci., Polym. Phys. Ed.*, 1976, **14**, 2119.
4. Stein, R. S., *J. Polym. Sci., Polym. Lett. Ed.*, 1969, **7**, 657.
5. Flory, P. J. and Rehner, J., *J. Chem. Phys.*, 1949, **11**, 521.
6. Flory, P. J., *Principles of Polymer Chemistry*, Cornell University Press, Ithaca, New York, 1953.
7. Flory, P. J., *J. Chem. Phys.*, 1942, **10**, 51.
8. Huggins, M. H., *J. Phys. Chem.*, 1942, **46**, 151; *Ann. NY Acad. Sci.*, 1942, **41**, 1.
9. Kuhn, W., *Koll. Zeitschr.*, 1936, **76**, 258; *Angew. Chem.*, 1938, **51**, 640.
10. James, H. M. and Guth, E., *J. Chem. Phys.*, 1947, **15**, 651.
11. Langley, N. R., *Macromolecules*, 1968, **1**, 348.
12. Edwards, S. F., *Brit. Polym. J.*, 1977, **9**, 140.
13. Flory, P. J. and Erman, B., *Macromolecules*, 1982, **15**, 800, 806.
14. Einstein, A., *Ann. de Physik*, 1910, **33**, 1275.
15. Mark, J. E., in *Elastomers and Rubber Elasticity*, Mark, J. E. and Lal, J. (Eds), American Chemical Society, Washington, DC, 1982.
16. Schultz, G. V., *Z. Physik. Chem.*, 1939, **B43**, 25.
17. Stein, R. S., Farris, R. J., Kumar, S. and Soni, V., in *Elastomers and Rubber Elasticity*, Mark, J. E. and Lal, J. (Eds), American Chemical Society, Washington, DC, 1982.
18. Soni, V. K., Optical studies of swollen polymer networks, PhD Dissertation, University of Massachusetts, Amherst, 1986.
19. Fischer, A. and Gotlieb, M., Poster at Networks '86, Elsinore, Denmark, September 1986.

20. Stein, R. S., Soni, V. K. and Yang, H. E., Optical Studies of Network Topology, in press.
21. Stein, R. S. and Keane, J. J., *J. Polym. Sci.*, 1955, **17**, 21.
22. Hrabowska, J. and Stein, R. S., Poster at Networks '86, Elsinore, Denmark, September 1986.
23. Bastide, J., Picot, C. and Candau, S., *J. Macromol. Sci., Phys.*, 1981, **B19**, 13.

SECTION 4

SWELLING OF POLYMER NETWORKS

27

SWELLING OF POLYMER NETWORKS

MOSHE GOTTLIEB

Chemical Engineering Department, Ben Gurion University,
Beer-Sheva 84105, Israel

ABSTRACT

The theory of polymer swelling has been formulated by Flory and Rehner based on three main assumptions: (1) the change in the free energy of a swollen system is the sum of the contributions due to mixing of polymer and solvent and elastic deformation of network strands; (2) the mixing term can be described by the mean-field Flory–Huggins theory; (3) the elastic deformation of the network is described by the affine-deformation model.

Using these three assumptions, one can obtain the relationship between network structure and equilibrium saturation swelling. The method is commonly used to determine network parameters such as crosslink density.

*In this chapter, the basic assumptions will be critically reviewed in view of recent work in rubber-like elasticity (Flory, P. J. and Erman, B., Macromolecules, 1982, **15**, 800), scaling theories and experimental results on network expansion during swelling, non-saturation swelling and swelling of networks by homopolymers.*

INTRODUCTION

The swelling of crosslinked polymer networks in the presence of a solvent is a phenomenon of great practical and theoretical interest. Its applications range from physiological processes to important mass transfer, membrane separation, extraction and biomedical applications. In the study of polymers, the swelling of a network has been used as a standard method for the determination of crosslink density.[1,2] Swelling has also been suggested as a method for the measurement of the polymer–solvent interaction parameter χ.[3] From the theoretical aspect, the study of swelling affords an important test of molecular theories of rubber elasticity in deformations not accessible by other means, and of fundamental concepts in polymer solution thermodynamics.[4-9] In this paper we review the theory used to analyze swelling behavior of polymers, outline the assumptions made in deriving it, and point out some of the problems associated with them.

The physical principles responsible for the swelling of networks are readily recognized.[4,7] When a dry polymer gel is placed in contact with a solvent (small molecular weight solvent, oligomer or polymer chain[9-12]), the tendency of the segments comprising the network to increase the system entropy by dispersion throughout the solvent is counterbalanced by the decrease in entropy as result of network deformation. The degree of swelling at any given time is defined as the network volume relative to its dry volume:

$$Q = V/V_d \tag{1}$$

A closely related measure of swelling is the volume fraction of polymer in the swollen network.

$$v_2 = V_2/V = V_d/V \tag{2}$$

where subscripts 1 and 2 refer to the solvent and crosslinked polymer, respectively. For isotropic swelling the linear deformation (or extension ratio) is given by

$$\lambda^3 = v_{2,0}/v_2 = V/V_0 = v_{2,0} + n_1 \tilde{v}_1/V_0 \tag{3}$$

where λ is the linear deformation of the network relative to its dimensions upon formation, the subscript 0 refers to the state at network formation, n is the number of molecules and $\tilde{v}$ the molar volume. If additional deformation is superimposed on that due to swelling, eqn (3) has to be modified to account for the different λ-values in the directions normal and parallel to the superimposed strain.

In general, the study of swelling will include any process in which a crosslinked polymer network is in contact with a solvent. A large number of swelling-related experiments have been carried out over the years. The most common ones may be classified as follows:

(1) *Saturation swelling.*[6,13-15] This is the most widely studied behavior of polymer networks. The network is placed in contact with a large excess of solvent and allowed to reach thermodynamic equilibrium. The volume fraction of solvent and polymer in the swollen gel at saturation, v_{1s} and v_{2s}, respectively, are related to the network structure, type of polymer and type of solvent.

(2) *Equilibrium swelling.*[16,17] The network is exposed to limited amounts of solvent such that v_1 is smaller than v_{1s} (or $Q < Q_s$). After equilibrium has been reached, the relative magnitude of the different thermodynamic quantities is obtained as function of deformation, network structure and polymer–solvent pair.

(3) *Deformation of swollen networks.* Networks containing a constant amount of solvent are subjected to different types (uniaxial, biaxial, shear, torsion) and magnitudes of deformation. The stress–strain relationship of the network is studied as a function of structure and degree of swelling.

(4) *Osmotic deswelling.*[18,19] Swollen networks are brought into contact with a solution of high molecular weight polymer. The latter cannot penetrate the network due to its size (or semipermeable membrane in some versions[19]), resulting in an osmotically driven depletion of solvent in the gel. The amount of swelling may be modified by variation of polymer concentration in the external solution.

(5) *Dynamics of swelling.*[20] The time evolution of swelling is studied in these experiments and related to polymer and solvent characteristics.

(6) *Scattering and resonance.* Light scattering,[21,22] neutron scattering[23] and NMR[24–26] of swollen and deformed networks (with labeled chains, crosslinks or solvent in some cases) enable study of molecular-scale deformations in the networks.

(7) *Critical phase transitions.*[20,27] Recently, the existence of network 'collapse' as result of critical phase transitions has been observed for the case of electrolyte solutions of ionic polymers.

Only a fraction of the large number of experiments described above will be discussed here, with emphasis placed on the equilibrium behavior of swollen networks. Initially, the commonly used theories will be reviewed, followed by a discussion of some pertinent experimental results.

GENERAL FORMULATION OF SWELLING THEORY

The amount of change in the Gibbs function of a polymer network as a result of swelling depends on the change due to mixing of polymer and solvent, and due to elastic network deformation:

$$\Delta G = f(\Delta G_{mix}, \Delta G_{el}) \tag{4}$$

In order to evaluate eqn (4), we have to specify the two contributions to the Gibbs function and assume a form for the function relating the two. Additional assumptions will involve the ideality of network structure and the volume additivity of solution components (i.e. $\Delta V_{mix} = 0$). Traditionally, the following three assumptions have been used.

(1) The contributions of mixing and elasticity to the overall change in the Gibbs function are additive. Thus

$$\Delta G = \Delta G_{mix} + \Delta G_{el} \qquad (5)$$

This is the so called Flory–Rehner assumption.[28] It has been widely used and only recently some questions regarding its validity have been raised.[29,30] In terms of a_1, the chemical activity of the solvent in the system

$$\ln a_1 = \ln a_{1,mix} + \ln a_{1,el} = [(\partial \Delta G_{mix}/\partial n_1) + (\partial \Delta G_{el}/\partial n_1)]/kT \qquad (6)$$

(2) The mixing term is described by the Flory–Huggins[31] lattice model for polymer solutions. Thus

$$\ln a_{1,mix} = \ln v_1 + (1 - 1/r)v_2 + \chi v_2^2 \qquad (7)$$

The value of r, the ratio of number of segments in a polymer molecule to that in a solvent molecule, is infinitely large in the case of a gel (i.e. $1/r = 0$). In addition, it is assumed that crosslinks do not contribute to the mixing term. This may not hold in the case of highly crosslinked systems or systems in which the crosslinks are chemically very different from the polymer chains. In the original Flory–Huggins model the interaction parameter χ was a constant for a given polymer–solvent pair, accounting for the enthalpic contribution to the mixing process. As result of experimental evidence, it has been modified to include additional entropic effects, free volume contribution and to depend on concentration. It may be treated as an empirical parameter or computed from more advanced solution theories.[32,33] It has been shown recently that χ may be evaluated *a priori* by solution-of-groups theories.[34]

(3) The last major assumption involves the changes in Gibbs function as result of the elastic deformation of the network upon swelling. The choice of the specific constitutive equation will determine the success of the treatment of swelling. In all these treatments of rubber elasticity, it is commonly assumed that deformation-related enthalpic effects may be ignored and that the network is devoid of defects such as dangling chains, internal loops and local crosslink density inhomogeneities. The presence of such defects will be of importance in any study of rubber elasticity, but more so in the case of swelling because elastically ineffective parts of the network will not contribute to its elastic behavior, yet will contribute to the mixing term in eqn (7).

THEORY OF ELASTICITY AND SATURATION SWELLING

The classical theory of swelling,[4] which has been used almost universally to obtain network structure from saturation swelling experiments, is based on the affine network model.[35] According to this model, the junctions (crosslinks) are firmly embedded in the polymer matrix, their relative positions are determined by the macroscopic deformation, and the strands between junctions possess no physical properties allowing them to cut across each other when required to do so as result of the imposed deformation. The change in Gibbs function as result of deformation for such a network is[36]

$$\Delta G_{el} = \tfrac{1}{2}\nu kT\sum(\lambda_i^2 - 1) - \mu kT\sum\ln\lambda_i \tag{8}$$

where ν and μ are, respectively, the number of elastically effective strands and junctions in the network, and λ_i are the extension ratios in the directions of the three principal axes. For a perfect network, ξ, the network cycle rank, is related to ν, μ and ϕ (the average junction functionality, i.e. number of elastically effective strands emanating from a junction) by

$$\xi = \nu - \mu \tag{9}$$

$$\mu/\xi = 2/(\phi - 2) \tag{10}$$

The contribution of elasticity to the chemical activity of the solvent in the swollen gel is given by

$$\ln a_{1,el} = (\partial\Delta G_{el}/\partial n_1)/kT$$
$$= (\partial\Delta G_{el}/\partial\lambda)(\partial\lambda/\partial n_1)/kT \tag{11}$$

Using the definition in eqn (3) we can easily obtain

$$\partial\lambda/\partial n_1 = \tilde{v}_1/(3V_0\lambda^2) \tag{12}$$

For isotropic swelling, λ is equal in all directions and eqns (8) and (12) may be used in eqn (11) to yield

$$\ln a_{1,el} = \tilde{v}_1(\nu - \mu\lambda^{-2})/\lambda V_0 \tag{13}$$

For saturation equilibrium $\Delta G = 0$, and with the aid of eqn (10) we can show that

$$\frac{\xi}{V_0} = -\frac{\ln v_{1s} + v_{2s} + \chi v_{2s}^2}{\tilde{v}_1(v_{2s}/v_{2,0})^{1/3}[1 + (2/\phi - 2)(1 - (v_{2s}/v_{2,0})^{2/3})]} \tag{14}$$

If the network functionality is known, the concentration of cycles (which in turn may be easily related to concentration of junctions or strands or molecular weight between junctions, with the help of eqns (9) and (10) is readily obtained from measurement of equilibrium swelling, given eqn (8) is the correct expression for elastic deformation. Yet in most practical situations a large degree of uncertainty is associated with the value of average functionality, and even in the case of 'model' networks incomplete reaction and non-idealities in structure will render the exact determination of functionality difficult. A somewhat different model for rubber elasticity is the one proposed by James and Guth.[37] In their model the junctions are not fixed but rather fluctuate around a mean position. The resulting expression for the change in Gibbs energy is

$$\Delta G_{el} = \tfrac{1}{2}\xi k T \sum (\lambda_i^2 - 1) \tag{15}$$

and for saturation swelling

$$\xi/V_0 = -[\ln v_{1s} + v_{2s} + \chi v_{2s}^2]/\{\tilde{v}_1(v_{2s}/v_{2,0})^{1/3}\} \tag{16}$$

In this case, knowledge of the functionality of network junctions is not required for the determination of the concentration of network cycles from saturation swelling measurements.

The expressions obtained for the dependence of swelling on network structure are quite simple and afford easily accessible means for the determination of network parameters. Unfortunately, it has long been known that the two models used in obtaining these relations, the affine and phantom models (Wall–Flory[35] and James–Guth,[37] respectively), are unable to account correctly for any type of deformation,[36,38,39] and thus cannot be expected to do so for swelling. This well-documented failure is due to the effect of network topology (chain entanglements) ignored by both models as a result of the 'phantom' nature of the chains. Unfortunately, despite their obvious inapplicability, the use of eqn (14) and (16) is still widespread.

To account for the real nature of polymer chains and to obtain better description of network behavior under deformation, a large number of models have been proposed over the last few years. These have been recently reviewed and their predictions for several types of deformations compared with experimental data.[38–40] We will mention only two of them: the widely used Mooney–Rivlin[41] phenomenological equation and the suppressed-junction-fluctuation model of Flory and Erman.[42] The expressions for network structure as function of degree of saturation swelling for several molecular theories are listed in Table 1.

The Mooney–Rivlin equation is the first term from a more general expression derived from continuum mechanics principles.[48] For a long

TABLE 1

Network Cycle Concentration from Saturation Swelling Values

Model	Y^a	Parameters
Phantom[37]	1	—
Affine[35]	$1 + 2[1 - \lambda^{-2}]/(\phi - 2)$	—
Flory–Erman[42]	$1 + 2K/(\phi - 2)$	κ, ζ
	$K = (B/1 + B)\partial B/\partial \lambda^2 + (Bg/1 + Bg)\partial Bg/\partial \lambda^2$	
	$B = (\lambda - 1)(1 + \lambda - \zeta\lambda^2)/(1 + g)^2$	
	$g = \lambda^2[\zeta(\lambda - 1) + 1/\kappa]$	
BDEW[43]	$(N_s/\xi)(1 + 2\eta + \eta^2\lambda^2)/(1 + \eta\lambda^2)^2 + 2/(\phi - 2)$	N_s, η
Edwards[44]	$\dfrac{1}{3}\left\{\dfrac{2\lambda(1 - \alpha) - \alpha}{2\lambda(1 - \alpha\lambda)} + \dfrac{\alpha\lambda(1 - \alpha)}{2(1 - \alpha\lambda)^2} + \dfrac{\beta}{2} + 2\right\}\dfrac{2}{\phi - 2}$	α, β
Graessley[45]	$1 + N(\phi/\phi - 2)[1/\lambda + (\lambda - 1)/3\lambda p]$	N, p
Marrucci[46]	$\{[2/(\phi - 2)]\lambda^{-2} + A(1 - r\lambda^{-4})\}/3$	A, r
Gaylord[47]	Same as for Marrucci	

^{a} Y is defined by $\xi/V_0 = -[\ln v_{1s} + v_{2s} + \chi v_{2s}^2]/\{[\tilde{v}_1(v_{2s}/v_{2,0})^{1/3}]Y\}$; $\lambda = (v_{2,0}/v_{2s})^{1/3}$.

time it has been accepted as a phenomenological equation with moderate success in certain types of deformations. Its simplicity should account for its widespread use in data analysis. Recently, it has been shown[49] that the equation is a special case of a more general molecular theory based on the deformed tube concept. The two model constants are related at saturation swelling equilibrium by

$$\frac{2C_1}{V_0} = -2(2C_2/V_0)(v_{2,0}/v_{2s})^{2/3} - \frac{[\ln v_{1s} + v_{2s} + \chi v_{2s}^2]}{\tilde{v}_1(v_{2s}/v_{2,0})^{1/3}} \quad (17)$$

When $2C_2$ is set equal to zero, eqn (16) is recovered. For this reason, this coefficient has been associated with the effect of network topology and a considerable amount of work has been performed to elucidate its nature (cf. Ref. 40 for a discussion of this point).

According to the Flory–Erman molecular theory,[42] the effect of the real properties of chains (as opposed to phantom chains) is to limit the amount of fluctuations performed by the junctions. These constraints are structure dependent (expressed by the structural parameter κ) and deformation dependent, with constraints becoming less severe as the network is deformed. The second parameter ζ is related to the deformation of the domain of constraints. If no constraints are present ($\kappa = 0$), or in the limit of very high deformations when all constraints have been removed, eqn (15) is recovered. For a highly entangled system ($\kappa = \infty$) at the limit of zero deformation, the behavior of the network is according to eqn (8). These two equations define the upper and lower bounds for network response. For saturation swelling, this model predicts

$$\frac{\xi}{V_0} = -\frac{\ln v_{1s} + v_{2s} + \chi v_{2s}^2}{\tilde{v}_1 (v_{2s}/v_{2,0})^{1/3} [1 + (2/\phi - 2) K(v_{2s}, \kappa, \zeta)]} \tag{18}$$

where K is a complicated function of deformation specified in Table 1. For $\kappa = 0$, $K = 0$ and eqn (16) is obtained. Since both κ and ζ depend on network structure, and with κ strongly dependent on crosslink density[50,51] (or molecular weight between crosslinks), it is impossible to obtain network structure based on swelling data and eqn (18) only.

SATURATION SWELLING EXPERIMENTS

To test the validity of any of the relations presented so far (eqns (14), (16), (18) or Table 1), we need to compare the predictions of these equations with v_{2s} values for networks of known structure. Due to the uncertainties still involving model networks,[38,52] an alternative approach is suggested. A given network may be swollen in different solvents of known $\chi(v_2)$. The swelling values are then used in the equation for the tested model which, if correct, should yield the same cycle concentration value for all solvents used. For a phantom or affine model, the process is straightforward. A slightly more complicated process is required in the case of the Flory–Erman model, since the relative results for the different solvents will depend on the choice of κ values.

Networks were obtained by end-linking in bulk, vinyl terminated PDMS chains (M_n ranging from 6000 to 20 000) with tri- and tetrafunctional silane crosslinkers. The networks were cast into 2-mm-thick films from which, upon curing, several pieces 1×1 cm in size were cut. Soluble materials were removed by extraction in cyclohexane for several weeks. After gradual deswelling by means of increasing amounts of ethanol and extensive drying, the specimens were weighed and measured by means of a travelling microscope. The networks were subsequently swollen in cyclohexane, toluene and n-octane, for all of which reliable χ-data as function of concentration are available. Saturation swelling was achieved by gradually increasing over a period of 10 days the amount of solvent in the solvent/non-solvent mixture in which the samples were immersed. After 10 additional days in the pure solvent, the degree of swelling was obtained by optical size determination and by gravimetric methods. Agreement between the two (assuming volume additivity) was in most cases about 5%. The samples were now gradually deswollen and weighed again, after drying, to check for possible loss of network material as a result of network damage and degradation. Five samples from each network formed were used, some of them swollen in all three solvents and others in only one of them, to assure swelling history had not affected the results. Variations between the samples from the same network were also monitored to detect any possible local inhomogeneities in network structure. None were found, which indicated that if such were present, as has been recently suggested,[22] they were averaged out due to the size of our samples.

The results were analyzed in terms of only three of the models listed in Table 1: the phantom, affine and Flory–Erman models. For the first model the analysis was straightforward; polymer saturation volume fraction values for the three solvents were used to calculate ξ/V_0 from eqn (16). For the affine model, the value of ϕ was varied within the bounds $0.75\phi_x < \phi < \phi_x$, where ϕ_x is the functionality of the crosslinker, such that the value of ϕ minimizing the difference between the three solvents' swelling results was used in eqn (14). In both cases the discrepancies between the results from the different solvents were between 10% and 120%. In the case of the phantom network, the agreement improved as the network became more lightly crosslinked, whereas an opposite trend was observed in the case of the affine model. Overall, the predictions of the affine model were considerably worse, in terms of the discrepancy between the solvents, than those of the phantom network. These may be explained by the fact that all networks were highly swollen ($v_{2s} < 0.45$), which corresponds to relatively high deformations. The immediate conclusion from these experiments is that none of these models may be used for the determination of network parameters from swelling measurements: only relative crosslink density on similar networks may be obtained.

For the analysis of the swelling data in terms of the Flory–Erman model, a more complicated scheme was required since the prediction of the model depends on κ, ζ and ϕ (cf. Table 1). In most instances the second parameter, ζ, is of negligible importance, but this is not necessarily the case for swelling.[48] We initially set the value of the latter to zero and optimized the values of κ and ϕ, under the condition that $0.75\phi_x < \phi < \phi_x$ and the value of κ increases with the increase in estimated amount of interpenetration,[50,51] such that the lowest difference between the predicted values from the three solvents for the same network is obtained. Subsequently, the value of ζ was allowed to change to improve the agreement. It was possible to obtain good agreement between the solvents (always better than 7%) with $7 < \kappa < 15$

and $\zeta < 0.1$. In most cases, with the exception of the higher crosslink densities, the value of the latter was of little significance. The values obtained for the parameters are in good agreement with values reported for similar networks.[50] No attempt was made to further test the model by, for example, using the parameters obtained to predict stress–strain behavior.[30] Attempts to analyze the data in terms of other models were found quite cumbersome and inconsistent in their results.

To summarize, the following conclusions have been reached from this study:

(1) Computation of network structure from swelling measurements with the aid of either the affine or phantom models is extremely unreliable. Only relative information, under some conditions, may be obtained.

(2) Consistent results are obtained with the Flory–Erman model but the procedure is quite cumbersome and not recommended as a 'routine' method for structure determination. Independent corroboration of network structure is highly desired to support the conclusions of this work.

(3) Additional models tested were found inconvenient and inconsistent for use in the determination of network structure by saturation swelling.

(4) The assumption of additivity of Gibbs energies and the contribution of chain–chain interactions were not tested by the described procedure, and thus their correctness or importance was not determined.

EQUILIBRIUM SWELLING

Only a limited amount of equilibrium swelling experiments have been reported.[16,17,29] These experiments are of major importance because they afford a new critical evaluation of molecular theories of rubber elasticity on the one hand, and an independent test of the additivity assumption discussed above, on the other. Unfortunately, the experiments are very sensitive, hard to perform and highly susceptible to experimental errors. The most complete set of these measurements was carried out by Eichinger and co-workers (Refs 17, 29 and references therein), without conclusively determining the correctness of the additivity assumption. Yet, based on the results obtained so far, there are strong indications that the assumption is in fact incorrect.

In these experiments, the solvent activity is measured in a swollen network $(v_1 < v_{1s})$ and in a solution of the same polymer (uncrosslinked) in the same volume fraction. If we denote by a_1^u and a_1^c the solvent activity in the uncrosslinked and crosslinked systems, respectively, we can readily obtain from eqns (6) and (7) for identical v_1

$$\ln(a_1^c/a_1^u) = \ln a_{1,el} \tag{19}$$

under the assumption that crosslinks do not contribute to the solvent activity.[29] We define the dilation modulus as

$$S = \lambda \ln a_{1,el} \tag{20}$$

and the reduced dilation modulus as the dilation modulus reduced by its value S_0 at $\lambda = 1$. For the phantom network (eqn (15)) we can easily show that $S/S_0 = 1$. The experimental data for different polymers and different solvents always show a maximum in the reduced dilation modulus when plotted against λ^2. Analysis of different molecular theories has been recently carried out.[53] This study has indicated that only four models out of the nine tested are capable of showing these experimentally observed maxima. Furthermore, even these four models require physically unacceptable parameter values to demonstrate these maxima. Whether the fault is in the inadequacy of all available molecular models or in the Flory–Rehner[28] additivity assumption is yet to be determined.

REFERENCES

1. Collins, E. A., Bares, J. and Billmeyer, F. W., *Experiments in Polymer Science*, John Wiley & Sons, New York, 1973.
2. Rabek, J. F., *Experimental Methods in Polymer Chemistry*, John Wiley & Sons, New York, 1980.
3. Orwoll, R. A., *Rubber Chem. Technol.*, 1976, **50**, 451.
4. Flory, P. J., *Principles of Polymer Chemistry*, Cornell University Press, Ithaca, New York, 1967.
5. de Gennes, P. G., *Scaling Concepts in Polymer Physics*, Cornell University Press, Ithaca, New York, 1979.
6. Flory, P. J. and Tatara, Y., *J. Polym. Sci., Polym. Phys. Ed.*, 1975, **13**, 683.
7. Dusek, K. and Prins, W., *Adv. Polym. Sci.*, 1969, **6**, 1.
8. Daoud, M., Bouchaud, E. and Jannink, G., *Macromolecules*, 1986, **19**, 1955.
9. Brochard, F., *J. Physique*, 1981, **42**, 505.
10. Gent, A. N. and Tobias, R. H., *J. Polym. Sci., Polym. Phys. Ed.*, 1982, **20**, 2317.
11. Garrido, L. and Mark, J. E., *J. Polym. Sci., Polym. Phys. Ed.*, 1985, **23**, 1933.
12. de Gennes, P. G., *Macromolecules*, 1986, **19**, 1245.
13. Lloyd, W. G. and Alfrey, T. Jr, *J. Polym. Sci.*, 1962, **62**, 301.
14. Mark, J. E. and Sullivan, J. L., *J. Chem. Phys.*, 1977, **66**, 1006.
15. Queslel, J. P. and Mark, J. E., *Eur. Polym. J.*, 1986, **22**, 273.
16. Gee, G., Herbert, J. B. M. and Roberts, R. C., *Polymer*, 1965, **6**, 541.

17. Yen, L. Y. and Eichinger, B. E., *J. Polym. Sci., Polym. Phys. Ed.*, 1978, **16**, 121.
18. Bastide, J., Candau, S. and Leibler, L., *Macromolecules*, 1981, **14**, 719.
19. Zrinyi, M. and Horkay, F., *J. Polym. Sci., Polym. Phys. Ed.*, 1982, **20**, 815.
20. Tanaka, T., *Phys. Rev. Lett.*, 1978, **40**, 820.
21. Munch, J. P., Candau, S., Herz, J. and Hild, G., *J. Physique*, 1977, **38**, 971.
22. Stein, R. S., Soni, V. K., Yang, H. and Erman, B., The scattering of light by swollen networks. *This volume*, p. 383.
23. Beltzung, M., Picot, C. and Herz, J., *Macromolecules*, 1984, **17**, 663, and references therein.
24. Deloche, B. and Samulski, E. T., *Macromolecules*, 1981, **14**, 575.
25. Gronski, W., Stadler, R. and Jacobi, M. M., *Macromolecules*, 1984, **17**, 741.
26. Cohen-Addad, J. P., Domard, M., Lorentz, G. and Herz, J., *J. Physique*, 1984, **45**, 575.
27. Ilavsky, M., *Polymer*, 1981, **22**, 1687.
28. Flory, P. J. and Rehner, J. Jr, *J. Chem. Phys.*, 1943, **11**, 521.
29. Brotzman, R. W. and Eichinger, B. E., *Macromolecules*, 1981, **14**, 1445; 1982, **15**, 531.
30. McKenna, G. B. and Hinkley, J. A., *Polymer*, 1986, **27**, 1368.
31. Flory, P. J., *J. Chem. Phys.*, 1942, **10**, 51; Huggins, M. H., *J. Phys. Chem.*, 1942, **46**, 151.
32. Flory, P. J., *Discuss. Faraday Soc.*, 1970, **49**, 7.
33. Lacombe, R. H. and Sanchez, I. C., *J. Phys. Chem.*, 1976, **80**, 2568.
34. Gottlieb, M. and Herskowitz, M., *Macromolecules*, 1981, **14**, 1468.
35. Wall, F. T. and Flory, P. J., *J. Chem. Phys.*, 1951, **19**, 1435.
36. Flory, P. J., *J. Chem. Phys.*, 1977, **66**, 5720.
37. James, H. M. and Guth, E., *J. Chem. Phys.*, 1947, **15**, 669.
38. Eichinger, B. E., *Annu. Rev. Phys. Chem.*, 1983, **34**, 359.
39. Mark, J. E., *Adv. Polym. Sci.*, 1982, **44**, 1.
40. Gottlieb, M. and Gaylord, R. J., *Macromolecules*, 1987, **20**, 130.
41. Mooney, M. J., *Appl. Phys.*, 1940, **11**, 582; Rivlin, R. S., *Philos. Trans. Royal Soc. London A.*, 1948, **A241**, 379.
42. Flory, P. J. and Erman, B., *Macromolecules*, 1982, **15**, 800.
43. Ball, R. C., Doi, M., Edwards, S. F. and Warner, M., *Polymer*, 1981, **22**, 1010.
44. Edwards, S. F., *Brit. Polym. J.*, 1977, **9**, 140.
45. Graessley, W. W., *Adv. Polym. Sci.*, 1982, **46**, 67.
46. Marrucci, G., *Macromolecules*, 1981, **14**, 434.
47. Gaylord, R. J., *Polym. Bull.*, 1982, **8**, 325.
48. Treloar, L. R. G., *The Physics of Rubber Elasticity*, Clarendon Press, Oxford, 1975.
49. Gaylord, R. J. and Douglas, J., *Bull. Am. Phys. Soc.*, 1987, **32**, 496.
50. Erman, B. and Flory, P. J., *Macromolecules*, 1982, **15**, 806.
51. Brotzman, R. W. and Mark, J. E., *Macromolecules*, 1986, **19**, 667.
52. Fischer, A. and Gottlieb, M., *Proc. Networks 86*, Elsinore, Denmark, 1986.
53. Gottlieb, M. and Gaylord, R. J., *Macromolecules*, 1984, **17**, 2024.

28

DIFFERENTIAL SWELLING OF ELASTOMERS

B. E. Eichinger and N. A. Neuburger

*Department of Chemistry, University of Washington,
Seattle, Washington 98195, USA*

ABSTRACT

Differential swelling measurements give directly the difference between the free energies of mixing a solvent with an elastomer and with the corresponding linear chain polymer. In these experiments, the difference between the amounts of solvent vapor that are absorbed by crosslinked and uncrosslinked, but otherwise identical, samples of a polymer is measured at various solvent activities. According to the Flory–Rehner theory, the mixing and elastic contributions to the swelling free energy are separable and additive, and the measurements should therefore give the elastic contribution to the swelling free energy. Differential sorption experiments provide sensitive tests both of the theory of elasticity and of the Flory–Rehner additivity assumption. We review the theory and the experiments, and present new results that bear on the validity of the crucial assumption in swelling theory.

INTRODUCTION

The analysis of elastomer swelling is based upon the theory of Flory and Rehner.[1-3] Kuhn, Pasternak and Kuhn,[4] and Hermans[5,6] contributed to the development of the theory as well, but the basic ideas derive from the Flory and Rehner papers. The essential assumption made in this theory is that the mixing and elastic free energies are separable and additive, which allows one to write

$$\Delta G = \Delta G_M + \Delta G_{el} \tag{1}$$

Here ΔG is the change in Gibbs free energy when the elastomer absorbs n_1 moles of solvent, ΔG_M is the mixing free energy to form an equivalent solution of linear, high molecular weight polymer, and ΔG_{el} is the elastic free energy change for dilation of the network. (In the original papers, the theory was primarily couched in terms of the entropy. That restriction is unnecessary, and it serves our purposes to consider the more general free energy changes that occur.)

Flory[3] made use of the following thermodynamic cycle to analyze the swelling behavior:

$$
\begin{array}{ccc}
E(0) + n_1 S & \xrightarrow{\Delta G} & E(\phi_1) \\
{\scriptstyle -\Delta G_x(0)} \Big\downarrow & & \Big\uparrow {\scriptstyle \Delta G_x(\phi_1)} \\
L(0) + n_1 S & \xrightarrow[\Delta G_M]{} & L(\phi_1)
\end{array}
\tag{2}
$$

It is clear from this cycle that

$$
\Delta G = \Delta G_M + [\Delta G_x(\phi_1) - \Delta G_x(0)]
\tag{3}
$$

and hence that

$$
\Delta G_{el} = \Delta G_x(\phi_1) - \Delta G_x(0)
\tag{4}
$$

In this rendition of the cycle, the dry elastomer, E(0), imbibes n_1 moles of solvent, S, and swells to a solvent volume fraction ϕ_1, with the free energy change ΔG. On the alternative path, the elastomer is uncrosslinked in the bulk with the free energy change $-\Delta G_x(0)$ to form linear chains, L(0), which then are dissolved in the solvent to form the solution $L(\phi_1)$. This process entails the mixing free energy ΔG_M. Finally, the linear chains in solution undergo an irreversible distortion to recover the same configurations as they have in the swollen elastomer, and then the crosslinks are formed. The free energy change accompanying the bond formation process is presumed to be the same in the bulk as in solution. Thus, only the distortional part of the free energy change for the last leg of the cycle, $\Delta G_x(\phi_1)$, remains as ΔG_{el} in eqn (4), when $\Delta G_x(0)$ is subtracted from $\Delta G_x(\phi_1)$ as required by the cycle. (If the network had been formed in solution, $\Delta G_x(\phi_1)$ would be reversible at the one particular solvent concentration that prevailed at the time of cure, but would be irreversible at all other concentrations, including $\phi_1 = 0$.)

In the original papers,[1-6] the elastic component ΔG_{el} was computed from the configurational entropy change that the individual chains experience when they are deformed so as to adopt the dimensions and positions that they are assumed to have in the elastomer. This was done in what has come to be known as the phantom network approximation. In somewhat more general terms, the ΔG_{el} component has been taken to arise from the statistical weight changes that accompany the deformation and constraint of the linear chains as they are forced into the configurations required to position them for bond formation.

The utility of swelling phenomena lies in the ability one has to characterize elastomers by relatively simple methods. According to the original theory, sketched above but discussed in detail below, swelling ratios measured in the presence of pure solvent give moduli of elasticity. This information is of considerable practical value. On the other hand, swelling can be exploited to explore the elastic equation of state. With somewhat more effort than that involved in the measurement of equilibrium swelling ratios, the solvent activity can be varied so as to swell an elastomer to different extents. This enables one to probe the elastic free energy in domains of extension that are not accessible in ordinary stress–strain measurements. In view of these facts, swelling deserves careful consideration, both as the basis for characterization techniques and as a means to test elasticity theory.

In what follows we will review the theory in more detail, describe the experimental apparatus, and discuss the method for data analysis. The published experimental results will then be reviewed, and, finally, our most recent measurements will be presented.

REVIEW

Theory

Differentiation of eqn (1) with respect to the number of moles of solvent, n_1, at constant temperature T and pressure p gives the chemical potential μ_1 of the solvent as

$$\mu_1 - \mu_1^\circ = (\mu_1 - \mu_1^\circ)_M + (\partial \Delta G_{el}/\partial n_1)_{T,p} \tag{5}$$

In terms of solvent activities $a_{1,c}$ above the elastomer and $a_{1,u}$ in the solution of linear chains, this becomes

$$RT \ln(a_{1,c}) = RT \ln(a_{1,u}) + (\partial \Delta G_{el}/\partial n_1)_{T,p} \tag{6}$$

For a linear polymer in solution at concentrations exceeding approximately 10% by volume, Flory–Huggins solution theory[7] gives the mixing term as

$$(\mu_1 - \mu_1^\circ)_M = RT \ln(a_{1,u}) = RT[\ln(\phi_1) + (1 - 1/r)\phi_2 + \chi\phi_2^2] \tag{7}$$

where the ratio of the molar volumes of polymer to solvent is r, and $\chi = \chi(\phi_2)$ is generally a function of the volume fraction ϕ_2 of polymer. For sufficiently high molecular weight linear chains, $r = \infty$ is a satisfactory approximation for analysis of swelling on the basis of the thermodynamic cycle. For the elastomer, $r = \infty$ is required as the $1/r$ term in eqn (7) derives from the translational degrees of freedom of the molecules in solution. The elastomer has only three translational modes, and these are of no consequence to solution thermodynamics.

For the sake of simplicity in the present discussion, and without any assertion of accuracy, we will assume that the elastic free energy ΔG_{el} is given by the James–Guth equation[8]

$$\Delta G_{el} = C(\lambda_1^2 + \lambda_2^2 + \lambda_3^2 - 3) \tag{8}$$

where the λ_i are the extension ratios along three mutually perpendicular Cartesian axes and C is the modulus. Other, more realistic and also more complicated, alternatives to the elastic free energy have been reviewed elsewhere.[9,10] (The James–Guth theory further predicts that C is directly proportional to the absolute temperature. This fact will be of use presently. The temperature dependence is susceptible to experimental test with swelling measurements.) For the swelling of an isotropic sample, $\lambda_1 = \lambda_2 = \lambda_3 = \lambda$, and eqn (8) reduces to

$$\Delta G_{el} = 3C(\lambda^2 - 1) \tag{9}$$

Since isotropy is assumed, we have $\lambda^3 = (V_0 + n_1 V_1)/V_0 = 1/\phi_2$, where V_0 is the volume of the dry elastomer and V_1 is the *apparent* molar volume of the solvent. Use of these relations gives

$$(\partial \Delta G_{el}/\partial n_1)_{T,p} = (V_1/3\lambda^2 V_0)(\partial \Delta G_{el}/\partial \lambda)_{T,p} \tag{10}$$

and with use of the James–Guth expression (eqn (9)), this becomes

$$(\partial \Delta G_{el}/\partial n_1)_{T,p} = 2V_1 C/\lambda V_0 \tag{11}$$

At swelling equilibrium, $a_1 = 1$, since the elastomer is in contact with pure solvent. Combining eqns (6), (7) and (11), and solving for χ, gives the locus of swelling equilibrium points as

$$\chi = -[\ln(\phi_1) + \phi_2 + A\phi_2^{1/3}]/\phi_2^2 \tag{12}$$

where $A = 2V_1 C/RTV_0$ and r in eqn (7) has been set to infinity. Illustrative calculations based on this equation with $A = 1/15$ are depicted as a χ vs. ϕ_2 plot in Fig. 1. (This value for A is considerably larger than one usually encounters with soft elastomers. It has been chosen to make clear the

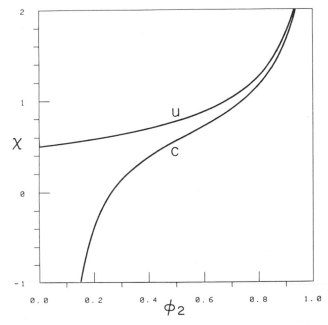

Fig. 1. Solubility limit for saturated solutions of linear chains of infinite molecular weight calculated from eqn (12) with $A = 0$ (upper curve, u) and swelling equilibrium curve for an elastomer with $A = 1/15$ (lower curve, c). Along each curve, the activity of the solvent is unity. The regions above and to the left of each curve are inaccessible at equilibrium.

activity differences to be discussed later.) Shown for comparison on the same plot is χ calculated from the Flory–Huggins equation, again with $r = \infty$. Any point, lying on, below or to the right of either curve represents an accessible state, whereas points above and to the left are inaccessible at equilibrium. If we had used another theoretical expression for ΔG_{el}, the swelling equilibrium curve would have been qualitatively similar to that shown in the figure at the scale used here.

It is of considerable importance to note that there are many assumptions lurking behind any calculation of the distortional free energy. Perhaps most important is the fact that, no matter what the cure conditions, there is an irreversible process in the Flory cycle. In fundamental thermodynamic terms, the entropy change that accompanies an irreversible process depends upon the detailed path. This is not a difficulty if the final state is perfectly specified. However, our current knowledge of elastic material is imprecise, and the choice of path is thereby crucial. This characteristic of irreversibility may have some bearing on the discrepancies that arose between the values that were calculated for the coefficient of the logarithmic term in the early theories.[1−6]

Experiments

According to the preceding theory, measurements of the difference $\Delta G - \Delta G_M$ should yield ΔG_{el} directly. An experiment to measure this difference, almost directly, was devised by Gee, Herbert and Roberts.[11] They determined the different amounts of benzene that were absorbed by samples of crosslinked and uncrosslinked natural rubber (NR) placed on the two pans of a balance. These $\Delta\phi$ vs. activity (a_1) data were then transformed into Δa_1 vs. ϕ, as described in detail below, so as to construct the elastic contribution to the solvent chemical potential. More precisely, substitution of eqn (10) into eqn (6) with rearrangement gives what Gottlieb[9] has called the dilation modulus, as

$$(RT\lambda/V_1)\ln(a_{1,c}/a_{1,u}) = (1/3\lambda V_0)(\partial\Delta G_{el}/\partial\lambda)_{T,p} \tag{13}$$

Gee *et al.*[11] found that neither James–Guth, Wall–Flory nor Mooney–Rivlin theory was able to account for their data. As shown in Fig. 2, the dilation modulus for the NR–benzene system exhibits a pronounced maximum at $\lambda^2 \simeq 1\cdot8$. The theories give dilation moduli that are either

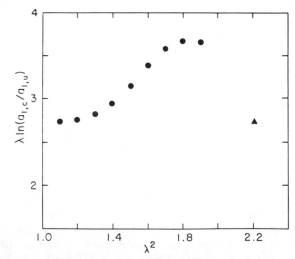

Fig. 2. Dilation modulus for natural rubber swollen with benzene as determined by Gee *et al.*[11] ●, vapor absorption measurements; ▲, swelling equilibrium point. The linear extension ratio is λ.

constant (James–Guth) or monotonically increasing (Wall–Flory and Mooney–Rivlin). Gee *et al.* were thus led to question whether the assumption of additivity is correct. As has been emphasized by Flory,[12] one experiment cannot settle this question, since it might be that additivity is correct while the theories of elasticity are incorrect. Relevant to this point, Flory and Erman [12-16] have developed the constrained junction theory to give a much better account of the observations, as shown by Flory,[12] Gottlieb[9] and Brotzman.[17] To assess the validity of the additivity assumption independent of the theory of elasticity requires an experimental protocol that can distinguish one contribution from the other.

We undertook a series of measurements[17-21] using the Gee *et al.* method, first to verify the general features of the swelling behavior that were observed, and then to investigate in more detail the dependence of the swelling upon network structure and solvent characteristics. Yen[19] found a pronounced maximum near $\lambda^2 \simeq 1.5$ for radiation-cured polydimethyl-siloxane (PDMS) swollen with benzene, which is similar to the NR–benzene result.[11] Results with peroxide-cured random poly(styrene-co-butadiene) (SBR) swollen in benzene and in *n*-heptane gave much less pronounced maxima; so much so that the data might have been fitted, nearly within experimental error, by the James–Guth equation. This curious variation in the general shape of the dilation modulus with system composition was puzzling at the time. Since then the Flory–Erman theory has been developed, and these differences can be acribed, at least qualitatively, to differences in the theoretical parameters for PDMS and SBR. More worrisome was the fact that the benzene and *n*-heptane data for the SBR sample did not coincide. This apparent solvent effect is beyond the realm of any theory of elasticity, and we tentatively agree with Gee *et al.* that the additivity assumption of Flory and Rehner is questionable.

Given these anomalies, we undertook a series of experiments on PDMS to explore the influence of network topology and solvent composition on the dilation modulus, in the hope that the behavior could be quantified. With the help of Professor J. E. Mark, who supplied PDMS samples, Brotzman investigated two radiation-cured networks [17,20] and three end-linked model networks.[21] Measurements on solution-cured and bulk-cured samples were found to be in qualitative agreement with the Flory–Erman theory,[17] but when the model network data[21] were analyzed together with the stress–strain data of Mark and co-workers[20,21] it became apparent that a single set of theoretical parameters would not suffice to fit all the results. But most importantly, when the bulk-cured sample was swollen with two different solvents, very different dilation moduli were obtained.[20]

In the latter experiments, a single sample of radiation-cured poly-dimethylsiloxane was swollen with benzene and then with cyclohexane. The dilation moduli measured in the two solvents differed from one another by more than 100% at the extreme.[20] This difference is far beyond any reasonable estimate of experimental uncertainty, and could only be ascribed to the breakdown of additivity.

There is one potential solution to the lack of additivity, as was first pointed out by Gee et al.[11] It might be that the Flory–Huggins parameters χ for the interaction of solvents with crosslinks are widely different from those with mid-chain segments. If so, there could be a contribution from the mixing term that does not cancel in the differences $\Delta G - \Delta G_M$, and that might be the source of the discrepancy. Calculations of the magnitude of the difference between such an interaction parameter for different solvents led us to conclude that this was an unlikely source of the anomaly.[20] To settle this issue definitively we have made networks with two different crosslinks, and these new measurements will be described later.

APPARATUS AND DATA ANALYSIS

The apparatus used to measure the differential swelling isotherms consists of a balance to record the difference in weight gained by the samples of crosslinked and uncrosslinked polymers placed on the two pans, a pressure gauge to measure the solvent vapor pressure and a good thermostatted bath. Our latest version of the apparatus is depicted in Fig. 3. We utilize two

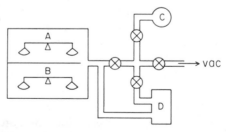

Fig. 3. Our present experimental apparatus consists of a stainless steel vacuum chamber containing two electronic balances, A and B. Balance A is loaded with a sample of the uncrosslinked polymer, and balance B has a crosslinked polymer on one pan and an uncrosslinked sample of the same polymer on the other pan. Integral sorption (balance A) is measured concurrently with the differential sorption (balance B). The solvent reservoir is labeled C and the capacitive diaphragm gauge is D.

electronic balances (Cahn Models RG and 2000), one of which is used for the differential measurements and the other for integral measurements of the solvent absorbed by a sample of uncrosslinked polymer. In this way we make simultaneous determinations of the elastic term and the χ parameter. The pressure gauge is of the capacitance diaphragm type (MKS Baratron). The entire apparatus, including valves, is submerged in a highly stable (± 1 mK) constant temperature bath (Tronac) equipped with a proportional heating element. The pressure gauge is contained in a water-tight submarine, and is maintained at a higher temperature than the bath to avoid condensation of vapor on cold spots.

The differential swelling experiments usually span the range $\sim 0.4 < \phi_2 < \sim 0.95$ for experimental reasons. In Fig. 4 we show a plot of activities calculated from the combination of eqns (6), (7) and (11), with $r = \infty$, $\chi = 0.5$, and with (i) $A = 0$ (linear chains) and (ii) $A = 1/15$ (network, see eqn (12) and Fig. 1). Even for such a highly crosslinked network there is relatively little difference between the two activity curves below $a_1 \simeq 0.6$. Measurements

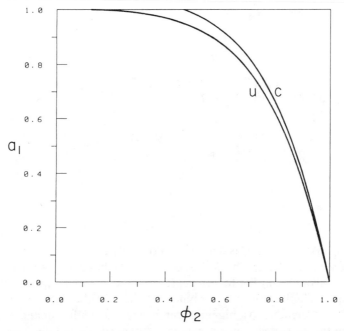

Fig. 4. Plots of solvent activities, a_1, calculated from eqns (6), (7) and (11) with $\chi = 0.5$. Curve u is for solutions of linear chains of infinite molecular weight, and curve c is for the elastomer with $A = 1/15$.

are limited on the side of large activities by the difficulty in measuring the partial pressure of the solvent in the vicinity of the equilibrium vapor pressure, and on the low side the mass measurements are subject to large uncertainties.

Figure 5 is a blow-up of Fig. 4, and will help us to describe the method of data analysis. Define $\Delta m_1 = m_{1,u} - m_{1,c}$: this is the directly measured difference in the mass of solvent absorbed by the two samples on the differential balance, which, respectively, weigh $m_{2,u}$ and $m_{2,c}$. These two samples are as close to identical masses as can be reasonably obtained. Let ρ_1 and ρ_2 be the densities of solvent and polymer, respectively (the densities of the lightly crosslinked samples that have been used in the experimental studies have been taken to be the same as those of the uncrosslinked samples). For the uncrosslinked sample on the second balance, the subscript u will be omitted. We have

$$\phi_1 = \frac{m_1/\rho_1}{m_1/\rho_1 + m_2/\rho_2} = \frac{1}{1 + (\rho_1/\rho_2)(m_2/m_1)} \tag{14}$$

and on defining $\Delta\phi = \phi_{1,u} - \phi_{1,c}$, we further have

$$\Delta\phi = \frac{1}{1 + (\rho_1/\rho_2)(m_{2,u}/m_{1,u})} - \frac{1}{1 + (\rho_1/\rho_2)(m_{2,c}/m_{1,c})} \tag{15}$$

Now, since $\phi_1 = \phi_{1,u}$, it follows that $m_2/m_1 = m_{2,u}/m_{1,u}$; these together with Δm_1 are all the relations required to construct the $\Delta\phi$ vs. $\phi_{1,c}$ curve. Specifically,

$$\Delta\phi = \frac{\phi_1\phi_2(\delta_1 - \delta_2)}{1 - \delta_1\phi_1 - \delta_2\phi_2} \tag{16}$$

and

$$\phi_{1,c} = \phi_1\left(\frac{1 - \delta_1}{1 - \delta_1\phi_1 - \delta_2\phi_2}\right) \tag{17}$$

where $\delta_i = \Delta m_i/m_{i,u}$ (Δm_2 is defined analogously to Δm_1: we strive to make δ_2 as small as possible). The $\Delta\phi$ vs. $\phi_{1,c}$ data are fitted with a least-squares polynomial of the third or fourth degree that is constrained to pass through the origin. The lowest order polynomial that fits the data, as judgment dictates, is preferable to a higher order polynomial with a smaller residual.

From the measured vapor pressure, p_1, one calculates $a_1 = f_1/f_1^\circ$, where the fugacities at pressure p_1 and at the equilibrium vapor pressure p_1°, f_1 and f_1°, respectively, are obtained from standard thermodynamic equations. In

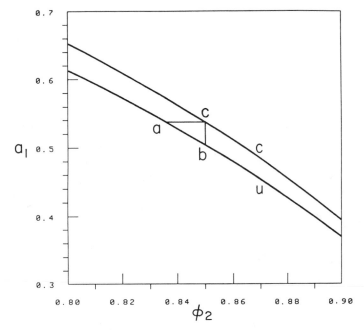

Fig. 5. Expanded region of Fig. 4. The measurements give points a and c, and the data analysis described in the text gives point b. The ratio of activities at points c and b is required to construct the dilation modulus. Curves for crosslinked and uncrosslinked polymers are labeled c and u, respectively.

most circumstances, the corrections for non-ideality of the vapor are negligible, and $a_1 = p_1/p_1^\circ$. The sorption data for the uncrosslinked sample are then rendered into χ vs. ϕ_2 form by use of eqn (7). The Flory–Huggins interaction parameter is thereby treated as an experimentally determined reduced residual chemical potential. The χ-values vary relatively slowly with composition, and it is usually possible to fit the χ vs. ϕ_2 data with a polynomial of low degree, and, if desired, to back-calculate a smoothed a_1 vs. ϕ_2 curve, as shown in Figs 4 and 5.

The data that are collected give directly the points a and c in Fig. 5. From these, one wishes to calculate the coordinates of point b. This is easy to do. At points a and c, $\ln(a_1) = \ln(a_{1,c}) = \ln(\phi_{1,u}) + \phi_{2,u} + \chi_u\phi_{2,u}^2$. At point c, $\phi_{1,c} = \phi_{1,u} - \Delta\phi$, where $\phi_{1,u}$ is the volume fraction of solvent at point a. However, $\phi_{1,u}$ at point b is the same as $\phi_{1,c}$ at point c, so that $\ln(a_{1,u}) = \ln(\phi_{1,c}) + \phi_{2,c} + \chi_c\phi_{2,c}^2$. Here $\chi_u = \chi(\phi_{2,u})$ and $\chi_c = \chi(\phi_{2,c})$. The raw data could be used directly to give points a and c, but to get point b the χ data must be smoothed. Our usual procedure, as noted above, is to smooth the $\Delta\phi$ data before attempting the data reduction.

NEW RESULTS

The best way to determine whether or not the Flory–Huggins interaction parameter for crosslinks is different from that for mid-chain segments is to prepare closely similar networks that have crosslinks with very different chemical structures.[22] Measurements on the two samples with various solvents would then presumably reveal different mixing contributions due to the crosslinks, should that be the source of the apparent breakdown of additivity. We have prepared two PDMS elastomers that meet this objective: they are described in detail in Table 1. The first network was prepared from silanol-terminated PDMS using tetraethoxysilane as the crosslinking agent; the second consists of vinyl terminated chains crosslinked with the cyclic tetramer of methyl siloxane. They have the same chain molecular weight, but the crosslinks are quite different.

The dilation moduli of these networks at 30°C that have been measured to date are shown in Figs 6 to 9. The data analysis was done using the physical properties of the polymer and solvents as given in Table 2. These swelling results are preliminary data which will be finalized upon calibration of our diaphragm pressure gauge with a dead-weight pressure tester. Since benzene and cyclohexane have closely similar equilibrium vapor pressures, the calibration will affect all data sets similarly, so that the conclusions that are reached here will not be vitiated in refinements.

As is clear from the figures, the two samples do not show identical behavior in a given solvent at a fixed temperature. This is doubtless a

TABLE 1
Characterization of Model Network Samples

End group	MW	Crosslinker	$[X]/[E]^a$	Sol fraction (%)
Silanol	26 000	$Si(OEt)_4^b$	1·0	4·70
Vinyl	26 000	$c—(HSiCH_3O)_4^c$	1·3	9·02

^{a} Ratio of concentrations of crosslink functional groups to chain ends.
^{b} Reaction conditions: room temperature, under vacuum, with 1 ppt tin as Sn(II) 2-ethylhexanoate catalyst.
^{c} Reaction conditions: 150°C in dry box with chloroplatinic acid as catalyst.

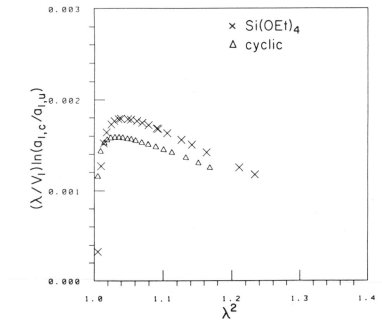

Fig. 6. Dilation moduli of tetraethoxysilane-crosslinked and 1,3,5,7-tetramethyl-cyclotetrasiloxane-crosslinked PDMS at 30°C for benzene.

TABLE 2
Physical Properties of Polymers and Solvents

Substance	Temperature (°C)	Density (g/cm³)	p°
PDMS[a]	20	0·974 2	—
	30	0·965 3	—
Benzene[b]	20	0·879 0	75·203
	30	0·868 5	119·338
Cyclohexane[c]	20	0·778 6	77·519
	30	0·769 3	121·726

[a] Ref. 23.
[b] Ref. 24a.
[c] Ref. 24b.

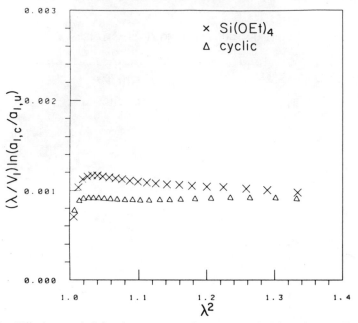

Fig. 7. Dilation moduli for the same two elastomers as in Fig. 6, but swollen with cyclohexane at 30°C.

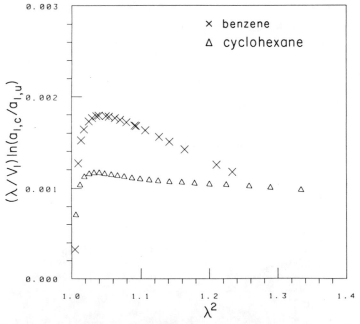

Fig. 8. Dilation moduli for the $Si(OEt)_4$-crosslinked elastomer swollen with benzene and with cyclohexane at 30°C.

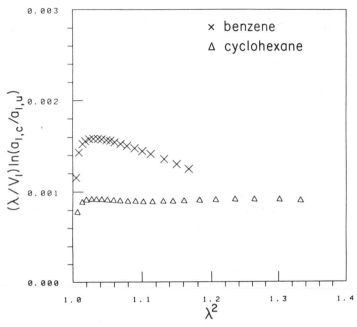

Fig. 9.   Dilation moduli for the cyclic siloxane-crosslinked sample swollen with benzene and cyclohexane at 30°C.

reflection of the different sol fractions found for the two samples, which in turn is likely due to the known side reactions[25,26] that accompany the hydrosilylation reaction. We found it necessary to unbalance the stoichiometry by 30% (Table 1) to make useful networks via the hydrosilylation reaction. It appears, nonetheless, that the curves could be scaled on the ordinates to become identical (for a given solvent) to within experimental uncertainty. This scale factor should be proportional to the modulus of elasticity, according to all known theories of elasticity. To investigate the truth of this assertion we are preparing to measure the stress–strain isotherms of the two elastomers so as to determine their elastic moduli. A comparison of the vertical shift factors with the moduli will provide a useful test of theory.

The fact that the behavior in benzene differs from that in cyclohexane in just the same way for the two samples proves, beyond all reasonable doubt, that a mixing term for crosslinks does not contribute to the breakdown of additivity. These results are in harmony with those obtained by Yen[19] and by Brotzman[20] on a different apparatus.[19] There are quantitative differences between the earlier data and these, but the trends are identical. These variations are, we believe, due to cure conditions. The previous PDMS sample[20] was radiation-crosslinked, which we now know to entail considerable chain scission;[27] there may be significant differences between that sample and the model networks used here. This uncertainty notwithstanding, we will attempt an analysis of the interrelation between the dilation and elastic moduli for all of these samples in the near future.

Figures 10 and 11 depict the temperature dependence of the dilation moduli for the two networks in benzene and in cyclohexane. Theories predict that the modulus is proportional to RT, and hence that $(\lambda/V_1)\ln(a_{1,c}/a_{1,u})$ is independent of temperature, except for the small temperature dependence of $\langle r^2 \rangle_o$.[28,29] The measured dilation moduli at 20°C are larger than those at 30°C, especially at the larger swelling ratios where the data are most reliable, and this suggests that $d\langle r^2 \rangle_o/dT$ is positive.[30] That is indeed the correct sign; the measured temperature coefficient of the unperturbed dimensions of PDMS is

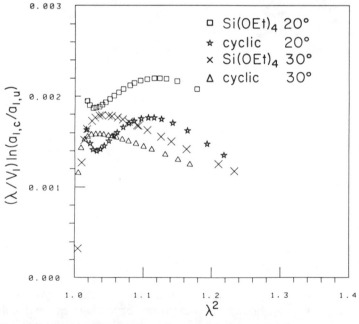

Fig. 10. Temperature dependence of the dilation moduli. The two elastomers (described in the text and in Fig. 6) were swollen in benzene at 20°C and 30°C. The data for 30°C are the same as in Fig. 6.

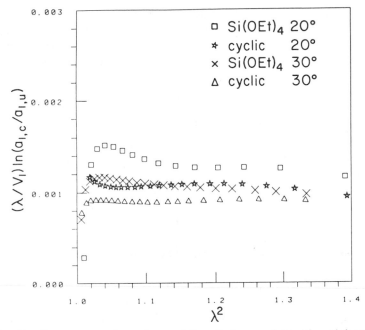

Fig. 11. Temperature dependence of the dilation moduli with cyclohexane.

positive.[30] However, the magnitude of the differences between the dilation moduli appear to be too large to be ascribed solely to $d\langle r^2 \rangle_o/dT$. These data will also be analyzed in greater detail in the near future, and we reserve further comments until that time.

CONCLUSIONS

The Flory–Rehner theory is in need of modification to account for an interaction between the elastic and mixing terms. The data that we have presented for two different samples with two different solvents effectively eliminate the possibility that a simple crosslink–solvent mixing term is responsible for the breakdown of additivity. The magnitude of the observed anomaly makes it appear unlikely that an explanation for the mixing–elastic cross-term can be made in terms of intrachain phenomena. Such a term would act multiplicatively on the dilation modulus, and it is difficult to see how such a term could ever attain the magnitude required to account for the observations. An interchain term, on the other hand, could act additively. Although the relative magnitude of the dilation moduli differences are large, the absolute values are small. For this reason, an additive term in ΔG appears to be the more likely source of the discrepancy. An interchain reorganization might be compatible with the unusual chain dimension changes that are seen with neutron scattering.[31,32]

We are currently engaged in additional measurements to explore the solvent dependence in more detail and to further characterize the mechanical properties of the samples used in this study. These results will be reported in the near future. Some theoretical possibilities are currently being explored in an attempt to come to a better understanding of the swelling of elastomers.

ACKNOWLEDGEMENT

This work was supported by grant DMR-8411523 from the Polymers Program of the Division of Materials Research, National Science Foundation. The instrument was partially constructed with funds from the Division of Materials Science, Office of Basic Energy Sciences, Department of Energy, Grant DE-FG06-84ER45123. The support of these agencies is gratefully acknowledged.

REFERENCES

1. Flory, P. J. and Rehner, J., Jr, *J. Chem. Phys.*, 1943, **11**, 512.
2. Flory, P. J. and Rehner, J., Jr, *J. Chem. Phys.*, 1943, **11**, 521.
3. Flory, P. J., *J. Chem. Phys.*, 1950, **18**, 108.
4. Kuhn, W., Pasternak, R. and Kuhn, H., *Helv. Chim. Acta*, 1947, **30**, 1705.
5. Hermans, J. J., *Trans. Faraday Soc.*, 1947, **43**, 591.
6. Hermans, J. J., *J. Polym. Sci.*, 1962, **59**, 191.
7. Flory, P. J., *Principles of Polymer Chemistry*, Cornell University Press, Ithaca, New York, 1953, Chapter 12.
8. James, H. M. and Guth, E., *J. Chem. Phys.*, 1947, **11**, 455.
9. Gottlieb, M., Swelling of polymer networks, *this volume*, p. 403.
10. Eichinger, B. E., *Ann. Rev. Phys. Chem.*, 1983, **34**, 359.
11. Gee, G., Herbert, J. B. M. and Roberts, R. C., *Polymer*, 1965, **6**, 541.
12. Flory, P. J., *Macromolecules*, 1979, **12**, 119.
13. Flory, P. J., *J. Chem. Phys.*, 1977, **66**, 5720.
14. Erman, B. and Flory, P. J., *J. Chem. Phys.*, 1978, **68**, 5363.
15. Erman, B., *J. Polym. Sci., Polym. Phys. Ed.*, 1981, **19**, 829.
16. Flory, P. J. and Erman, B., *Macromolecules*, 1982, **15**, 800.
17. Brotzman, R. W. and Eichinger, B. E., *Macromolecules*, 1981, **14**, 1445.

18. Yen, L. Y. and Eichinger, B. E., *J. Polym. Sci., Polym. Phys. Ed.*, 1978, **16**, 117.
19. Yen, L. Y. and Eichinger, B. E., *J. Polym. Sci., Polym. Phys. Ed.*, 1978, **16**, 121.
20. Brotzman, R. W. and Eichinger, B. E., *Macromolecules*, 1982, **15**, 531.
21. Brotzman, R. W., Jr and Eichinger, B. E., *Macromolecules*, 1983, **16**, 1131.
22. Kresge, E. N. and Eichinger, B. E., private discussion.
23. Shih, H. and Flory, P. J., *Macromolecules*, 1972, **5**, 761.
24. Rossini, F. D., Pitzer, K. S., Arnett, R. L., Braun, R. M. and Pimentel, G. C., *Selected Values of Physical and Thermodynamic Properties of Hydrocarbons and Related Compounds*, API Project 44, Carnegie Press, Pittsburgh, Pennsylvania, 1953: (a) Table 5d; (b) Table 23a.
25. Speier, J. L., *Adv. Organomet. Chem.*, 1979, **17**, 407.
26. Gustavson, W. A., Epstein, P. S. and Curtis, M. D., *J. Organomet. Chem.*, 1982, **238**, 87.
27. Shy, L. Y. and Eichinger, B. E., *Macromolecules*, 1986, **19**, 2787.
28. Flory, P. J., Hoeve, C. A. J. and Ciferri, A., *J. Polym. Sci.*, 1959, **34**, 337.
29. Flory, P. J., Hoeve, C. A. J. and Ciferri, A., *J. Polym. Sci.*, 1960, **45**, 235.
30. Flory, P. J., *Statistical Mechanics of Chain Molecules*, Interscience, New York, 1969, p. 39.
31. Bastide, J., Duplessix, R., Picot, C. and Candau, S., *Macromolecules*, 1984, **17**, 83.
32. Davidson, N. S. and Richards, R. W., *Macromolecules*, 1986, **19**, 2576.

29

PHASE TRANSITION IN SWOLLEN GELS. 10. EFFECT OF THE POSITIVE CHARGE AND ITS POSITION IN THE SIDE CHAIN ON THE COLLAPSE AND MECHANICAL BEHAVIOUR OF POLY(ACRYLAMIDE) NETWORKS

MICHAL ILAVSKÝ and KAREL BOUCHAL

Institute of Macromolecular Chemistry, Czechoslovak Academy of Sciences, 162 06 Prague 6, Czechoslovakia

ABSTRACT

The swelling and mechanical behaviour of ionized networks of copolymers of acrylamide (AAm) with four quaternary salts, which differ in the position of the positive centres in side chains (mole fraction of salts $x_s = 0$–0.2) in water–acetone mixtures, was investigated. In the range $x_s \geq 0.03$, phase transition was observed with increasing acetone content; both the extent of the transition, Δ, and the critical acetone concentration at which transition takes place, a_c, increased with increasing x_s. While the structure of salts does not affect Δ, the value of a_c increases with increasing distance of the position of the positive centre from the main chain. The jumpwise change in the gel volume is accompanied by a similar jump in the equilibrium modulus.

INTRODUCTION

With loosely crosslinked poly(acrylamide) (PAAm) networks prepared at high dilution at network formation and containing a small number of charges on the chain (1–5 mol%), first-order phase transition (collapse) was observed during the transition from a good to a poor solvent.[1-3] The charges could be introduced into the chain either by a spontaneous hydrolysis of AAm groups[1,2,4] or by copolymerizing AAm with a suitable monomer (e.g. sodium methacrylate[3]). The collapse could also be brought about by a change in external conditions, such as electric field[5] or temperature,[6] or by change in network parameters, such as the degree of ionization,[1-4] network density[7] or dilution at network formation.[8] It was found that the theory of polyelectrolyte networks[9] (which considers the effect of electrostatic interactions of charges on the chain irrespective of their polarity) gives a semiquantitative description of the formation and extent of collapse of PAAm networks in water–acetone mixtures.[3,4,7,8] In all these cases the negative charge was situated in close proximity to the main PAAm chain. First results showed that the behaviour of PAAm networks with a small number of positive charges also undergoes phase transition.[10-12]

In this study we investigate the effect of concentration of the positive charge on the swelling and mechanical equilibria of PAAm networks obtained by copolymerization with four quaternary salts. The advantage of such systems consists in that the degree of ionization is virtually pH-independent, and the positive centres are situated in various positions in side chains of the quaternary salts.

EXPERIMENTAL

Sample Preparation

The samples were prepared from 100 ml of a mixture which contained 5 g acrylamide (AAm), 0·135 g N,N'-methylenebisacrylamide (MBAAm), 20 mg ammonium persulphate and 20 mg sodium pyrosulphite. Four series of networks with various quaternary salts were prepared—N,N,N-trimethyl-N-2-methacryloyloxyethylammonium chloride (**I**), N,N,N-trimethyl-N-4-methacrylolyoxybutylammonium chloride (**II**), N,N-dimethyl-N-methoxycarbonylmethyl-N-2-methacryloyloxyethylammonium chloride (**III**) and N,N-dimethyl-N-butoxycarbonylmethyl-N-2-methacryloyloxyethyl-ammonium chloride (**IV**):

$$CH_2{=}\underset{\underset{CH_3}{|}}{C}{-}COO{-}CH_2{-}CH_2{-}\underset{\underset{CH_3}{|}}{\overset{\overset{CH_3\,Cl^{\ominus}}{|}}{N^{\oplus}}}{-}CH_3$$

I

$$CH_2{=}\underset{\underset{CH_3}{|}}{C}{-}COO{-}CH_2{-}CH_2{-}CH_2{-}CH_2{-}\underset{\underset{CH_3}{|}}{\overset{\overset{CH_3\,Cl^{\ominus}}{|}}{N^{\oplus}}}{-}CH_3$$

II

$$CH_2\!=\!\underset{\underset{\displaystyle CH_3}{|}}{C}\!-\!COO\!-\!CH_2\!-\!CH_2\!-\!\underset{\underset{\displaystyle CH_3}{|}}{\overset{\overset{\displaystyle CH_3Cl^{\ominus}}{|}}{N^{\oplus}}}\!-\!CH_2\!-\!COOCH_3$$

III

$$CH_2\!=\!\underset{\underset{\displaystyle CH_3}{|}}{C}\!-\!COO\!-\!CH_2\!-\!CH_2\!-\!\underset{\underset{\displaystyle CH_3}{|}}{\overset{\overset{\displaystyle CH_3Cl^{\ominus}}{|}}{N^{\oplus}}}\!-\!CH_2\!-\!COOCH_2\!-\!CH_2\!-\!CH_2\!-\!CH_3$$

IV

In each series six networks were prepared with a varying mole fraction of the salt, x_s, from 0 to 0·2 (Table 1). The polymerization proceeded at room temperature for ~ 5 h: after that, the gels were extracted in water for 7 days.

Swelling

Water–acetone mixtures were prepared from redistilled water in the range 0–95 vol.% acetone. The samples were swollen in mixtures at room temperature. After swelling for ~ 28 days, we determined the swelling ratio X related to the state of network formation from

$$X = (D^*/D)^3 = V^*/V \tag{1}$$

where D^* and D are sample diameters after the preparation and swelling in mixtures, respectively, and V^* and V are the respective gel volumes after the preparation and swelling. The measurements were performed with an Abbé comparator (Zeiss Jena, accuracy $\pm 0·002$ mm): the X-values in Fig. 1(a)–(d) are the means from at least three samples.

Mechanical Characteristics

Deformational measurements were carried out with apparatus described earlier,[3] in which a cylindrical specimen was compressed to $\lambda\,(=l/l_0$, where l and l_0 are the deformed and the initial specimen height, respectively). After relaxation for 30 s the force f was measured, and the whole procedure was repeated; ten λ_i and f_i values were measured (in the range $0·7 \leq \lambda < 1$), and from them the shear modulus G was determined using the relation

$$G = f/S_0(\lambda^{-2} - \lambda) \tag{2}$$

where S_0 is the initial cross-section of the specimen (Fig. 1(a)–(d)).

TABLE 1
Basic Network Characteristics and Phase Transition Parameters[a]

x_s	G_1 $(g\ cm^{-2})$	$10^5\ v_d$ $(mol\ cm^{-3})$	$\Delta \log X$	$\Delta \log G$	a_c	ϕ
Networks with quaternary salt **I**						
0	13·5	1·5	—	—	—	—
0·010	19·1	2·0	—	—	—	0·13
0·029	30·2	3·2	0·80	0·3	41	0·12
0·047	42·7	4·6	1·10	0·45	44	0·12
0·090	77·6	8·3	1·35	0·55	55	0·14
0·165	123·0	13·2	1·50	0·60	63	0·15
Networks with quaternary salt **II**						
0·015	32·1	3·4	—	—	—	0·15
0·030	36·5	3·9	1·0	0·35	50	0·15
0·060	50·1	5·4	1·3	0·40	63	0·15
0·100	52·8	5·7	1·4	0·42	67	0·11
0·150	72·7	7·8	1·5	0·44	70	0·11
0·200	64·7	6·9	1·5	0·47	75	0·08
Networks with quaternary salt **III**						
0·015	23·9	2·6	—	—	—	0·14
0·030	32·5	3·5	0·9	0·35	45	0·13
0·060	46·5	5·0	1·2	0·40	55	0·11
0·100	50·1	5·4	1·3	0·40	62	0·09
0·150	46·8	5·0	1·3	0·43	67	0·07
0·200	49·5	5·3	1·4	0·45	70	0·05
Networks with quaternary salt **IV**						
0·015	39·6	4·2	—	—	—	0·12
0·030	38·8	4·2	1·0	0·35	50	0·12
0·060	50·0	5·4	1·10	0·35	65	0·14
0·100	55·5	5·9	—	—	—	0·10
0·150	55·3	5·9	—	—	—	0·07
0·200	58·0	6·2	—	—	—	0·06

[a] x_s is the molar fraction of individual salts, G_1 is the shear modulus after the network formation, v_d is network density, $\Delta \log X$ is a change in the gel volume at collapse, $\Delta \log G$ is a change in the gel modulus at collapse, a_c is the critical acetone concentration at collapse, and ϕ is the correction factor ($\alpha = \phi x_s$).

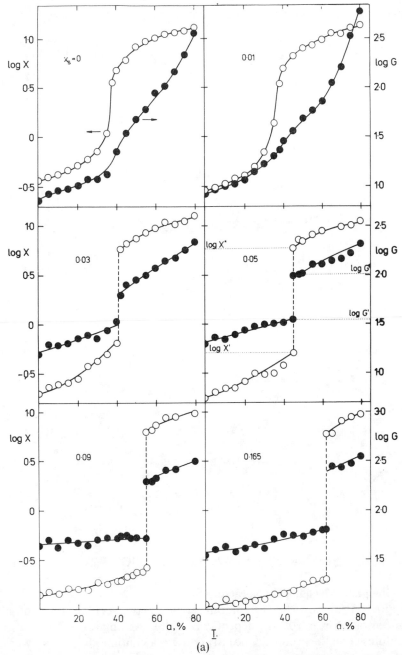

Fig. 1. Dependence of the swelling ratio X and of the modulus G ($g\,cm^{-2}$) on the acetone content a (vol.%) for networks with: (a) quaternary salt **I**; (b) quaternary salt **II**; (c) quaternary salt **III**; and (d) quaternary salt **IV**. $\bigcirc$, X; $\bullet$, G. Numbers correspond to the mole fraction of salt x_s.

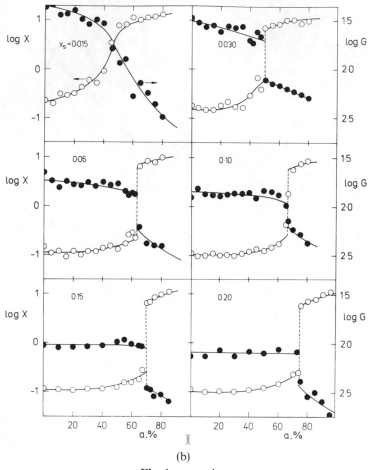

(b)

Fig. 1.—contd.

RESULTS AND DISCUSSION

Swelling and Mechanical Characteristics

Figure 1(a)–(d) shows that for all four series of networks in the range of salt concentration $x_s < 0.03$, the dependence of the swelling ratio X and of the modulus G on the acetone concentration a is a continuous one. Networks with $x_s \geq 0.3$ undergo a phase transition, reflected in a jumpwise change in volume and in the modulus of the gel. It is obvious that the extent of the

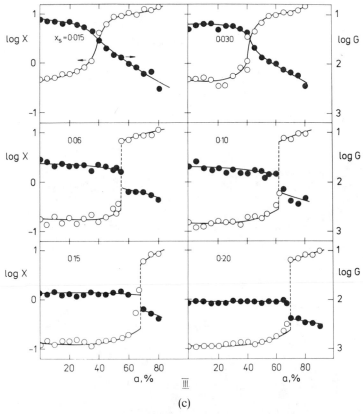

(c)

Fig. 1.—*contd.*

collapse $\Delta \log X$ ($=\log X'' - \log X'$) increases with increasing salt content; so does the critical acetone concentration a_c in the mixture at which the collapse takes place (Table 1). An exception is made by a series of networks with quaternary salt **IV**, in which the collapse occurs only for $x_s = 0.03$ and 0.06; for $x_s \geq 0.10$ the dependence of the swelling ratio X and of the modulus G on concentration is again continuous. As expected, the formation of the collapse does not depend on the charge polarity, but it can be brought about only if the concentration of quaternary salts is ~ 5 times higher compared with networks of copolymers of AAm with sodium methacrylate.[3]

A jumpwise change in the gel modulus $\Delta \log G$ ($=\log G'' - \log G'$) correlates well with a change in the gel volume $\Delta \log X$ ($\Delta \log G = 0.4 \times \Delta \log X$, Table 1): a somewhat higher slope was found earlier for PAAm

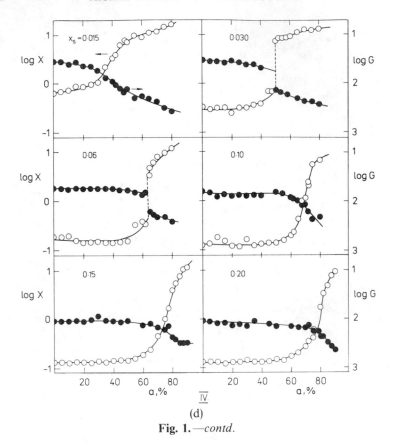

(d)

Fig. 1.—*contd.*

networks bearing a negative charge.[1-4] The dependence of the modulus $\log G$ on $\log X$ in the range of small X is linear (cf., for example, Fig. 2). Similarly, as has been done earlier,[3] deviations from linearity for $X \geq 10$ may be interpreted by the effect of the main transition region. From the values of the modulus G_1 measured after network preparation, the concentrations of elastically active chains were calculated with respect to the dry state, $v_d = G_1/RT\phi°$, where $\phi° = 0.037$ is the volume fraction of the polymer at network formation, R is the gas constant and T is temperature (Table 1). The low v_d values, and hence the low efficiency of the crosslinking reaction, indicate a high cyclization, which is a consequence of high dilution in the crosslinking reaction.

The dependence of the extent of collapse $\Delta \log X$ and of the critical value a_c on the molar salt concentration x_s is given in Fig. 3. Within the limits of

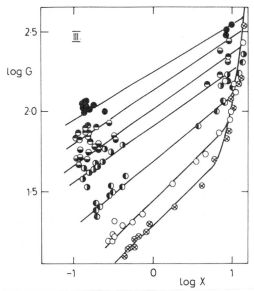

Fig. 2. An example of the dependence of the modulus G ($g\,cm^{-2}$) on the swelling ratio X for networks with quaternary salt **III**: ○, $x_s = 0.015$; ◐, $x_s = 0.03$; ◑, $x_s = 0.06$; ◓, $x_s = 0.10$; ◑, $x_s = 0.15$; ●, $x_s = 0.20$; ⊗, $x_s = 0$.

experimental error, $\Delta \log X$ is independent of x_s for all quaternary salts used; the effect of the concentration of quaternary salts on the extent of collapse is, however, much weaker than that of sodium methacrylate. The effect of the structure of side chains of salts on the critical acetone concentration at collapse, a_c, is somewhat unexpected. For networks of copolymers of AAm with salts **II** and **IV** containing a large amount of hydrophobic groups in the side chains, the a_c values (at the given x_s) increase compared with a_c of networks of salts **I** and **III**. This means that hydrophobic interactions stabilize the expanded state of the gel, and that an elevated acetone concentration in the water–acetone mixture is needed in order to bring about the collapse. It may be assumed that a_c increases as a result of the preferential sorption of acetone by hydrophobic regions of side chains of salts **II** and **IV**. This reduces the acetone content in the mixed water–acetone solvent, which interacts with AAm units.

Comparison Between Theory and Experiment
By including the effect of electrostatic interactions between chain charges in the theory of rubber-like elasticity, a relation

$$P = P_m + P_{os} + P_{el} + P_{els} \tag{3}$$

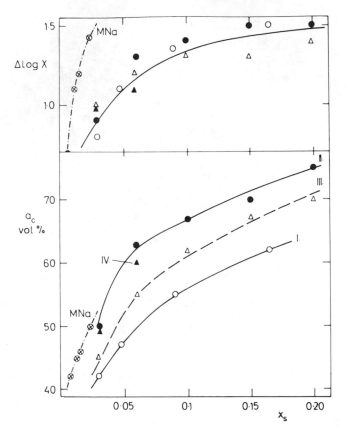

Fig. 3. Dependence of the extent of collapse $\Delta \log X$ and of the critical acetone concentration at collapse a_c on the mole fraction of salt x_s: $\bigcirc$, networks with salt **I**; $\bullet$, networks with salt **II**; $\triangle$, networks with salt **III**; $\blacktriangle$, networks with salt **IV**; $\otimes$, networks of copolymers AAm with sodium methacrylate.[3]

was obtained for the swelling pressure P (cf. Ref. 9). In eqn (3), the contribution P_m is given by the mixing of chain segments with the solvent (the Flory–Huggins relation with the interaction parameter χ), P_{os} is given by the mixing of gel ions with the solvent, P_{el} is given by a change in the elastic free energy with swelling, and P_{els} is determined by a change in the free energy of electrostatic interactions with swelling. P_i is expressed in detail in molecular parameters in eqns (2)–(5).[9] Using the known molecular parameters (density of the dry polymer $\rho = 1 \cdot 35\, \mathrm{g\, cm^{-3}}$, molecular volumes of acetone–water mixtures V_1, network density ν_d, degree of ionization x_s

and experimental values of volume fractions of the polymer in the swollen state $\phi_2 = \phi^\circ X$), the dependences of the interaction parameters χ of networks on ϕ_2 were calculated (as a consequence of free swelling, $P = 0$ in eqn (3)). The same procedure was used earlier for PAAm networks with a negative charge.[3,4]

The use of eqn (3) for a PAAm network without a quaternary salt ($x_s = 0$) swollen in water led to the expected value of the interaction parameter $\chi = 0.48$. Similar to the earlier case of variously aged PAAm networks,[4] in our case too the use of the degree of ionization x_s gave unrealistically high χ values for ionized networks of all four series swollen in water. Since χ is a measure of the polymer–solvent interaction, when the charges are screened (the effect of charges in eqn (3) is included in the contributions P_{os} and P_{els}), one may require that $\chi = 0.48$ also for ionized networks of all four series swollen in water. This requirement may be satisfied by assuming that the effective degree of ionization α is lower than the salt concentration used in the copolymerization x_s, i.e. $\alpha = \phi x_s$ where ϕ is the empirical correction factor (Table 1).

The ϕ values thus calculated, and ranging between 0·05 and 0·3, are much lower than those found earlier for PAAm networks with a negative charge.[4] There are probably several reasons for the low value of the factor. (a) Positively charged centres are localized at a larger distance from the main chain, which is probably reflected in a smaller influence on its conformation. (b) The bulkiness of the ammonium group has as a result that the electrostatic field of the positively charged nitrogen atom is weaker compared with the electrostatic field of the carboxylic anion. Hence the degree of hydration of the ammonium group is lower. (c) The degree of hydration of the centres of electrostic charges is affected by counterions. In the case of quaternary salts, the anion has the average number of hydration two, while that of the sodium cation is three. Due to this, a higher degree of hydration of electronegative centres compared with positive centres may be anticipated. (d) One may expect differences in the association of counterions for negatively and positively charged centres, and thus also participation of ionic pairs. While points (a)–(c) are reflected in a decrease in ϕ, the effect of the association of counterions is difficult to discuss.

CONCLUSIONS

From measurements of the swelling and mechanical equilibria of polyacrylamide networks with a low content of quaternary salts (mole fraction of salts, x_s, ranging from 0 to 0·2) in the mixed water–acetone solvent, we have the following:

(a) In the range $x_s \geq 0\cdot03$, phase transition takes place with all four salts. Both the extent of transition $\Delta \log X$ and the critical acetone concentration in the mixed solvent at which the transition takes place, a_c, increase with increasing x_s. In networks of quaternary salt **IV** with the longest side chain, in the range $x_s \geq 0\cdot1$ the phase transition disappears.

(b) While the structure of salts does not affect the extent of transition $\Delta \log X$, the critical acetone concentration a_c increases with the increasing distance between the position of the positive centre from the main chain, and mainly with the increasing content of hydrophobic groups in the side chain of the salt. Hence, hydrophobic interactions stabilize the expanded state of the gel.

(c) The effect of the positive charge of the salt on the formation and extent of transition is smaller by 5 to 10 times than the effect of the negative charge observed with copolymers of AAm and sodium methacrylate.

(d) In all cases the jumpwise change in the gel volume observed in the transition is accompanied by a jumpwise change in the equilibrium modulus of the gel.

REFERENCES

1. Tanaka, T., Phase transitions in gels and a single polymer, *Polymer*, 1979, **20**, 1404–11.
2. Stejskal, J., Gordon, M. and Torkington, J. A., Collapse of polyacrylamide gels, *Polym. Bull.*, 1980, **3**, 621–5.
3. Ilavský, M., Phase transition in swollen gels. 2. Effect of charge concentration on the collapse and mechanical behavior of polyacrylamide networks, *Macromolecules*, 1982, **15**, 782–8.
4. Ilavský, M., Hrouz, J., Stejskal, J. and Bouchal, K., Phase transition in swollen gels. 6. Effect of ageing on the extent of hydrolysis of aqueous polyacrylamide solutions and on the collapse of gels, *Macromolecules*, 1984, **17**, 2868–74.
5. Tanaka, T., Nishio, I., Sun, T. S. and Ueno-Nishio, S., Collapse of gels in an electric field, *Science*, 1982, **218**, 467–71.
6. Ilavský, M., Hrouz, J. and Havlíček, I., Phase transition in swollen gels. 7. Effect of charge concentration on the temperature collapse of poly(N,N-diethyl-acrylamide) networks in water, *Polymer*, 1985, **26**, 1514–18.
7. Ilavský, M. and Hrouz, J., Phase transition in swollen gels. 4. Effect of concentration of the crosslinking agent at network formation on the collapse and mechanical behaviour of polyacrylamide gels, *Polym. Bull.*, 1982, **8**, 387–94.
8. Ilavský, M. and Hrouz, J., Phase transition in swollen gels. 5. Effect of the amount of diluent at network formation on the collapse and mechanical behaviour of polyacrylamide networks, *Polym. Bull.*, 1983, **9**, 159–66.

9. Ilavský, M., Effect of electrostatic interactions on phase transition in the swollen polymeric network, *Polymer*, 1981, **22**, 1687–91.

10. Ilavský, M., Hrouz, J. and Bouchal, K., Phase transition in swollen gels. 9. Effect of the concentration of quaternary salt on the collapse and mechanical behaviour of poly(acryl amide) networks, *Polym. Bull.*, 1985, **14**, 301–5.

11. Kotayama, S. and Ohata, A., Phase transition of a cationic gel, *Macromolecules*, 1985, **18**, 2781–2.

12. Hirokawa, Y., Tanaka, T. and Sato, E., Phase transition of positively ionized gels, *Macromolecules*, 1985, **18**, 2782–4.

30

DESWELLING OF GELS INDUCED BY UNIDIRECTIONAL COMPRESSION

Ferenc Horkay*

*Research Laboratory for Inorganic Chemistry,
Hungarian Academy of Sciences, Budaörsi ut.45, H-1112 Budapest, Hungary*

and

Miklós Zrinyi

*Department of Colloid Science, Loránd Eötvös University,
Puskin ut. 11–13, H-1088 Budapest, Hungary*

ABSTRACT

Elastic and swelling properties of chemically crosslinked poly(vinyl acetate) gel homologues swollen by good and θ-diluents were studied. Deswelling was induced by unidirectional compression of the gels, on the one hand, and by lowering the chemical potential of the diluent in the surrounding liquid phase, on the other. An attempt was made to take into account the effect of unidirectional deformation on the equilibrium concentration of the gel. The results were compared with experimental findings. Satisfactory agreement was found both in good and θ-diluents. The equivalence of the response of the network to istropic shrinking and to unidirectional compression was confirmed experimentally.

INTRODUCTION

The effect of deformation on the swelling equilibrium concentration of a swollen polymer network has long been recognized and was discussed by Flory and Rehner.[1,2] The theory was further developed by Treloar,[3,4] who discussed the general case of a homogeneous strain of any type and derived equations for a number of particular types of strain, including single

* Permanent address: Department of Colloid Science, Loránd Eötvös University, Puskin ut. 11–13, H-1088 Budapest, Hungary.

elongation, two-dimensional extension and unidirectional compression. Previous experimental results of Flory, Rehner, Gee and Treloar[1−5] confirmed the validity of the proposed relationships for gels swollen by good diluents. In poor solvents the agreement between predictions and experimental findings was only qualitative.[5]

If a swollen gel equilibrated with a diluent is deformed, the chemical potential of the diluent inside the gel is altered, through the influence of the isotropic component of the stress arising from the deformation of the network. Thus the swelling degree of the gel changes, although the activity of the diluent in the equilibrium liquid or vapour phase remains the same. In the case of single elongation the swelling degree increases, while in the case of unidirectional compression the swelling degree decreases.

Deswelling can also be achieved by lowering the chemical potential of the diluent around the freely swollen network. This can be carried out either by dissolving a polymer in the equilibrium liquid phase,[6] or through the vapour phase by reducing the partial pressure of the diluent.[7,8] According to these methods, the volume change is not accompanied by distortion of the isotropic gel. Due to the lowering of the activity of the swelling agent in the surrounding media, the gel loses a certain amount of diluent but its shape remains unchanged.

Although several methods are available to decrease the activity of the diluent, no experimental work has been reported yet in which the dependence of the swelling degree on both the activity of the diluent and the extent of deformation was investigated.

The main purposes of our work are:

(i)   to determine the swelling degree as a function of the deformation ratio for gels equilibrated with a pure swelling agent;
(ii)   to determine the swelling degree as a function of the activity of the diluent;
(iii) to study the correspondence between anisotropic deswelling induced by unidirectional deformation of the gel, and isotropic deswelling due to lowering the chemical potential of the diluent around the swollen network.

EXPERIMENTAL

Mechanical and equilibrium deswelling experiments were performed on chemically crosslinked poly(vinyl acetate) (PVAc) gels swollen by toluene and isopropyl alcohol, respectively. Toluene can be considered as a good solvent for PVAc, while for PVAc/isopropyl alcohol gels the θ-temperature was found to be 52°C. PVAc gels were prepared by acetylation of poly(vinyl alcohol) samples according to a previously described method.[9] Poly(vinyl alcohol) was crosslinked with glutaric aldehyde in a mixture of 90% (v/v) dimethylsulphoxide and 10% (v/v) water. Cylindrical-form gel specimens of equal height and diameter (approximately 1 cm) were obtained. Several series of gel homologues were prepared differing in degree of crosslinking and in polymer concentration when crosslinks were introduced. Table 1 shows the list of PVAc gels used for the experiments.

Deswelling induced by unidirectional compression was performed in a home-made apparatus.[10] Gels were deformed between two parallel flat plates, whose distance of separation could be adjusted in order to impose the desired compression on the sample. During the experiments the whole apparatus was immersed in pure diluent. The time required to reach equilibrium was found to be between 3 and 5 days. In several cases, the measurements were performed at successively increasing compression ratios following the sequences of measurements carried out with decreasing compression ratios. The two sets of experimental data showed no

TABLE 1
List of Poly(vinyl acetate) Gels

Symbol[a]	PVAc/toluene	PVAc/isopropyl alcohol
3/50	+ [b]	+
3/100	+	−
6/50	+	+
6/100	+	−
6/200	+	+
6/400	+	−
9/50	+	+
9/100	+	+
9/200	+	+
9/400	+	+
12/50	+	+
12/100	+	−
12/200	+	+
12/400	+	−

[a] Polymer concentration at which crosslinks were introduced/degree of crosslinking.
[b] +, gels investigated; −, gels not studied in the swelling agent.

appreciable differences. Finally, the gels were dried and measured. Detectable loss of mass was not observed as a result of the swelling and drying cycle.

The swelling degree (q) was calculated from height and diameter data of swollen and dry cylinders on the one hand, and from the masses of swollen and dry networks on the other. Table 2 shows that the agreement between swelling degrees calculated from independent data is within experimental error, indicating the additivity of specific volumes of polymer and solvent.

In Fig. 1 the swelling degree is presented as a function of the deformation ratio (Λ_e) for PVAc gels swollen by toluene. (Λ_e is the ratio of the length of the deformed gel to that of the freely swollen undeformed sample. DC is the degree of crosslinking, which means the average number of monomer units between neighbouring crosslinks.) The effect of unidirectional compression is apparent: gels deswell under the applied pressure—the effect is more pronounced for slightly crosslinked networks.

Deswelling was also achieved by the equilibrium deswelling method described in Ref. 11. This method differs from other deswelling techniques, since the swollen gels are not directly immersed in the equilibrium liquid phase. The gels are separated from the polymer solution by a semipermeable membrane which prevents the penetration of the mobile macromolecules used for lowering the activity of the diluent into the gel. The activity of the diluent was calculated from the osmotic pressure of the surrounding polymer solution.

In Fig. 2 the swelling degree is presented against the activity of the diluent for the same PVAc gels as shown in Fig. 1. The swelling degree considerably decreases with decreasing diluent activity.

TABLE 2

Swelling Data of 9/400 Poly(vinyl acetate) Gel Cylinder at Different Stages of Dilution

Height (mm)	Mass (g)	Volume swelling degree	
		From mass	From dimension
5·271	0·135	1·00	1·00
12·669	1·399	13·89	13·87
12·850	1·458	14·49	14·50
13·109	1·547	15·38	15·39
13·463	1·675	16·67	16·69
13·776	1·794	17·86	17·87
14·032	1·895	18·87	18·86

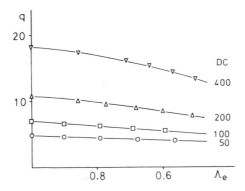

Fig. 1. Dependence of the volume swelling degree (q) on the deformation ratio (Λ_e) for PVAc gels swollen by toluene ($T = 25°C$). Polymer concentration when crosslinks were introduced, 9% (m/m).

Figure 3 shows the kinetics of deswelling in both cases. Figure 3(a) refers to deformation-induced deswelling when the gels are directly immersed in the liquid phase. In Fig. 3(b) the swelling degree is presented as a function of time for PVAc/toluene gels separated from the polymer solution by a semipermeable membrane. In the latter case the time required to attain equilibrium is considerably longer.

The reversibility of the deswelling process was also controlled by transferring the deswollen gels into polymer solutions with solvent activities increasing up to that of the pure swelling agent. No deviation was

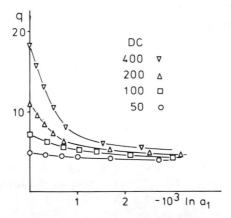

Fig. 2. Dependence of the volume degree of swelling (q) on the activity of the swelling agent (a_1) for PVAc gels swollen by toluene ($T = 25°C$).

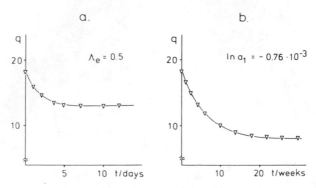

Fig. 3. Deswelling of PVAc/toluene gel as a function of time: (a) deswelling induced by unidirectional compression of the gel; (b) deswelling induced by lowering the chemical potential of the diluent in the equilibrium liquid phase. Symbol of the gel: 9/400.

found between swelling ratios obtained by increasing or decreasing the solvent activity. The height and diameter data of partially deswollen gels indicated that the deswelling process was isotropic.

Stress–compression measurements were also performed on the gels at different stages of dilution.[12] The shear moduli of the gels were calculated on the basis of the Mooney–Rivlin equation

$$\sigma = C_1(\Lambda - \Lambda^{-2}) + C_2(\Lambda - \Lambda^{-2})\Lambda^{-1}$$

where σ is the nominal stress, Λ is the experimental deformation ratio, and C_1 and C_2 are constants. C_2 was found to be equal to zero independently of the swelling ratio. The C_1 term was identified with the shear modulus.

The accuracy of the measurements of equilibrium swelling degree and deformation ratio was better than 0.5%. The reproducibility of activity measurements proved to be within 2%, while that of modulus measurements did not exceed 5%.

DISCUSSION

The chemical potential of the diluent in a swollen gel can be given as the sum of an elastic contribution $(\Delta\mu_1)_{el}$ associated with the deformation of the network and a mixing term $(\Delta\mu_1)_{mix}$ which takes into account the thermodynamics of mixing of polymer and solvent:[13]

$$\Delta\mu_1 = (\Delta\mu_1)_{el} + (\Delta\mu_1)_{mix} \tag{1}$$

The elastic contribution is related to the shear modulus[14]

$$(\Delta\mu_1)_{el} = G(\phi)\bar{V}_1 \tag{2}$$

where $G(\phi)$ is the shear modulus of the gel and $\bar{V}_1$ is the partial molar volume of the diluent. The modulus obeys a simple power-law dependence on the concentration given by eqn (3):

$$G(\phi) = K_1\phi^l \tag{3}$$

where K_1 is a constant and ϕ is the volume fraction of the network polymer. The exponent, l, equals 1/3 according to Flory and Treloar,[13,14] or 0·5 according to Bastide *et al.*[15]

For a gel equilibrated with a pure diluent, we get

$$G(\phi_e) = K_1\phi_e^l \tag{4}$$

where $G(\phi_e)$ is the shear modulus of the freely swollen gel and ϕ_e is the equilibrium volume fraction of the polymer in the gel.

From eqns (3) and (4) follows

$$\frac{G(\phi)}{G(\phi_e)} = \left(\frac{\phi}{\phi_e}\right)^l \tag{5}$$

i.e. the ratio of the moduli is the only function of ϕ/ϕ_e.

In Fig. 4 experimental data are presented according to eqn (5) for PVAc gels swollen by toluene and isopropyl alcohol, respectively. Experimental points scatter around the same straight line for both kinds of systems. The

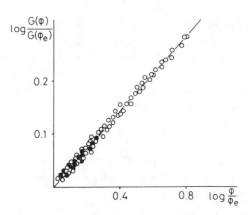

Fig. 4. Log–log representation of the ratio of $G(\phi)/G(\phi_e)$ vs. ϕ/ϕ_e for PVAc/ toluene gels at $T = 25°C$ (○) and for PVAc/isopropyl alcohol gels at $T = 52°C$ (●).

slope of the straight line is equal to 0·34 in good agreement with the theoretical exponent, 1/3.

Taking into account that for freely swollen gels $\Delta\mu_1 = 0$, we get

$$(\Delta\mu_1)_{el} = -(\Delta\mu_1)_{mix} = \pi_n \bar{V}_1 \tag{6}$$

where π_n is the osmotic pressure of network chains. There is reason to assume that π_n obeys the same power-law dependence on the concentration as the osmotic pressure of polymer solutions in the semidilute regime:

$$\pi_n = K\phi_e^m \tag{7}$$

The exponent, m, equals 2 according to the classical Flory–Huggins theory. In good solvent conditions the scaling theory predicts the exponent $m = 9/4$, while in the θ-state $m = 3$.[16] Combination of eqns (2), (4), (6) and (7) yields the relationship between K_1 and K:

$$K_1 = K\phi_e^{m-l} \tag{8}$$

Substituting eqn (8) into eqn (4) gives the concentration dependence of the elastic moduli of gel homologues equilibrated with a pure swelling agent

$$G(\phi_e) = K\phi_e^m \tag{9}$$

The validity of eqn (9) has been supported by several experimental findings.[17-21]

In Fig. 5 the elastic moduli are presented against the volume fraction of the network polymer for PVAc gel homologues swollen by good and θ-solvent, respectively. The exponents are close to those predicted by the scaling theory.

Until now we have discussed the concentration dependence of the elastic modulus. From now on we are interested in the effect of deformation on the equilibrium concentration. In this case, the equation describing the swelling equilibrium of an isotropic gel should be modified[22] by introducing the deformation ratio, Λ_e, defined as the ratio of the length of the deformed gel to that of the freely swollen undeformed sample:

$$\Delta\mu_1 = K_1 \bar{V}_1 \phi_e^l \Lambda_e^{-1} - K\bar{V}_1 \phi^m \tag{10}$$

For gels equilibrated with a pure diluent, the two terms of eqn (10) compensate each other and the deformation ratio can be expressed by the volume fractions

$$\Lambda_e = \left(\frac{\phi_e}{\phi}\right)^m \tag{11}$$

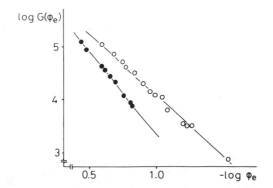

Fig. 5. Double logarithmic representation of variations in shear moduli, $G(\phi_e)$, with the volume fraction of the polymer for PVAc gel homologues swollen by toluene ($T = 25°C$) ($\bigcirc$) and isopropyl alcohol ($T = 52°C$) ($\bullet$). Solid lines were calculated by least-squares fitting through the experimental points. Slopes: 2·33 (PVAc/toluene); 2·97 (PVAc/isopropyl alcohol).

In Fig. 6 experimental results are presented according to eqn (11). In the figure, results of Treloar[3] obtained for natural rubber samples swollen by benzene and heptane are also shown. In these cases, both elongation and compression measurements were performed. The solid curve was calculated with exponent $m = 9/4$, while the dotted line with exponent $m = 2$. Experimental points scatter around the solid curve for both types of deformation. In Table 3 deformation ratio and volume fraction data taken

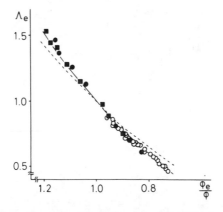

Fig. 6. Plot of Λ_e as a function of ϕ_e/ϕ for natural rubber/benzene[3] ($\blacksquare$), natural rubber/heptane[3] ($\bullet$) and PVAc/toluene ($\bigcirc$) gel systems. The solid curve was calculated with $m = 9/4$, the dotted line with $m = 2$.

TABLE 3
Volume Fraction Data for Natural Rubber/Benzene
Gels: Comparison of Experimental Results of Treloar[3]
with Calculations

Experimental data		Calculated ϕ values $(\phi = \phi_e \Lambda_e^{-4/9})$
Λ_e	ϕ	
0·753	0·208	0·210
0·882	0·196	0·197
0·977	0·191	0·188
1·000	0·186 $(= \phi_e)$	0·186
1·144	0·180	0·180
1·277	0·173	0·171
1·410	0·167	0·164
1·451	0·163	0·162
1·549	0·160	0·157
1·642	0·157	0·153

from Treloar's paper[3] are summarized. The volume fraction data given in
the third column were calculated from experimental Λ_e and ϕ_e values. The
agreement between experiments and calculation is satisfactory.

In Fig. 7, prediction of eqn (11) is compared with experimental findings
for PVAc gel homologues swollen by a θ-diluent. The solid curve was
calculated with exponent $m = 3$. PVAc/isopropyl alcohol gels fit the
theoretical curve rather well.

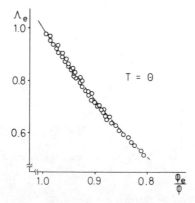

Fig. 7. Plot of Λ_e as a function of ϕ_e/ϕ for PVAc gel homologues swollen by
isopropyl alcohol ($T = \theta = 52°C$). The solid curve was calculated with $m = 3$.

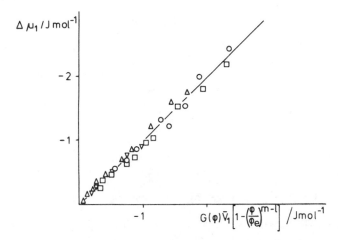

Fig. 8. The chemical potential change of the diluent ($\Delta\mu_1$) plotted against $G(\phi)\bar{V}_1[1 - (\phi/\phi_e)^{m-l}]$. The diagonal straight line corresponds to the equivalence of the response of the network to isotropic shrinking and to unidirectional compression. Experimental points refer to PVAc/toluene gels ($T = 25°C$). Polymer concentrations when crosslinks were introduced: ○, 12; □, 9; △, 6; ▽, 3% (m/m).

From eqn (10) follows the correspondence between anisotropic deswelling induced by unidirectional deformation and isotropic deswelling due to lowering the chemical potential of the diluent. Equation (12) gives the chemical potential difference required to shift the volume fraction of the freely swollen gel from ϕ_e to ϕ:

$$\Delta\mu_1 = G(\phi)\bar{V}_1\left[1 - \left(\frac{\phi}{\phi_e}\right)^{m-l}\right] = G(\phi_e)\bar{V}_1[\Lambda_e^{-l/m} - \Lambda_e^{-1}] \tag{12}$$

$\Delta\mu_1$ is known from equilibrium deswelling experiments and $G(\phi)$ can be calculated from mechanical measurements. In Fig. 8 experimental results can be seen. The points scatter around the theoretical straight line, indicating the equivalence of the response of the network to isotropic shrinking and to unidirectional compression.

ACKNOWLEDGEMENT

This work was supported by the Hungarian Academy of Sciences under contract AKA No. 1-36-86-229.

REFERENCES

1. Flory, P. J. and Rehner, J. Jr, Statistical mechanics of cross-linked polymer networks. II. Swelling, *J. Chem. Phys.*, 1943, **11**, 521–6.
2. Flory, P. J. and Rehner, J. Jr, Effect of deformation on the swelling capacity of rubber, *J. Chem. Phys.*, 1944, **12**, 412–14.
3. Treloar, L. R. G., The swelling of cross-linked amorphous polymers under strain, *Trans. Faraday Soc.*, 1950, **50**, 783–9.
4. Treloar, L. R. G., The equilibrium swelling of crosslinked amorphous polymers, *Proc. Roy. Soc.*, 1950, **A200**, 176–83.
5. Gee, G., The interaction between rubber and liquids. X. Some new experimental tests of a statistical thermodynamic theory of rubber–liquid systems, *Trans. Faraday Soc.*, 1946, **42B**, 33–50.
6. Boyer, R. F., Deswelling of gels by high polymer solutions, *J. Chem. Phys.*, 1945, **13**, 363–72.
7. Gee, G., Herbert, J. B. M. and Roberts, R. C., The vapour pressure of a swollen cross-linked elastomer, *Polymer*, 1965, **6**, 541–8.
8. Yen, L. Y. and Eichinger, B., Volume dependence of the elastic equation of state, *J. Polym. Sci., Polym. Phys. Ed.*, 1978, **16**, 121–30.
9. Horkay, F. and Nagy, M., Elasticity of swollen poly(vinyl alcohol) and poly(vinyl acetate) networks, *Polym. Bull.*, 1980, **3**, 457–63.
10. Horkay, F. and Zrinyi, M., Effect of deformation on the swelling equilibrium concentration of gels, *Macromolecules*, submitted.
11. Nagy, M. and Horkay, F., A simple and accurate method for the determination of solvent activity in swollen gels, *Acta Chim. Acad. Sci. Hung.*, 1980, **104**, 49–61.
12. Horkay, F., Nagy, M. and Zrinyi, M., Apparatus for the measurement of the mechanical–rheological properties of gels, *Acta Chim. Acad. Sci. Hung.*, 1980, **103**, 387–95.
13. Flory, P. J., *Principles of Polymer Chemistry*, Cornell University Press, Ithaca, New York, 1953.
14. Treloar, L. R. G., *The Physics of Rubber Elasticity*, Clarendon Press, Oxford, 1976.
15. Bastide, J., Candau, S. and Leibler, L., Osmotic deswelling of gels by polymer solutions, *Macromolecules*, 1981, **14**, 719–26.
16. de Gennes, P. G., *Scaling Concepts in Polymer Physics*, Cornell University Press, Ithaca, New York, 1979.
17. Candau, S., Bastide, J. and Delsanti, M., Structural, elastic and dynamic properties of swollen polymer networks, *Adv. Polym. Sci.*, 1982, **44**, 27–71.
18. Horkay, F. and Zrinyi, M., Test of the scaling law for the elastic modulus of chemically different network systems, *Polym. Bull.*, 1981, **4**, 361–8.
19. Horkay, F. and Zrinyi, M., Extension of the scaling approach to gels swollen to equilibrium in a diluent of arbitrary activity, *Macromolecules*, 1982, **15**, 1306–10.
20. Zrinyi, M. and Horkay, F., Crossover effects above and below the θ temperature, *Macromolecules*, 1984, **17**, 2805–11.
21. Geissler, E. and Hecht, A. M., The Poisson ratio in polymer gels, *Macromolecules*, 1980, **13**, 1276–80.
22. Dusek, K. and Prins, W., Structure and elasticity of non-crystalline polymer networks, *Adv. Polym. Sci.*, 1969, **6**, 1–102.

31

ON THE TEMPERATURE DEPENDENCE OF EQUILIBRIUM CONCENTRATION OF SLIGHTLY CROSSLINKED GELS

Miklós Zrinyi

Department of Colloid Science, Loránd Eötvös University,
Puskin ut. 11-13, H-1088 Budapest, Hungary

and

Ferenc Horkay*

Research Laboratory for Inorganic Chemistry, Hungarian Academy of
Sciences, Budaörsi ut. 45, H-1112 Budapest, Hungary

ABSTRACT

The temperature dependence of equilibrium concentration of chemically crosslinked swollen polymer networks was investigated. The predictions of the classical James–Guth theory and the scaling theory were compared with each other as well as with the experimental findings.

It was shown that the scaled concentration is a function of the scaled temperature only, according to both theories. Experimental results obtained for poly(vinyl acetate) gel homologues swollen by isopropyl alcohol supported the validity of the theoretical predictions.

INTRODUCTION

Considerable attention has been paid recently to the theoretical and experimental investigation of the influence of solvent power on network properties. In particular, the effect of temperature (or solvent composition) as well as the crosslinking density on the equilibrium swelling degree has attracted many discussions.[1-4] New concepts like scaling and universality

* Permanent address: Department of Colloid Science, Loránd Eötvös University, Puskin ut. 11–13, H-1088 Budapest, Hungary.

have been proved to be highly appropriate tools in the understanding of gel properties.[2,5]

The purpose of the present work is to show that the concept of scaling can be successfully used even in the framework of classical James–Guth theory. The dependence of equilibrium concentration of slightly crosslinked swollen network systems will be considered here, in order to compare the prediction of scaling and James–Guth approaches with each other as well as with the experimental results.

THEORETICAL CONSIDERATIONS

The scaling theory can be simply postulated mathematically. Experience with phase transition,[6] percolation[7] and chain conformation[5] has indicated that most functions $F(x, y)$ of two variables (x and y) can be written in a simpler form

$$F(x, y) = x^a f(y/x^b) \tag{1}$$

if both x and y approach zero. In eqn (1), a and b are constants and f is a function of a new variable y/x^b. Despite the fact that eqn (1) is not a general mathematical theory, most simple functions have this property. If the values for x and y are sufficiently small to make the correction terms unimportant, one can test eqn (1) by plotting the scaled quantity $Q = F(x,y)/x^a$ versus the scaled variable $z = y/x^b$. If eqn (1) is valid, the points belonging to different x and y values all collapse into a single curve, the function $f(z)$. Thus scaling means that if $F(x, y)$ is scaled by some power of x, and y by some other power of x, then the resulting scaled quantity Q is a function of only one scaled variable z, instead of being a function of two independent variables x and y.

Now let us apply this idea for slightly crosslinked networks (large N) swollen to equilibrium in a poor solvent (the temperature T is near the θ-temperature). Let the function $F(x, y)$ equal the equilibrium concentration ϕ, x belong to the reciprocal value of N (we need small x) and y stand for the reduced temperature $\tau = (T - \theta)/T$. Making these substitutions into eqn (1), one obtains

$$\phi(N^{-1}, \tau) = N^{-a} \times f(\tau/N^b) \tag{2}$$

Now the main task is to determine the value of exponents a and b and the function $f(z)$. To achieve this, the 'c^* theorem' may help.[2,5] It says that the equilibrium polymer concentration of a swollen network can be expressed in terms of overlap concentration, ϕ^*, of the solution of a polymer having the same molecular mass as the network chains:

$$\phi \propto \phi^* \propto \frac{N}{R^3} \propto N^{1-3v} \qquad (3)$$

where R is the radius of gyration of the equivalent polymer chain in dilute solution, which scales with N^v, where v is the excluded volume exponent. This idea is strongly supported by recent small-angle neutron scattering measurements.[8]

The exponent a can be determined easily if we consider the θ-state. In this case $\tau = 0$, $f(0) = $ constant, and the relation $\phi \propto N^{-1/2}$ (see eqn (3) with $v = 1/2$) must hold. As a result of this $a = 1/2$.

The other exponent, b, can be determined by the following scaling consideration. According to Daoud and Jannink,[9] the temperature dependence of the radius of gyration of a flexible chain is $R(\tau) \propto |\tau|^{(2v-1)}$. This results in $\phi(\tau) \propto |\tau|^{-3(2v-1)}$ (see eqn (3)), and consequently the relation

$$f(\tau/N^b) \propto (\tau/N^b)^{-3(2v-1)} \qquad (4)$$

must hold.

We know from eqn (3) that the left-hand side of eqn (2) must be proportional to N^{1-3v}; therefore we have to choose a value for b such that the N dependence of the right-hand side of eqn (2) results in N^{1-3v}. In other words, the exponents of N must be equal on both sides, i.e. the solution of the equation

$$1 - 3v = 3b(2v-1) - 1/2 \qquad (5)$$

gives the value of b. Solving eqn (5) for b, one obtains $b = -1/2$ independently of the value of v. Thus if eqn (2) is valid, the equilibrium concentration divided by $N^{-1/2}$ is a single variable function of $\tau/N^{-1/2}$.

To check the validity of eqn (2) one needs to know the polymerization index of the network chains. However, the experimental determination of N is a rather complicated task, and therefore it is much more convenient to use the equilibrium concentration ϕ_θ measured at the θ-temperature instead of N. The correspondence between ϕ_θ and N is given by eqn (3). At the θ-temperature, $v = 1/2$ and $\phi_\theta \propto N^{-1/2}$, and eqn (2) can be written as follows:

$$\phi = \phi_\theta \times f\left(\frac{\tau}{\phi_\theta}\right) \qquad (6)$$

This scaling relation has already been tested previously.[3,10] It was verified that the ratio ϕ/ϕ_θ is the only function of τ/ϕ_θ for poly(vinyl acetate) network homologues swollen in isopropyl alcohol at either below or above the θ-temperature.

To give the form of the scaling function, f, one needs models. The results concerning the thermal blob theory can be found elsewhere.[3,11]

Here we want to show that eqn (5) can also be derived from the classical James–Guth theory,[12] and we want to determine the scaling function on this basis.

Dependence of the Equilibrium Concentration on the Temperature According to the James–Guth Theory (JG)

The condition of equilibrium for free swelling is given by

$$Av^*q_0^{-2/3}V\phi^{1/3} + \ln(1-\phi) + \phi + \chi\phi^2 = 0 \tag{7}$$

where v^* is the amount (expressed in moles) of network chains per unit dry polymer volume, V is the partial molar volume of the diluent, χ is the Huggins interaction parameter and q_0 is the isotopic deformation factor. The 'front factor' A includes the possible corrections due to elastically ineffective chain ends, loops and entanglements. It is often taken as unity if, instead of network chains, the elastically active chains are regarded.[1]

Since we are considering loosely crosslinked networks ($\phi \ll 1$), it is more straightforward to use the virial form of eqn (7). In this approach the logarithmic term is substituted by its third-order series and generalized in order to take into account not only the binary, but also the ternary, monomer interactions:

$$e\phi^{1/3} - u\phi^2 - w\phi^3 = 0 \tag{8}$$

where $e = Av^*q_0^{-2/3}$, and u and w are dimensionless second and third virial coefficients of monomer interactions, respectively. In the literature, u is usually identified as $u = 1/2 - \chi$. The third virial coefficient w is supposed to be temperature independent and dominated by the chain flexibility. In eqn (8) the only quantity which depends on the temperature is u, as far as the difference in heat expansion between the dry network and the solvent can be neglected. In our analysis we neglect the possible temperature dependence of the product $v^*Vq_0^{-2/3}$.

Near the Critical Solution Temperatures, and consequently near the θ-temperature, the second virial coefficient is linearly proportional to the reduced temperature:

$$u = \frac{d_n}{\theta} \times \tau \tag{9}$$

Equation (9) comes from the following choice of the Huggins parameter:[13]

$$\chi = a_s + \frac{d_n}{T} \tag{10}$$

where a_s and d_n are related to the temperature-independent entropy term (a_s) and enthalpy term (d_n).

At the θ-state, $\tau = 0$, $u = 0$ and $\chi = 1/2$, and eqn (8) can be written as follows:

$$e\phi_\theta^{1/3} = w\phi_\theta^3 \tag{11}$$

Combination of eqns (8), (9) and (11) results in

$$\phi^{1/3} - c\tau\phi_\theta^{-8/3}\phi^2 - \phi_\theta^{-8/3}\phi^3 = 0 \tag{12}$$

where

$$c = \frac{d_n}{w\theta}$$

One can see that the equilibrium concentration depends on the reduced temperature; the crosslinking density, which is taken into consideration via ϕ_θ; and the numerical factor c, which includes the effects of chain stiffness and interaction between the network polymer and the swelling agent.

For network homologues swollen in the same liquid, c is constant and consequently the equilibrium concentration is determined by τ and ϕ_θ. At first sight one may think that ϕ is a two-variable, implicit function of τ and ϕ_θ; however, eqn (12) has simple scaling properties with respect to the reduced temperature, and as a result of this the number of variables can be decreased.

Introducing scaled concentration $\tilde{\phi}$ and scaled temperature $\tilde{T}$, defined as

$$\tilde{\phi} = \frac{\phi}{\phi_\theta} \quad \text{and} \quad \tilde{T} = \frac{\tau}{\phi_\theta} \tag{13}$$

one can see that τ is eliminated from eqn (12) and the equilibrium concentration can be given by a single variable function:

$$(\tilde{\phi})^{-5/3} - \tilde{\phi} = c\tilde{T} \tag{14}$$

Equation (14) predicts that, in the case of network homologues ($c = $ constant), the $\tilde{\phi}$ vs. $\tilde{T}$ curve is independent of structural details, such as the crosslinking density v^* and the isotropic deformation factor q_0. Now the qualitative agreement between eqns (6) and 14 is obvious.

EXPERIMENTAL RESULTS AND DISCUSSION

For the experiments, gel systems of poly(vinyl acetate) (PVAc) swollen to equilibrium in isopropyl alcohol were used. The θ-temperatures of PVAc/isopropyl alcohol gels were found to be 54·7°C in our previous work.[3,10] A more critical analysis of data has shown that $\theta = 52 \pm 2$°C. Thus for the calculations $\theta = 325$ K is used.

The preparation of gel samples can be found in our other papers.[3,14,15]

Networks with different crosslinking density were prepared. The crosslinking density x_c, which is the mole fraction of crosslinking agent in the dry network, was varied from 2×10^{-3} to 2×10^{-2}. Not only the crosslinking density, but also the concentration, c_0, at which the crosslinks were introduced, were varied. Some characteristics of the PVAc networks can be seen in Table 1.

The equilibrium concentrations of gels were determined by a weighing technique. At each of the temperatures 25, 30, 37, 45, 50, 55, 60 and 70°C the gels were stored for not less than 1 month to attain equilibrium. The density of the dry networks and that of the isopropyl alcohol were also measured as a function of temperature. The volume fraction of the polymer in the equilibrated gel was calculated from the measured mass and density values, supposing the additivity of the specific volumes. The error of concentration measurements did not exceed 0·2%.

In Fig. 1 the concentration versus temperature curves are shown.

In order to compare the experimental results with the prediction of eqns (6) and (14), the scaled concentration is plotted against the scaled

TABLE 1
Characteristics of the PVAc Gels[a]

Symbol	C_0 (wt%)	$10^2 \times x_c$
□	3	2
×	6	2
■	6	0·5
○	9	2
●	9	1
△	9	0·5
▲	9	0·25
▽	12	2
▼	12	0·5

[a] The same symbols are used on Figs 1 and 2.

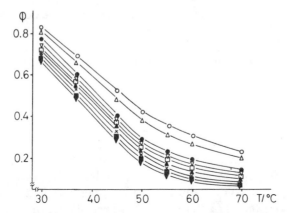

Fig. 1. Concentration vs. temperature diagram for poly(vinyl acetate)/isopropyl alcohol gels. For identification of symbols see Table 1.

temperature in Fig. 2. It can be seen that, within the experimental accuracy, all points are on the same curve. This result supports the validity of eqns (6) and (14), since ϕ seems to depend only on one variable, $\tilde{T}$.

To compare the experimental results quantitatively with eqn (14), one needs the value of c. This parameter has to be determined by direct comparison between theory and experiment. For the PVAc/isopropyl alcohol gels the best fit was obtained at $c = 0.1$.

Figure 2 shows how eqn (14) gives the temperature dependence of

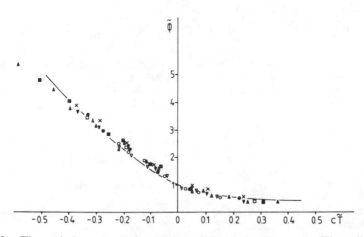

Fig. 2. The scaled concentration against the scaled temperature. The solid line was calculated by eqn (14).

equilibrium concentration. It can be seen that the agreement is quite good: only a small deviation can be observed in the vicinity of the θ-point.

A more detailed analysis of experimental data, as well as the comparison of scaling functions obtained by other network theories, can be found in Ref. 11.

CONCLUSION

We have studied the temperature dependence of equilibrium concentration of loosely crosslinked swollen networks. Two basically different approaches (the classical James–Guth and the scaling approach) were considered in order to obtain the $\tilde{\phi}(\tilde{T})$ relation. It has been shown that the JG theory results in a dependence which is qualitatively consistent with the prediction of the scaling approach, i.e. both give a 'universal' dependence if scaled concentration and scaled temperature are used.

Experiments performed on poly(vinyl acetate) gels support these findings.

ACKNOWLEDGEMENT

The authors gratefully acknowledge the support of AKA and OTKA.

REFERENCES

1. Dusek, K. and Prins, W., Structure and elasticity of non-crystalline polymer networks, *Advances in Polymer Science*, 1969, **6**, 1–102.
2. Candau, S., Bastide, J. and Delsanti, M., Structural, elastic and dynamic properties of swollen polymer networks, *Advances in Polymer Science*, 1982, **44**, 27–72.
3. Zrinyi, M. and Horkay, F., Crossover effects above and below the θ temperature, *Macromolecules*, 1984, **17**, 2805–11.
4. Erman, B. and Baysal, B., Temperature dependence of swelling of polystyrene networks, *Macromolecules*, 1985, **18**, 1696–9.
5. de Gennes, P. G., *Scaling Concepts in Polymer Physics*, Cornell University Press, Ithaca, New York, 1979.
6. Stanley, H. E., *Introduction to Phase Transitions and Critical Phenomena*, Clarendon Press, Oxford, 1971.
7. Stauffer, D., *Introduction to Percolation Theory*, Taylor and Francis, London and Philadelphia, 1985.

8. Beltzung, M., Herz, J. and Picot, C., Investigation of the conformation of elastic chains in poly dimethylsiloxane networks by small-angle neutron scattering, *Macromolecules*, 1983, **15**, 1594–1604.

9. Daoud, M. and Jannink, G., Temperature–concentration diagram for polymer solutions, *Journal de Physique*, 1976, **37**, 973–7.

10. Zrinyi, M. and Wolfram, E., Temperature–concentration dependence of polyvinyl acetate gels with respect to the collapse phenomenon, *Progress in Colloid and Polymer Science*, 1985, **71**, 20–5.

11. Zrinyi, M. and Horkay, F., A Comparative study on the meanfield and scaling theories of temperature–concentration dependence of slightly crosslinked network systems, *Macromolecules*, submitted.

12. James, H. M. and Guth, E., Theory of the elastic properties of rubber, *Journal of Chemical Physics*, **11**, 455–81.

13. Flory, P. J., *Principles of Polymer Chemistry*, Cornell University Press, Ithaca, New York, 1953.

14. Horkay, F. and Zrinyi, M., Extension of the scaling approach to gels swollen to equilibrium in a diluent of arbitrary activity, *Macromolecules*, 1982, **15**, 1306–10.

15. Horkay, F. and Zrinyi, M., Gels immersed in polymer solutions, *Journal of Macromolecular Science, Phys.*, 1986, **B25**(3), 307–34.

32

NMR APPROACH TO THE SWELLING PROCESS OF PDMS NETWORKS AND SILICA-FILLED SILOXANES

J. P. COHEN-ADDAD and A. VIALLAT

Laboratoire de Spectrométrie Physique associé au CNRS, Université Scientifique, Technologique et Médicale de Grenoble, BP 87, 38402 St Martin d'Hères Cedex, France

ABSTRACT

This work deals with the NMR observation of the progressive swelling of siloxane systems. Gels resulting from covalent crosslinks, on the one hand, and network-like structures formed from silica-filled siloxane, on the other hand, were examined. The transverse magnetization of protons was used to probe the unfolding mechanism of elementary chains. The residual energy of spin–spin interactions governing the magnetization dynamics was found to obey static scaling laws upon variation of the solvent concentration. In calibrated gels, these properties reflect both a packing condition and screening effects due to trapped entanglements, whereas neither any affine property nor a packing condition could account for scaling properties observed in silica-filled siloxane systems.

The spin–lattice relaxation rate was found to be independent of the silica concentration, on the one hand, and of the polymer weight fraction w_2 (for $w_2 < 0.5$), on the other hand. These two results were assumed to give evidence for the presence of dynamic screening domains, analogous to those existing in any entangled chain solution.

1 INTRODUCTION

The purpose of the present work was to illustrate the NMR approach to the swelling process of siloxane gels: calibrated gels formed from end-linked chains and network-like structures obtained from random silica–siloxane

471

mixtures. The magnetic relaxation of protons bound to siloxane chains was used to investigate unfolding mechanisms of elementary chains, induced by a progressive swelling of these systems.

Elementary chains are known to obey Gaussian statistics in dry gels. When these are swollen at equilibrium in a good solvent, elementary chains obey both a packing condition and excluded volume statistics: the maximum swelling is described by the c^*-theorem.[1,2] Statistical properties of elementary chains in a gel at intermediate swelling states have not yet been given a full theoretical description, even though some approaches have been recently proposed.[3]

The present NMR approach to the observation of the progressive unfolding of elementary chains is based on the existence of a residual energy of spin–spin interactions responsible for the irreversible dynamics of the transverse nuclear magnetization.[4] This energy reflects an average orientational order of monomeric units induced by topological constraints exerted on chain segments and resulting from entanglements or crosslinks. The progressive swelling of covalent gels formed from end-linked polydimethylsiloxane (PDMS) chains has been already described from NMR, according to a two-step process.[5] Starting from a dry gel, a disinterspersion of chains first occurs: it is analogous to that observed in an uncrosslinked system by adding small amounts of solvent. This effect is associated with a decrease in the magnetic relaxation rate δ. In the second step, elementary chains are progressively swollen. This unfolding keeps junctions away from one another: elementary chains are slightly stretched and the relaxation rate δ is progressively increased. In this study, effects of the polymer concentration before gel formation upon swelling properties are examined over a wide range of values, using two swelling agents.

The present work also concerns gel-like structures formed from siloxane chains linked to silica particles through hydrogen bonding. Random mixtures resulting from a saturated adsorption of chains onto the silica surface may behave like permanent gels: they can be reversibly swollen, using a good solvent. Mineral filler particles serve the function of interlinkages.[6]

2 NMR APPROACH

2.1 The Spin System
NMR probes used to investigate properties of the network structures are three-proton methyl groups. Random motions of any methyl group

attached to long polymer chains are known to occur according to a time scale split into two well-separated intervals. A fast diffusional rotation of the methyl group about its $\vec{c}$-axis of symmetry occurs during the first interval of the time scale. The second time interval is associated with the slow orientational diffusion of the $\vec{c}$-axis, induced by random motions of the chain skeleton. Accordingly, dipole–dipole interactions existing within a methyl group are averaged according to a two-step process. Furthermore, because of symmetry properties of the three-spin system, these interactions are divided into two parts which induce two separated responses of the spin system:[7] the transverse component $m(t)$ consists of two terms

$$m(t) = \mu_1(t) + \mu_2(t)$$

$\mu_1(t)$ only depends upon fast random rotations of a methyl group about its $\vec{c}$-axis—it usually corresponds to a long relaxation time ($\simeq 1$ s); $\mu_2(t)$ only depends upon time fluctuations of the angle $\theta_c(t)$ which the $\vec{c}$-axis makes with the direction of the steady magnetic field; $\mu_1(0)$ and $\mu_2(0)$ have equal amplitudes.

2.2 Residual Energy of Spin–Spin Interactions

In polymeric gels, relative fluctuations in space of two junctions connected by a chain segment are not isotropic. From the NMR point of view, it is considered that locations of junctions are frozen: they correspond to their positions averaged in space. Any elementary chain connecting two successive junctions or two trapped entanglements is supposed to have a fixed end-to-end vector $\vec{r}_e$. Accordingly, skeletal bonds cannot undergo isotropic rotations, and spin–spin tensorial interactions of nuclei which depend upon the angle $\theta_c(t)$ are not averaged to zero. Thus a residual energy ε_e of interaction arises from these non-isotropic motions: it governs the part $\mu_2(t)$ of the spin system response. The residual energy ε_e has been already calculated:

$$\varepsilon_e = \beta(3\cos^2\theta_e - 1)\frac{r_e^2}{N_e^2 a^2} \tag{1}$$

where a is the mean skeletal bond length; β accounts for flexibility properties of the chain segment; θ_e is the angle which $\vec{r}_e$ makes with the steady magnetic field direction; and N_e is the number of skeletal bonds in the chain segment. Neglecting all magnetic interactions between different methyl groups, the transverse magnetic relaxation function is expressed as

$$M_2(t) = \langle \mu_2(t) \rangle = \tfrac{1}{2}\langle \cos(\varepsilon_e(\vec{r}_e))t \rangle_{\vec{r}_e} \tag{2}$$

The average is calculated over all orientations of $\vec{r}_e$ vectors in space, assuming an isotropic distribution. The averaged contribution $M_1(t)$ originating from $\mu_1(t)$ must be added to $M_2(t)$ to describe the whole spin-system response. In semidilute or dilute solutions, $M_1(t)$ is expected to decay within a long time scale (1 s).

2.3 Concentrated Solutions

In the case of real chains in concentrated solutions, interactions between different methyl groups cannot be neglected An additional residual energy contributes to the relaxation mechanism of the part $\mu_1(t)$ of the magnetization. It can appreciably reduce the relaxation time of this contribution.

3 EXPERIMENTAL

3.1 Materials

Covalent polydimethylsiloxane gels were synthesized according to an experimental procedure previously reported.[8] Silica–siloxane systems were prepared by mechanically mixing high molecular weight siloxane chains with fused silica particles.[9] Free polymer fractions were removed from random mixtures by exhaustive extraction, using a good solvent; the corresponding experimental procedure has been already described elsewhere.[9] Every sample was kept at equilibrium with solvent vapor in a sealed tube. All measurements were performed at 313 K.

3.2 NMR Measurements

All NMR measurements were performed by using a CXP 100 Bruker spectrometer, operating at 60 MHz. Relaxation functions of the transverse magnetization were determined from spin–echo sequences. They were numerically analyzed from the sum of two exponential time functions of equal initial amplitudes.

4 CALIBRATED GELS: SCALING PROPERTIES

The PDMS networks considered are tetrafunctional gels. The weight-average chain molecular weight between two covalent bridges is 9.7×10^3. The polymer concentration v_c of gel synthesis was given five values: 0·56, 0·664, 0·722, 0·823 and 0·884. Two swelling agents were used: benzene and toluene. Observed maximum swelling ratio values are reported in Table 1.

TABLE 1
Calibrated Gels: Maximum Swelling Ratio Values

v_c	Q_{max} (toluene)	Q_{max} (benzene)
0·56	7·4	6·7
0·664	6·3	5·6
0·722	6·1	5·4
0·823	5·5	5·4
0·884	5·5	5·5

4.1 Progressive Swelling: A Packing Condition

Variations of the relaxation rate δ_2 of the $\mu_2(t)$ part of the transverse magnetization are reported as a function of the swelling ratio Q, according to a logarithmic plot represented in Fig. 1. The relaxation rate δ_2 is found to increase upon addition of solvent. All curves exhibit a linear variation characterized by a slope equal to 0.65 ± 0.05. Similar results have been

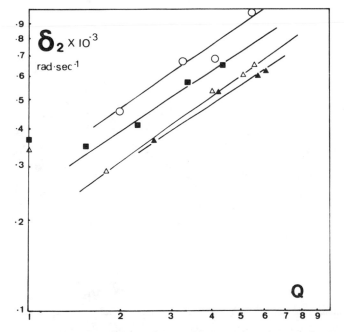

Fig. 1. Log–log plot of variations of the transverse relaxation rate δ_2 observed on calibrated gels as a function of the swelling ratio Q. $\bigcirc$, $v_c = 0.884$; $\blacktriangle$, $v_c = 0.56$, the swelling agent is deuterated toluene; $\triangle$, $v_c = 0.664$; $\blacksquare$, $v_c = 0.823$, the swelling agent is deuterated benzene.

previously observed in calibrated gels considering different polymer concentrations of gel synthesis.[5] The power law

$$\delta_2 \propto Q^{2/3}$$

is assumed to reflect both a packing condition and an affine deformation of elementary chains, starting from

$$Q \propto r_e^3/a^3 N_e$$

and using eqn (1)

$$\delta_2 \propto Q^{2/3}/N_e^{4/3} \tag{3}$$

Experimental results clearly show that average distances between two consecutive junctions, compared with overall dimensions of a gel, present affine variations. This property applies to intermediate swelling at

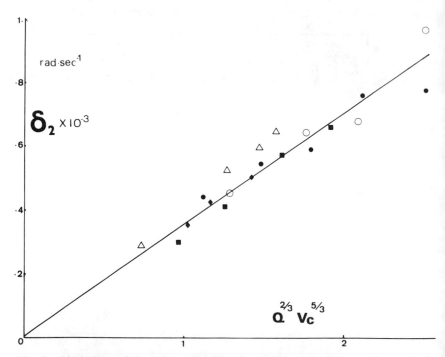

Fig. 2. Variations of the transverse relaxation rate δ_2 observed on calibrated gels as a function of the reduced variable $Q^{2/3}v_c^{5/3}$: $\triangle$, $v_c = 0.664$; $\blacklozenge$, $v_c = 0.722$; $\blacksquare$, $v_c = 0.823$; $\bullet$, $v_c = 0.884$, the swelling agent is deuterated benzene; $\bigcirc$, $v_c = 0.884$, the swelling agent is deuterated toluene.

equilibrium states as well as to the maximum swelling at equilibrium with the solvent. Correspondingly, eqn (3) indicates that amplitudes of fluctuations of junction points in space are small compared with mean chain dimensions.

4.2 Trapped Entanglements

Effects of trapped entanglements can be observed by varying the polymer concentration v_c of gel formation; then the average number of monomeric units between entanglements is determined from

$$N_e \propto v_c^{-5/4} \tag{4}$$

Variations of δ_2 are reported as a function of the reduced variable $Q^{2/3}v_c^{5/3}$ in Fig. 2. All curves are superposed with one another, within reasonable agreement. This result shows that the effective mesh size of a gel is determined by dynamic screening effects occurring in the solution before the gelation process. Trapped topological constraints act as covalent junctions. They induce a cut in correlations of space fluctuations of elementary chains.

5 RANDOM SILICA–SILOXANE MIXTURES

The size and the concentration of silica particles, on the one hand, and the average chain molecular weight, on the other hand, were chosen in a way appropriate to the formation of a network structure: a multiple-aggregate chain binding process occurs. It is worth noting that these gel-like systems are characterized by a high functionality and a large number of dangling chains. Furthermore, entanglements originating from the pure melt are probably trapped after the mechanical mixing. Properties of elementary chains joining silica aggregates have been investigated in three silica-filled siloxane network structures. They are described in Table 2 and will be hereafter referred to as a, b and c.

The progressive unfolding of elementary chains induced by adding small controlled amounts of solvent to dry systems is reflected by variations of the relaxation rate δ_2. These are reported as a function of the linear gel dimension q in Figs 3, 4 and 5. δ_2 was found to obey the following power laws:

$$\delta_2 - \delta_2(0) \propto q^{3.15} \quad \text{for sample a}$$
$$\delta_2 - \delta_2(0) \propto q^{6.75} \quad \text{for sample b}$$
$$\delta_2 - \delta_2(0) \propto q^{1.1} \quad \text{for sample c}$$

TABLE 2
Siloxane–Silica Mixtures

$M_w{}^a$	$I_p{}^b$	$C_{Si}{}^c$ (w/w)	$Q_r{}^d$ (w/w)	Q_{max} (C_7H_{14})
(a) $6\cdot1 \times 10^5$	1·7	0·29	1·7	6·5
(b) $6\cdot1 \times 10^5$	1·7	0·17	2·4	8·2
(c) $2\cdot4 \times 10^5$	1·8	0·23	1·2	5·2

^{a} M_w = weight-average molecular weight.
^{b} I_p = polydispersity index.
^{c} C_{Si} = initial silica concentration.
^{d} Q_r = residual amount of polymer bound to silica after exhaustive extraction.

The purpose of this NMR study was to investigate statistical properties of elementary chains which do not necessarily obey an affine property because of the complex structure of the polymeric medium. The presence of screening domains characterizes any polymer–solvent system. Likewise, it is assumed that the addition of solvent to random mixtures gives rise to small domains. Then, considering eqn (1), both the end-separation vector $\vec{r}_e$

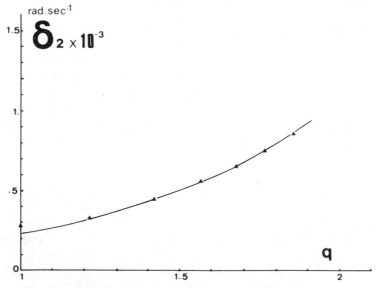

Fig. 3. Variations of the relaxation rate δ_2 observed on sample a as a function of the linear gel dimension q. The swelling agent is deuterated methylcyclohexane.

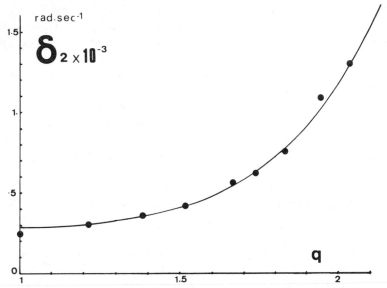

Fig. 4. Variations of the relaxation rate δ_2 observed on sample b as a function of the linear gel dimension q. The swelling agent is deuterated methylcyclohexane.

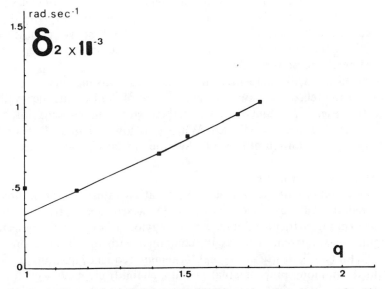

Fig. 5. Variations of the relaxation rate δ_2 observed on sample c as a function of the linear gel dimension q. The swelling agent is deuterated methylcyclohexane.

and the average number N_e of monomeric units determining a domain can be dependent upon the swelling ratio Q. Accordingly, even though small domains obey a packing condition, the relaxation rate δ_2 does not reflect affine properties. An analogous situation has been already described concerning the elastic modulus of vulcanized systems.[3] It may be worth emphasizing that the role of dangling chains in any polymeric gel has not yet been clearly elucidated. Their contribution to the swelling mechanism of silica–siloxane mixtures may also partly govern the δ_2 dependence upon the swelling ratio Q. Semi-local investigations clearly show that the silica concentration is closely involved in the progressive swelling mechanism.

6 LOCAL PROPERTIES: EXISTENCE OF CORRELATION DOMAINS

Dynamical local properties of silica–siloxane mixtures were investigated from measurements of both the spin–lattice relaxation rate T_1^{-1} and the relaxation rate δ_1 characterizing the second part $M_1(t)$ of the transverse magnetic relaxation function.

The parameter T_1^{-1} reflects both the rotational diffusion of a methyl group about its $\vec{c}$-axis and high relaxational frequency motions of short-chain segments. The relaxation rate δ_1 is determined from dipole–dipole interactions between protons located on different methyl groups. The strength of these interactions varies with the distance b between nuclei according to the power law b^{-3}.

All measurements were performed on silica–siloxane mixtures. Variations of the relaxation rates T_1^{-1} and δ_1 are plotted versus the swelling ratio Q in Fig. 6: T_1^{-1} and δ_1 were found to exhibit similar behavior. Two concentration ranges are defined. When Q is lower than 2, T_1^{-1} and δ_1 decrease upon addition of solvent; then they reach plateau values.

6.1 Dynamic Screening Effects

Similar variations of the spin–lattice relaxation rates have been already observed in polymer–solvent systems. Data from Ref. 10 have been reported in Fig. 6; they exhibit the same behavior although chain molecular weights were different. This result, combined with the observation of a plateau for $Q \geq 2$, clearly shows that local high frequency motions are fully shielded from long-range fluctuations of a chain. It gives evidence for the existence of dynamical correlation domains in silica–siloxane mixtures; they probably originate from entanglements existing in the melt.

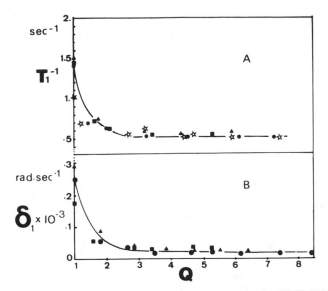

Fig. 6.   (A) Variations of the longitudinal relaxation rate T_1^{-1} as a function of the swelling ratio Q (60 MHz). (B) Variations of the relaxation rate δ_1 as a function of the swelling ratio Q. Key: ▲, sample a; ●, sample b; ■, sample c; ☆, data from Ref. 10 ($\bar{M}_w = 7·7 \times 10^4$ and $13·8 \times 10^4$)—for these data Q is the dilution ratio.

6.2 Orientational Interactions of Methyl Groups

The parameter δ_1 mainly characterizes dipole–dipole interactions of protons attached to different methyl groups. The decrease of δ_1 upon addition of solvent reveals both a decrease of the non-isotropic character of relative diffusional motions of chain segments and an increase of the average distance between interacting nuclei.

7 CONCLUSION

This work was aimed at comparing swelling behaviors of calibrated gels and silica–siloxane random mixtures. NMR static scaling laws were observed in both cases as a function of the swelling ratio. From this approach, the role of trapped topological constraints in calibrated gels is emphasized; they induce a cut in correlation of fluctuations along chain segments; these obey a packing condition and exhibit affine deformations. Although the unfolding of chain segments in silica-filled siloxane cannot be described according to a simple scheme, experimental results reveal the crucial role played by the silica concentration in these systems. However, high relaxational frequency motions of short segments are independent of the presence of silica. Consequently, the high frequency viscoelastic response of these systems should not vary with the silica concentration, whereas the elastic modulus is increased and the terminal relaxation spectrum is shifted towards a longer time scale.[11]

REFERENCES

1. de Gennes, P. G., *Scaling Concepts in Polymer Physics*, Cornell University Press, Ithaca, New York, 1979.
2. Bastide, J., Picot, C. and Candau, S., *J. Macromol. Sci. Phys.*, 1981, **B19**, 13.
3. Daoud, M., Bouchaud, E. and Jannink, G., *Macromolecules*, 1986, **19**, 1955.
4. Cohen-Addad, J. P., *J. Physique*, 1982, **43**, 1509.
5. Cohen-Addad, J. P., Domard, M., Lorentz, G. and Herz, J., *J. Phys.*, 1984, **45**, 575.
6. Viallat, A., Cohen-Addad, J. P. and Puchelon, A., *Polymer*, 1986, **27**, 843.
7. Andrew, E. R. and Bersohn, R., *J. Chem. Phys.*, 1950, **18**, 159.
8. Belkebir-Mrani, A., Beinert, G., Herz, J. and Rempp, P., *Eur. Polym. J.*, 1973, **13**, 277.
9. Cohen-Addad, J. P., Roby, C. and Sauviat, M., *Polymer*, 1985, **26**, 1231.
10. Cuniberti, C., *J. Polymer Science, Part A*2, 1970, **8**, 2051.
11. Cohen-Addad, J. P., Viallat, A. and Huchot, P., *Macromolecules*, 1987, **20**, 2146.

33

INVESTIGATION ON POLYSTYRENE NETWORKS CONTAINING PENDENT POLY(ETHYLENE OXIDE) CHAINS

Z. Mouflou, J. G. Zilliox, G. Beinert,
Ph. Chaumont and J. Herz

*Institut Charles Sadron (CRM-EAHP) (CNRS-ULP),
6 rue Boussingault, 67083 Strasbourg Cedex, France*

ABSTRACT

Polystyrene networks grafted with pendent poly(ethylene oxide) chains were prepared in pure THF or in THF–heptane mixtures by anionic block copolymerization. In a good solvent of both polymer components such networks are highly swollen. In water, deswelling occurs; however, a rather high residual swelling ratio is observed. The equilibrium swelling and mechanical behaviour of these samples, when swollen in THF or in water, have been investigated.

1 INTRODUCTION

The mechanical properties of hydrophilic polymer networks swollen in water are generally very poor. Such a lack of mechanical resistance was observed, for example, in crosslinked dextrans or agaroses.[1-3]

An improvement of these properties can be obtained from an association of a hydrophilic and a hydrophobic component;[4] due to the hydrophilic constituent, the residual swelling degree in water of such a 'copolymer' network can be appreciable, whereas the hydrophobic part constitutes a rigid structure which provides for good mechanical resistance of the system.

Such materials have a potential interest as column fillings for

chromatography in aqueous media, where not only a large porous volume
in water but also a good resistance to high driving pressures are required.

The 'copolymer' networks used for the present investigation are formed
by elastic polystyrene chains (PS) and by pendent poly(ethylene oxide)
chains (PEO) fitted by one end to the crosslinks.

The present paper will be focused on the influence of these pendent
chains, the chemical nature of which is very different from that of the elastic
network chains, on the behaviour of the system.

After a short description of the network synthesis, we will examine the
swelling behaviour of these gels in a good solvent of both the elastic chains
and the PEO grafts, as well as in pure water. The last section of the paper
will be devoted to the mechanical properties of these systems in the
equilibrium-swollen state in tetrahydrofuran (THF) or when deswollen in
water.

2 SYNTHESIS

Polystyrene networks with well-defined characteristics can be prepared by
anionic block copolymerization of styrene and divinylbenzene.[5] The
synthesis of polystyrene networks containing PEO grafts attached to the
crosslinks is achieved by the same method[4] (Fig. 1). In a first step, an α, ω-
dicarbanionic precursor polystyrene of known molecular weight is
prepared in THF solution at low temperature ($\sim -70°C$) using the
dipotassium derivative of α-methylstyrene tetramer or naphthalene–potassium as a difunctional initiator.

A small amount of divinylbenzene (DVB) is then added to the solution of
'living' polymer under efficient stirring (approximately 5 units DVB per
anionic site). Polymerization of the divinyl monomer, initiated by the
anionic chain ends of the precursor polymer, yields a network in which the
crosslinks are constituted of small poly-DVB nodules: the precursor chains
have become elastic network chains linked to these nodules. In the absence
of a proton donor, the anionic sites, now located within the poly-DVB
nodules, remain active and can be used to initiate polymerization of
another monomer. In a last step, ethylene oxide is introduced into the
reactor, at low temperature ($\sim -70°C$), and allowed to diffuse into the
'living' network, 'swollen' in THF. Under these conditions, it reacts with the
anionic sites to form alkoxide groups but it does not polymerize. After
complete penetration of the cyclic monomer into the network, the
temperature is raised to approximately 20°C. At this temperature, a rather

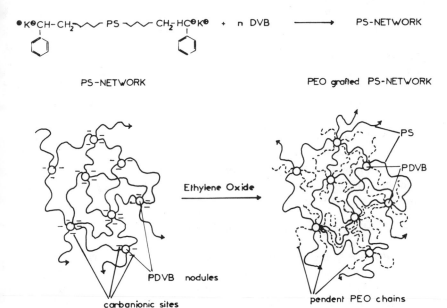

Fig. 1. Schematic representation of the 'copolymer' network synthesis.

fast polymerization yields PEO grafts fitted by one chain end to the poly-DVB nodules. Finally, the alkoxide chain ends are protonated.

The networks prepared generally contain only a small fraction of soluble polymers of the order of 1% which can be extracted with THF and characterized separately.

The fraction of PEO actually linked to the network has been established very accurately by elemental analysis of the networks free of soluble components. This allows also an estimation of the average length of the PEO grafts, under the assumption that each anionic site initially present in the reaction medium generates a pendent PEO chain.

The average mesh size of the networks is defined by the average molecular weight of the precursor chains (PSP), systematically characterized by size exclusion chromatography (SEC).

Network syntheses were also carried out in THF/heptane mixtures of various composition, however at volume ratios THF/heptane > 55/45, in order to avoid precipitation of the PS chains before crosslinking.

The reduced solvent quality favours solvent expulsion, as minute droplets, during the crosslinking process and might induce formation of micropores in the system (microsyneresis).

3 SWELLING BEHAVIOUR

The 'copolymer' networks, first swollen to equilibrium in excess THF, a good solvent of PS and PEO, deswell on progressive addition of water. After complete exchange of THF these networks do not collapse completely, but still exhibit an important residual volume swelling degree in water. The swollen state in water, which is obtained probably through a rather complicated mechanism, corresponds to a pseudo-equilibrium state which can be reached only by the above deswelling, where THF acts as a plasticizer of the PS phase. When completely dried, such a network can no longer be swollen in water since the PS phase, far below its glass transition temperature, is too rigid.

3.1 Networks Swollen to Equilibrium in THF

3.1.1 Influence of the Solvent Medium Used for Network Synthesis
One can expect that the quality of the solvent used for the synthesis, pure THF or THF/heptane mixtures, influences the topology of the networks (micropores) and modifies their swelling behaviour. This has been checked on both ungrafted networks (Table 1) and on PEO-containing samples. Table 1 shows the equilibrium swelling degrees in THF (Q_{THF}) of ungrafted gels prepared in THF and in 55/45 THF–heptane mixtures (in volume). One observes a slightly higher swelling degree for the samples prepared in the solvent mixture.

The same can be observed in the case of networks containing PEO grafts (Table 2): in spite of the shorter elastic chains of the network Mz1 prepared in the solvent mixture, the latter presents a higher equilibrium swelling degree in THF.

TABLE 1
Quality of the Solvent Used for Synthesis: Influence on the Swelling Behaviour in THF of Ungrafted Networks

Network	THF/heptane	$\bar{M}_n{}^a$ PSP (g/mol)	$Q_{THF}{}^b$
Mz4	55/45	19 600	16·4
Mz5	100/0	20 400	15·1
MJ7	55/45	18 100	16·1
MJ9	100/0	18 800	13·4

[a] Number-average molecular weight of precursor polymer.
[b] Equilibrium swelling degree in THF.

TABLE 2

Quality of the Solvent Used for Synthesis: Influence on the Swelling
Behaviour in THF and in Water of PEO Grafted Networks

Network	THF/heptane	$\bar{M}_n$ PSP (g/mol)	% PEO	Q_{THF}	$Q_{H_2O}{}^a$
Mz1	55/45	19 700	32·5	17·5	6
MJ10	100/0	25 600	29	15·6	2·7

[a] Residual swelling degree in water.

3.1.2 Influence of the Weight Fraction of PEO in the Network

One could expect an increase of the equilibrium swelling ratio of the
networks with increasing weight fraction of PEO in the system, due to a
repulsion between the incompatible PS network chains and the pendent
PEO chains. The equilibrium swelling ratios of the grafted networks Mz1
and Mz2 (Table 3) are slightly higher than those of the ungrafted samples
Mz4 and MJ7 (Table 1), which seems to indicate a small effect. However, no
increase of the equilibrium swelling degree in THF is observed with
increasing weight fraction of PEO (Table 3).

3.2 Networks Swollen in Water

The residual swelling degrees in water of the PEO grafted networks are
always quite high: the lowest values are close to 3 (examples in Tables 2 and
3). However, no evidence exists for a direct relation between the swelling
degree in water and the network characteristics or the weight fraction PEO,
as previously observed for systems swollen in a good solvent. Here the
behaviour is not governed by the classical thermodynamics.

Fig. 2, however, shows that, surprisingly, for copolymer networks
prepared in 'good' solvent media (also mixtures rich in THF), the ratio of
the swelling degrees in THF and in water Q_{THF}/Q_{H_2O} decreases linearly with
increasing weight fraction PEO. The experimental points corresponding to

TABLE 3

Influence of Weight Fraction of PEO on the Swelling Behaviour in
THF and in Water of Grafted Networks

Network	THF/heptane	$\bar{M}_n$ PSP (g/mol)	% PEO	Q_{THF}	Q_{H_2O}
Mz1	55/45	19 700	32·5	17·5	6
Mz2	55/45	18 800	21	17·5	3·7

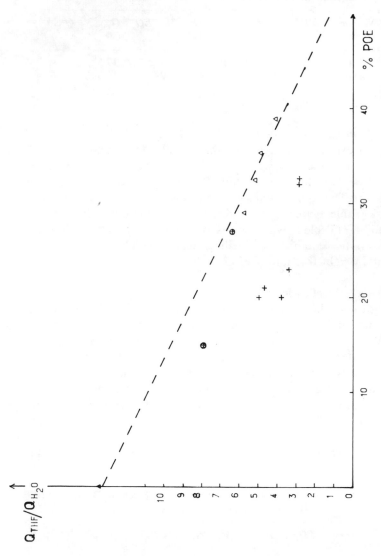

Fig. 2. Variation of the ratio Q_{THF}/Q_{H_2O} versus the weight fraction PEO in the networks. $+$, networks prepared in a 55/45 THF/heptane solvent mixture.

the networks prepared in 55/45 THF–heptane mixtures all lie below this line.

The general trend of the variation of Q_{THF}/Q_{H_2O} can be explained only qualitatively: in the absence of PEO, $Q_{H_2O} = 1$, thus Q_{THF}/Q_{H_2O} increases with decreasing weight fraction of PEO to reach the equilibrium swelling degree in THF of the ungrafted network.

When the weight fraction of PEO increases, the swelling power of water for the crosslinked system improves: Q_{H_2O} increases whereas Q_{THF} should not vary very much. Thus Q_{THF}/Q_{H_2O} decreases with increasing % POE. A linear variation seems plausible only for networks with comparable elastic chain length, as in the case here. However, more experiments are required to determine the influence of parameters such as mesh size, functionality or deswelling treatment.

4 MECHANICAL BEHAVIOUR

4.1 Networks Swollen in THF

As previously, before considering the case of grafted networks, it was necessary to check whether the quality of the solvent used for network synthesis influences the elastic modulus, as could be expected from the swelling behaviour.

This is illustrated in Table 4, where the uniaxial compression moduli E of two ungrafted samples prepared in 55/45 THF–heptane mixtures are compared with that of a similiar network prepared in pure THF: the modulus of the latter is more than twice that of the two other samples.

Similar behaviour is observed for networks containing PEO grafts (Table 5): the moduli of samples prepared in THF–heptane mixtures are

TABLE 4

Quality of the Solvent Used for Synthesis: Influence on the Uniaxial Compression Modulus of Ungrafted PS Networks Swollen in THF

Network	THF/heptane	$\bar{M}_n$ PSP (g/mol)	$E_{THF}{}^a$ ($Pa \times 10^{-3}$)
Mz4	55/45	19 600	2·1
MJ7	55/45	18 100	2·3
MJ9	100/0	18 800	5·0

^{a} Uniaxial compression modulus normalized to unity swelling degree.

systematically lower than those of the networks prepared in THF. The observed influence of solvent quality probably reflects the existence of heterogeneities resulting from microsyneresis during the crosslinking reaction.

The influence of the weight fraction of PEO grafts in the networks was investigated on samples prepared in the 55/45 THF–heptane mixtures. Table 6 demonstrates that the presence of pendent PEO chains is of no noticeable influence on the uniaxial compression modulus, which depends only upon the length of the elastic chains and fluctuates very little.

In order to compare all grafted networks with ungrafted samples, prepared in solvent mixtures as well as in pure THF, we have plotted the compression moduli versus the corresponding equilibrium concentration C_{eq} (at equilibrium swelling) on a double logarithmic scale (Fig. 3). The experimental points including those corresponding to the ungrafted networks are scattered around a straight line. The slope $(s \sim 3)$ is much higher than the theoretical value of 2·25, as already observed in the case of networks with defects.[6] The above presentation confirms that no noticeable difference in behaviour exists between grafted and ungrafted samples when swollen in a good solvent.

4.2 Networks Swollen in Water

Compression experiments were carried out on samples deswollen very progressively in water in order to avoid crazes. However, the accuracy of these measurements is rather poor and does not allow the mechanical behaviour to be related to the characteristics of the networks, such as length and number of PEO grafts, or to the experimental conditions used for the

TABLE 5

Quality of the Solvent Used for Synthesis: Influence on the Uniaxial Compression Modulus of Grafted PS Networks Swollen in THF

Network	THF/heptane	$\bar{M}_n$ PSP (g/mol)	% PEO	$E_{THF}{}^a$ $(Pa \times 10^{-3})$
MJ9	100/0	18 800	0	5·0
2 790	100/0	22 200	39	4·4
MJ2	70/30	19 800	27	3·2
MJ3	60/40	21 400	15	3·2
MJ8	55/45	21 100	20	2·3

[a] Uniaxial compression modulus normalized to unity swelling degree.

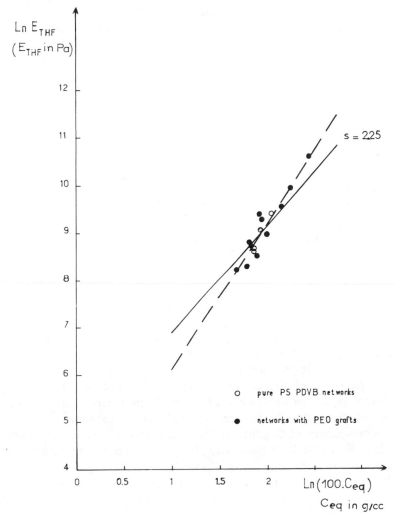

Fig. 3. Double logarithmic plot of the compression moduli versus concentration at equilibrium swelling.

crosslinking reaction, which can be responsible for heterogeneities in the system.

Examples of compression moduli E_{H_2O} of networks swollen in water are shown in Table 6. The values are of the same order for all the grafted samples. The compression moduli of these gels are always several orders of magnitude larger than those of the networks swollen in THF ($E_{H_2O}/E_{THF} >$

TABLE 6
Uniaxial Compression Moduli of Grafted and Ungrafted Networks

Network	THF/heptane	$\bar{M}_n$ PSP (g/mol)	% PEO	$E_{THF}{}^a$ $(Pa \times 10^{-3})$	$E_{H_2O}{}^b$ $(Pa \times 10^{-3})$
MJ7	55/45	18 100	0	2·3	—
MJ8	55/45	21 100	20	2·3	6 000
Mz4	55/45	19 600	0	2·1	—
Mz3	55/45	19 200	32	2·0	1 800
Mz1	55/45	19 700	32·5	2·6	2 700

^{a} Uniaxial compression modulus normalized to unity swelling degree; networks swollen in THF.
^{b} Uniaxial compression modulus of networks swollen in water.

10^3), since they depend exclusively upon the rigid PS phase which is below the glass transition temperature.

5 CONCLUSION

The behaviour of PS networks fitted with PEO grafts has been characterized by equilibrium swelling and by uniaxial compression measurements in the swollen state.

The samples investigated were prepared by anionic block copolymerization of styrene and DVB, followed by the polymerization of ethylene oxide initiated by the 'living' sites located in the crosslinks.

The presence of PEO grafts in the system seems to have only a small influence on its swelling behaviour, in spite of the incompatible nature between PS network chains and PEO grafts.

The residual swelling ratio in water increases slightly with the fraction of pendent chains in the system, but general behaviour, common to all gels, relating swelling degree and PEO content was not observed. However, for samples prepared in a good solvent the ratio Q_{THF}/Q_{H_2O} decreases linearly with increasing weight per cent of grafted PEO.

The uniaxial compression modulus of networks swollen in THF is practically independent of the presence of pendent PEO chains (within the limit of experimental accuracy), and depends on the effectively elastic network chains. For networks prepared in THF–heptane mixtures the modulus is drastically lowered, an effect due probably to an inhomogeneous network structure.

The modulus of the same networks swollen in water is always several orders of magnitude larger than for the systems swollen in THF.

The high mechanical resistance of these water-swollen materials makes them suitable for possible applications in size exclusion chromatography in aqueous media.

REFERENCES

1. Barth, H. G., A practical approach to steric exclusion chromatography of water soluble polymers, *J. Chromat. Sci.*, 1980, **18**, 409–29.
2. Alfredson, Th. V., Wehr, C. T., Tallman, L. and Klink, F., Evaluation of new microparticulate packings for aqueous steric exclusion chromatography, *J. Liq. Chromat.*, 1982, **5**, 489–524.
3. Rinaudo, M., Application de la chromatographie de permeation sur gel dans le domaine des polysaccharides, *Bull. Soc. Chim. Fr.*, 1974, **11**, 2285–94.
4. Beinert, G., Chaumont, Ph., Rempp, P. and Herz, J., Hydrophylic non–ionic polymer networks, *Europ. Polym. J.*, 1984, **20**, 837–40.
5. Weiss, P., Hild, G., Herz, J. and Rempp, P., Préparation de gels réticulés par copolymérisation séquencée anionique du styrène avec du divinylbenzène, *Makromol. Chem.*, 1970, **135**, 249–61.
6. Hild, G., Okasha, R., Macret, M. and Gnanou, Y., Relationship between elastic modulus and volume swelling degree of networks swollen to equilibrium in good diluents. Interpretation of experimental results on the basis of scaling concepts, *Makromol. Chem.*, 1986, **187**, 2271–88.

SECTION 5

RUBBER ELASTICITY

34

ANISOTROPY OF RUBBER NETWORKS CROSSLINKED IN STATES OF STRAIN

Burak Erman

*School of Engineering, Bogazici University,
Bebek 80815, Istanbul, Turkey*

ABSTRACT

Anisotropy of amorphous polymer networks crosslinked first in the isotropic state and subsequently in the stretched state is analyzed using the junction constraint theory. According to the theory, the anisotropy results from the anisotropy of elastic constraint domains trapped in the network structure during the two stages of crosslinking. The calculated differences in elastic moduli parallel and perpendicular to the direction of stretch during crosslinking are observed to compare favourably with experimental data on natural rubber and poly(ethylacrylate) networks. The theory also predicts a small anisotropy in swelling where the linear dilation along the parallel direction to crosslinking differs from those in the lateral directions.

INTRODUCTION

The reduced force, $[f^*]$, for a network in uniaxial tension is defined as

$$[f^*] = f/A^*(\alpha - \alpha^{-2}) \tag{1}$$

Here, f is the force acting on the sample, A^* is the cross-sectional area in the reference state, and α is the ratio of the final length to the initial length.[1] The reduced force defined in this manner may suitably be used to compare the behavior of real networks with the predictions of molecular models. Uniaxial stress experiments on networks indicate that the reduced force in general decreases with increasing extension, and approaches the value for the phantom network, given according to the theory[2] as

$$[f^*] = \xi kT/V^\circ \tag{2}$$

where ξ is the cycle rank, k is the Boltzmann constant, T is the temperature and $V°$ is the network volume in the state of reference.

According to the theory,[3] fluctuations of junctions from their mean locations in a phantom network are substantial. For a tetrafunctional phantom network, for example, the mean-squared fluctuations, $\langle(\Delta R)^2\rangle$, of junctions equal $(3/8)\langle r^2\rangle_0$, where $\langle r^2\rangle_0$ is the mean-squared end-to-end distance of unperturbed chains.[3] The mean positions of junctions in a phantom network transform affinely with macroscopic deformation. Fluctuations from these mean positions are independent of the macroscopic state of strain. This property is the result of the phantom-like structure in which chains may pass through each other without steric hindrances. Thus, a phantom network describes a hypothetical structure in which effects of intermolecular constraints are absent. In real networks, intermolecular constraints arising from the dense interspersion of chains with their environments contribute further to the reduced force.[2] This contribution is in general dependent on the level of strain. The theory describing the effects of constraints has been formulated[3,4] over the past few years, and is now referred to as the 'constraint theory'. According to this theory, the fluctuations of junctions are affected by the surrounding *domains* of constraint. These domains transform with macroscopic deformation. In simple tensions, for example, a spherical domain transforms into an ellipsoid, the major axis of which lies in the direction of stretch and the minor axes in the lateral directions. Fluctuations of a junction are enhanced along the direction of stretch and decreased in the lateral directions. The decrease of the reduced force with extension (the C_2 effect in the Mooney–Rivlin formulation) is a consequence of the extension of the constraint domains along the direction of stretch.

A network is expected to exhibit anisotropic elasticity if the constraint domains are anisotropic in the unstressed structure. Anisotropy of constraint domains results from crosslinking the system in two stages, where at least one of the stages is performed in the deformed state.

The behavior of a network crosslinked in states of strain has previously been analyzed.[5-7] The network models used in these studies corresponded essentially to the phantom model. In as much as the concept of constraint domains does not exist in the phantom network model, it was not possible to obtain anisotropic elasticity in these works. Experiments, however, show[8,9] that the reduced force for the networks crosslinked in the stretched state differs in parallel and perpendicular directions to the direction of stretch during crosslinking. These experiments also show[9] that swelling of networks differs along different directions.

In the following analysis, the anisotropy of networks crosslinked in the isotropic and the deformed states is investigated using the molecular model of a network with constrained junctions. The anisotropy of the constraint domains resulting from the second-stage crosslinking is seen to lead to results which are in quantitative agreement with experiments on natural rubber and poly(ethylacrylate) networks.[8]

THEORY

The State of Deformation

The following formulation is restricted to the study of (i) simple extension and (ii) free anisotropic dilation of a network. The sample is assumed to be crosslinked in the state of isotropy, followed by a second-stage crosslinking in the stretched state. The reference state is taken as the state of ease reached after the second-stage crosslinking. We let L_0 and L_0' denote the length of the sample during the first and second stages of crosslinking, respectively. The length at the state of ease following the second-stage crosslinking is denoted by L_1. We let $L_\parallel$ denote the final length of the sample if tension is parallel to the direction of stretch during crosslinking. Similarly, $L_\perp$ denotes the final length if tension is perpendicular to the direction of stretch during second-stage crosslinking.

Various extension ratios may be defined in terms of the lengths L_0, L_0', L_1, $L_\parallel$ and $L_\perp$. Thus we let $\lambda_1 = L_0'/L_0$, $\lambda_2 = L/L_0'$ and $\varepsilon = L_1/L_0$. Additionally, $\lambda = L/L_0$ and $\alpha = L/L_1$, where L may be taken as $L_\parallel$ or $L_\perp$, depending on the direction of stretch. Using these definitions for the extension ratios, the following displacement gradient tensors may be introduced.

(a) Stretching parallel to the direction of stretch during second-stage crosslinking:

$$\lambda_1 = \text{diag}\,(\lambda_1,\, \lambda_1^{-1/2},\, \lambda_1^{-1/2})$$
$$\alpha = \text{diag}\,(\alpha,\, \alpha^{-1/2},\, \alpha^{-1/2})$$
$$\varepsilon = \text{diag}\,(\varepsilon,\, \varepsilon^{-1/2},\, \varepsilon^{-1/2}) \tag{3}$$
$$\lambda \equiv \alpha\varepsilon = \text{diag}\,[\alpha\varepsilon,\, (\alpha\varepsilon)^{-1/2},\, (\alpha\varepsilon)^{-1/2}]$$
$$\lambda_2 \equiv \lambda\lambda_1^{-1} = \text{diag}\,[\varepsilon\alpha\lambda_1^{-1},\, (\lambda_1/\varepsilon\alpha)^{1/2},\, (\lambda_1/\varepsilon\alpha)^{1/2}]$$

The tensors in eqn (3) are defined such that the direction of stretch is along the x-axis of a laboratory-fixed coordinate system $Oxyz$.

(b) Stretching perpendicular to the direction of stretch during crosslinking:

$$\lambda_1 = \text{diag}(\lambda_1^{-1/2}, \lambda_1, \lambda_1^{-1/2})$$
$$\alpha = \text{diag}[\alpha, (\gamma/\alpha)^{1/2}, (\gamma\alpha)^{-1/2}]$$
$$\varepsilon = \text{diag}(\varepsilon^{-1/2}, \varepsilon, \varepsilon^{-1/2}) \qquad (4)$$
$$\lambda \equiv \alpha\varepsilon = \text{diag}[\alpha\varepsilon^{-1/2}, (\gamma/\alpha)^{1/2}\varepsilon, (\gamma\alpha)^{-1/2}\varepsilon]$$
$$\lambda_2 \equiv \lambda\lambda_1^{-1} = \text{diag}[\alpha\varepsilon^{-1/2}\lambda_1^{1/2}, (\gamma/\alpha)^{1/2}\varepsilon\lambda_1^{-1}, (\gamma\alpha)^{-1/2}\varepsilon\lambda_1^{1/2}]$$

The direction of stretch in eqn (4) is represented by the x-axis. The direction of stretch during crosslinking coincides with the y-axis. The factor $\gamma = \alpha_y/\alpha_z$ arises from the fact that the y and z directions are not equivalent.

(c) Swelling: letting the x-axis represent the direction of stretch during crosslinking in the second stage, we have

$$\lambda_1 = \text{diag}(\lambda_1, \lambda_1^{-1/2}, \lambda_1^{-1/2})$$
$$\alpha = \text{diag}(\alpha, \delta\alpha, \delta\alpha)$$
$$\varepsilon = \text{diag}(\varepsilon, \varepsilon^{-1/2}, \varepsilon^{-1/2}) \qquad (5)$$
$$\lambda = \text{diag}[\delta^{-2/3}\varepsilon(V/V^\circ)^{1/3}, \delta^{1/3}\varepsilon^{-1/2}(V/V^\circ)^{1/3}, \delta^{1/3}\varepsilon^{-1/2}(V/V^\circ)^{1/3}]$$
$$\lambda_2 = \text{diag}[\delta^{-2/3}\varepsilon(V/V^\circ)^{1/3}\lambda_1^{-1}, \delta^{1/3}\varepsilon^{-1/2}(V/V^\circ)^{1/3}\lambda_1, \delta^{1/3}\varepsilon^{-1/2}(V/V^\circ)^{1/3}\lambda_1]$$

where $\delta = \alpha_y/\alpha_x$ is the swelling anisotropy and $\alpha_x = \delta^{-2/3}(V/V^\circ)^{1/3}$. The latter follows from the equality $\det(\alpha) = (V/V^\circ)$, where V and V° are the swollen and dry volumes of the network, respectively.

The Elastic Free Energy

Following the previous theory of anisotropic networks,[6] we assume the total elastic free energy, ΔA, of the network to be the sum of contributions, ΔA_1, from the first-stage network and ΔA_2 from the second-stage network:

$$\Delta A = \Delta A_1 + \Delta A_2 \qquad (6)$$

where

$$\Delta A_1 = \tfrac{1}{2}\xi_1 kT\left\{\sum_t\left[(\lambda_t^2 - 3) + \frac{\mu_1}{\xi_1}(B_t - \ln(B_t + 1) + D_t - \ln(D_t + 1))\right]\right\} \qquad (7)$$

$$\Delta A_2 = \tfrac{1}{2}\xi_2 kT\left\{\sum_t\left[(\lambda_{2,t}^2 - 3) + \frac{\mu_2}{\xi_2}(B_{2,t} - \ln(B_{2,t} + 1) + D_{2,t} - \ln(D_{2,t} + 1))\right]\right\} \qquad (8)$$

where $t = x$, y or z, and

$$B_t = (\lambda_t - 1)(1 + \lambda_t - \zeta_1\lambda_t^2)/(1 + g_t)^2$$
$$D_t = g_t B_t \qquad\qquad (9)$$
$$g_t = \lambda_t^2[\kappa_1^{-1} + \zeta_1(\lambda_t - 1)]$$

$B_{2,t}$ and $D_{2,t}$ are obtained from eqn (9) by replacing components of λ by λ_2, and κ_1, ζ_1 by κ_2, ζ_2. Equations (7) to (9) follow from the constraint theory[4] developed for isotropic networks. ξ_1 and μ_1 denote the cycle rank and the number of junctions obtained in the first stage of crosslinking, respectively. ξ_2 and μ_2 represent the corresponding parameters of the second-stage crosslinking. κ_1 and κ_2 represent the extent of constraints on the fluctuations of junctions originating from the first and second stages of crosslinking, respectively. ζ_1 and ζ_2 are the additional parameters of the network representing the effects of nonhomogeneities of networks.[4]

It should be noted that the superposition acknowledged by eqn (6) is an assumption of the present analysis. For a phantom network ($\kappa_1 = \kappa_2 = 0$) or an affine network ($\kappa_1 = \kappa_2 = \infty$), the validity of eqn (6) follows rigorously from theory, as was shown previously by Flory.[6]

The Reduced Force

The uniaxial force f acting on the network is obtained from the elastic free energy, according to the relation $f = \partial\Delta A/\partial L$. In experiments, extension ratios are measured relative to the state of ease following the second-stage crosslinking. Thus, expressing f in terms of α as $f = \partial\Delta A/L_1\partial\alpha$ and using eqns (1) and (6), the reduced force may be obtained as

$$[f^*] = f/A^*(\alpha - \alpha^{-2}) = \frac{1}{V_1(\alpha - \alpha^{-2})}\left[\sum_t\left(\frac{\partial\Delta A_1}{\partial\lambda_t^2}\frac{\partial\lambda_t^2}{\partial\alpha} + \frac{\partial\Delta A_2}{\partial\lambda_{2,t}^2}\frac{\partial\lambda_{2,t}^2}{\partial\alpha}\right)\right]$$

$$(10)$$

where V_1 is the volume in the reference state. For networks crosslinked in the bulk state, V_1 equals V°.

For extension parallel to the direction of stretch during crosslinking, using eqns (3) in (7), (8) and substituting into eqn (10), we obtain, for a tetrafunctional network ($\mu_1 = \xi_1$, $\mu_2 = \xi_2$):

$$[f^*]_\| = \frac{\xi_1 kT}{V_1(\alpha - \alpha^{-2})}[P(\lambda_x^2)\varepsilon^2\alpha - P(\lambda_y^2)/\varepsilon\alpha^2]$$

$$+ \frac{\xi_2 kT}{V_1(\alpha - \alpha^{-2})}[P(\lambda_{2,x}^2)\varepsilon^2\alpha/\lambda_1^2 - P(\lambda_{2,y}^2)\lambda_1/\varepsilon\alpha^2] \qquad (11)$$

where

$$P(\lambda^2) = 1 + \frac{\mu}{\xi}\left(\frac{B\,\partial B/\partial \lambda^2}{1+B} + \frac{D\,\partial D/\partial \lambda^2}{1+D}\right)$$

$$\partial B/\partial \lambda^2 = B[1/2\lambda(\lambda - 1) + (1 - 2\zeta\lambda)/2\lambda(1 + \lambda - \zeta\lambda^2) - 2(\partial g/\partial \lambda^2)/(1 + g)]$$
$$\tag{12}$$

$$\partial D/\partial \lambda^2 = g\,\partial B/\partial \lambda^2 + B\,\partial g/\partial \lambda^2$$

$$\partial g/\partial \lambda^2 = \kappa^{-1} - \zeta(1 - 3\lambda/2)$$

The components of λ and λ_2 appearing in eqn (11) are given in eqn (3). In as much as the elastic free energy has to be a minimum in the state of rest, the value of the extension ratio has to be obtained from

$$\partial \Delta A/\partial \varepsilon = \sum_t\left[\frac{\partial \Delta A_1}{\partial \lambda_t^2}\frac{\partial \lambda_t^2}{\partial \varepsilon} + \frac{\partial \Delta A_2}{\partial \lambda_{2,t}^2}\frac{\partial \lambda_{2,t}^2}{\partial \varepsilon}\right] = 0 \tag{13}$$

Using eqn (3) (with α equal to unity) in eqn (13) and performing the indicated differentiations, the condition of minimum ΔA is obtained for stretching parallel to the direction of stretch as

$$P(\lambda_x^2)\varepsilon - P(\lambda_y^2)/\varepsilon^2 + \frac{\xi_2}{\xi_1}[P(\lambda_{2,x}^2)\varepsilon/\lambda_1^2 - P(\lambda_{2,y}^2)\lambda_1/\varepsilon^2] = 0 \tag{14}$$

The value of ε may be obtained numerically from eqn (14).

The expression for the reduced force for extension perpendicular to the direction of stretch during second-stage crosslinking may similarly be obtained as

$$[f^*]_\perp = \frac{\xi_1 kT}{2V_1(\alpha - \alpha^{-2})}[2P(\lambda_x^2)\alpha/\varepsilon - P(\lambda_y^2)\gamma\varepsilon^2/\alpha^2 - P(\lambda_z^2)/\gamma\varepsilon\alpha^2]$$
$$+ \frac{\xi_2 kT}{2V_1(\alpha - \alpha^{-2})}[2P(\lambda_{2,x}^2)\alpha\lambda_1/\varepsilon - P(\lambda_{2,y}^2)\gamma(\varepsilon/\alpha\lambda_1)^2 - P(\lambda_{2,z}^2)\lambda_1/\gamma\varepsilon\alpha^2]$$
$$\tag{15}$$

where various extension ratios are given in eqn (4) and the value of γ is obtained by minimizing ΔA at each α. The condition for minimizing ΔA at fixed α is given by eqn (13), where ε is replaced by γ. Substituting from eqn (4) into eqns (7) and (8), and performing the differentiations indicated in eqn (13), one obtains

$$P(\lambda_y^2)\varepsilon^2/\alpha - P(\lambda_z^2)/\gamma^2\varepsilon\alpha + \frac{\xi_2}{\xi_1}[P(\lambda_{2,y}^2)\varepsilon^2/\lambda_1^2\alpha - P(\lambda_{2,z}^2)\lambda_1/\gamma^2\varepsilon\alpha] = 0 \quad (16)$$

Solution of eqn (16) leads to the value of γ for each α.

In the case of swelling, the elastic free energy has to be a minimum with respect to δ. Letting $\partial \Delta A / \partial \delta = 0$ leads to eqn (13), where ε is replaced by δ. Using eqn (5) for the extension ratios in swelling and performing the indicated differentiations, we obtain

$$\partial \Delta A / \partial \delta = \frac{\xi_1 kT}{2V_1} \left[P(\lambda_x^2) \frac{\partial \lambda_x^2}{\partial \delta} + 2P(\lambda_y^2) \frac{\partial \lambda_y^2}{\partial \delta} \right]$$

$$+ \frac{\xi_2 kT}{2V_1} \left[P(\lambda_{2,x}^2) \frac{\partial \lambda_{2,x}^2}{\partial \delta} + 2P(\lambda_{2,y}^2) \frac{\partial \lambda_{2,y}^2}{\partial \delta} \right]$$

$$= \frac{\xi_1 KT}{V_1} \left\{ P(\lambda_x^2)[-\delta^{-7/3} \varepsilon^2 (V/V^\circ)^{2/3}] P(\lambda_y^2)[\delta^{-1/3}(V/V^\circ)^{2/3}/\varepsilon] \right\}$$

$$+ \frac{\xi_2 kT}{V_1} \left\{ P(\lambda_{2,x}^2)[-\delta^{-7/3} \varepsilon^2 (V/V^\circ)^{2/3}/\lambda_1] \right.$$

$$\left. + P(\lambda_{2,y}^2)[\delta^{-1/3}(V/V^\circ)^{2/3} \lambda_1^{1/2}/\varepsilon] \right\} = 0 \qquad (17)$$

Numerical solution of eqn (17) for a given V/V° gives the swelling anisotropy δ.

NUMERICAL CALCULATIONS

Values of the reduced force in anisotropic networks are presented as a function of α^{-1} in Figs 1 and 2. The curves for $[f^*]_\parallel$ are obtained by using eqn (11). Those for $[f^*]_\perp$ are obtained from eqn (15). The direction of stretch is indicated on each curve. The solid curves in Fig. 1 are obtained for $\lambda_1 = 4$, with $\kappa_1 = \kappa_2 = 10$, $\zeta_1 = \zeta_2 = 0$, $\xi_1 kT/V_1 = 0.1 \, \text{Nmm}^{-2}$ and $\xi_2 kT/V_1 = 0.2 \, \text{Nmm}^{-2}$. The extension ratio during the state of ease, ε, is obtained numerically as 2.094 from eqn (14). The dashed curves are calculated for the same material parameters but with $\lambda_1 = 2$. ε is calculated for this case as 1.519. In Fig. 2 similar calculations are repeated by taking $\zeta_1 = \zeta_2 = 0.3$, all other parameters being the same as in Fig. 1. Calculated values of ε are 2.178 and 1.561 for $\lambda_1 = 4$ and 2, respectively. Comparison of Figs 1 and 2 shows that the parameter ζ has a significant effect on network anisotropy. Values of γ obtained numerically from eqn (16) for

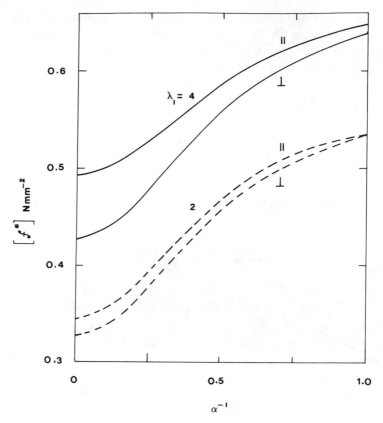

Fig. 1. Reduced force as a function of α^{-1}. Calculations are made for $\xi_1 kT/V_1 = 0.1\ \mathrm{Nmm}^{-2}$, $\xi_2 kT/V_1 = 0.2\ \mathrm{Nmm}^{-2}$, $\kappa_1 = \kappa_2 = 10$, $\zeta_1 = \zeta_2 = 0$, and for $\lambda_1 = 4$ and 2 as shown on the curves. Calculated values of ε are 2·094 and 1·519 for $\lambda_1 = 4$ and 2, respectively.

perpendicular stretching were very close to unity for the networks treated for all values of α. Exploratory calculations of $[f^*]_\perp$, by assuming $\gamma = 1$, resulted in virtually the same curves as shown in Figs 1 and 2.

Calculations of anisotropy of swelling are shown in Fig. 3. Values of δ obtained numerically from eqn (17) are presented as a function of linear dilation ratio, $(V/V^\circ)^{1/3}$. Calculations are presented for the two networks used for Fig. 1. Solid lines are for $\zeta_1 = \zeta_2 = 0.3$ and the dashed lines are for $\zeta_1 = \zeta_2 = 0$. Values of λ_1 are indicated on each curve. The value of ε for $\lambda_1 = 6$ was obtained as 2·504.

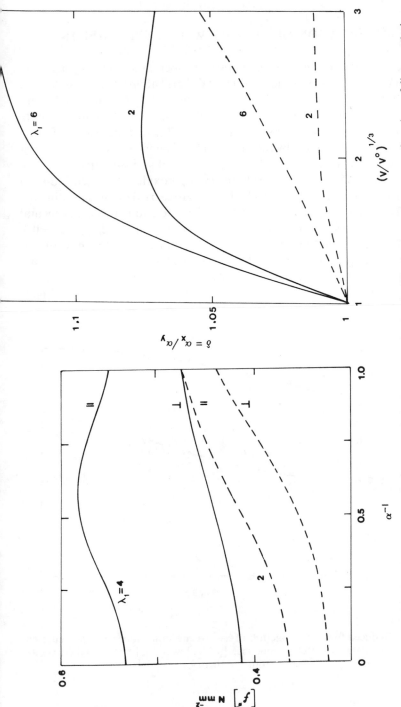

Fig. 2. Reduced force as a function of α^{-1}. The data of Fig. 1 were used except $\zeta_1 = \zeta_2 = 0.3$. Calculated values of ε are 2·178 and 1·561 for $\lambda_1 = 4$ and 2, respectively.

Fig. 3. Swelling anisotropy, δ, as a function of linear dilation ratio, $(V/V^\circ)^{1/3}$. Calculations are for $\lambda_1 = 6$ ($\varepsilon = 2.504$) and $\lambda_1 = 2$ ($\varepsilon = 1.561$). Solid lines are for $\zeta_1 = \zeta_2 = 0.3$. Dashed lines are for $\zeta_1 = \zeta_2 = 0$. Other parameters are as reported in Fig. 1.

COMPARISON OF THEORY WITH EXPERIMENTS

Simple tension experiments on natural rubber and poly(ethylacrylate) networks reported by Greene, Smith and Ciferri[8] afford direct comparison of experiments with theory. In Fig. 4, values of $[f^*]_\parallel$ and $[f^*]_\perp$ are presented as a function of α^{-1} for a natural rubber network. Circles denote experimental data, the upper set referring to $[f^*]_\parallel$ and the lower to $[f^*]_\perp$. Solid curves are obtained by the theory. The network was crosslinked first in the isotropic state by a Co^{60} source at 4 Mrad and subsequently in the stretched state ($\lambda_1 = 2\cdot3$) with 2·5% dicumyl peroxide. The extension ratio in the state of ease was $\varepsilon = 2\cdot0$. Calculations are carried out by adopting the reported values of λ_1 and ε and choosing ξ, κ and ζ to match experimental data. The curves obtained for a choice of $\kappa_1 = \kappa_2 = 12$, $\zeta_1 = \zeta_2 = 0\cdot3$, $\xi_1 kT/V_1 = 0\cdot8\,\text{Nmm}^{-2}$ and $\xi_2 kT/V_1 = 1\cdot75\,\text{Nmm}^{-2}$ are shown in Fig. 4.

The poly(ethylacrylate) network was obtained by crosslinking in the first isotropic state by Co^{60} radiation at 4 Mrad, and subsequently at $\lambda_1 = 2\cdot6$ at 8 Mrad. ε was measured as 2·0. Circles denote experimental data in Fig. 5.

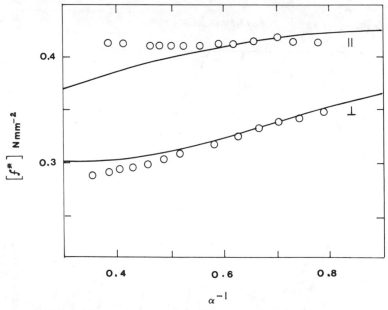

Fig. 4. Reduced force as a function of α^{-1} for natural rubber. Circles represent experimental data from Ref. 8: upper set, $[f^*]_\parallel$; lower set, $[f^*]_\perp$. Solid curves are from the theory.

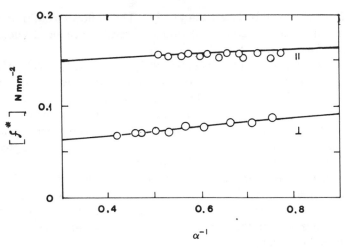

Fig. 5. Reduced force as a function of α^{-1} for poly(ethylacrylate). See caption to Fig. 4.

The solid curves are obtained for $\kappa_1 = 10$, $\kappa_2 = 5$, $\zeta_1 = \zeta_2 = 0{\cdot}3$, $\xi_1 kT/V_1 = 0{\cdot}315\,\mathrm{Nmm}^{-2}$ and $\xi_2 kT/V_1 = 0{\cdot}715\,\mathrm{Nmm}^{-2}$.

DISCUSSION AND CONCLUSION

The state of anisotropy of networks crosslinked in the anisotropic state followed by second-stage crosslinking in the stretched state is analyzed using the constraint theory. According to the theoretical model and the present treatment, network anisotropy results from the anisotropy of constraint domains trapped in the network structure during crosslinking. Domains of constraint in the isotropic state are spherical. During the second stage of crosslinking, these spheres are ellipsoids with their major axes along the direction of λ_1. The domains of constraint introduced in the second stage of crosslinking are spherical. When the network is released to its state of ease after the second stage of crosslinking, the major axes of the ellipsoidal constraint domains of the first-stage crosslinking shorten, and the second stage spheres become ellipsoids with their major axes normal to the direction of λ_1. This argument follows from the assumption that these domains act in an elastic spring-like manner in constraining the fluctuations of the junctions. In the final structure, if the average domain dimension along the parallel direction is shorter than that in the lateral direction, $[f^*]_{\parallel}$ will be larger than $[f^*]_{\perp}$. This follows from the fact that smaller dimensions of constraint domains result in reduced fluctuations and therefore in larger $[f^*]$.

There are six parameters of the theory: $\xi_1, \xi_2, \kappa_1, \kappa_2, \zeta_1$ and ζ_2. Although ξ and κ are expressed in terms of molecular parameters in the original theory[4] of isotropic networks, the correspondence to anisotrpic networks is not obvious at this time. However, qualitatively, it may be stated that if the second crosslinking is more dense than the first stage, then ξ_2 should be larger than ξ_1 and κ_2 should be less than κ_1. This argument agrees with results of calculations for the poly(ethylacrylate) network, for which the second-stage crosslinking was more than that in the first stage.

Finally, it should be noted that the values of ξ, κ and ζ chosen for the description of experimental data on natural rubber and poly(ethylacrylate) networks are within the ranges observed in the analysis of previous experiments on a large number of isotropic networks.

REFERENCES

1. Queslel, J. P. and Mark, J. E., Rubber elasticity and characterization of networks, *Comprehensive Polymer Science*, Vol. 1, Allen, G. (Ed.), Pergamon Press, Oxford, 1986, p. 41.
2. Erman, B. and Flory, P. J., *Macromolecules*, 1982, **15**, 806.
3. Flory, P. J., *J. Chem. Phys.*, 1977, **66**, 5720.
4. Flory, P. J. and Erman, B., *Macromolecules*, 1982, **15**, 800.
5. Berry, L. P., Scanlan, J. and Watson, W. F., *Trans. Faraday Soc.*, 1956, **52**, 1137.
6. Oosawa, F., *J. Polym. Sci.*, 1958, **32**, 229.
7. Flory, P. J., *Trans. Faraday Soc.*, 1960, **56**, 722.
8. Greene, A., Smith, K. J. and Ciferri, A., *Trans. Faraday Soc.*, 1965, **61**, 2772.
9. Batsberg, W., Hvidt, S. and Kramer, O., *J. Polym. Sci., Polym. Lett. Ed.*, 1982, **20**, 341.

35

HIGH-VINYL POLYBUTADIENE CROSSLINKED IN THE STRAINED STATE TO DIFFERENT DEGREES OF CROSSLINKING

W. Batsberg

Department of Chemistry, Risø National Laboratory,
DK-4000 Roskilde, Denmark

S. Hvidt

Department of Chemistry, University of Roskilde,
DK-4000 Roskilde, Denmark

O. Kramer

Department of Chemistry, University of Copenhagen,
DK-2100 Copenhagen, Denmark

and

L. J. Fetters

Exxon Research and Engineering Co., Annandale,
New Jersey 08801, USA

ABSTRACT

It is demonstrated that chain entangling in a well crosslinked polybutadiene network results in an equilibrium elastic contribution, which equals the rubber plateau modulus of the same uncrosslinked polymer. It is found to be true for crosslink densities varying by as much as a factor of four. This result is obtained by introducing the crosslinks in the strained state. No theory and no assumptions are required.

INTRODUCTION

The role of chain entangling in crosslinked elastomers is a question that still needs to be settled. However, there is general agreement that chain entangling gives rise to the rubber-like properties which are observed for uncrosslinked linear polymers of very high molecular weight. Given enough time, the long polymer chains will disengage completely from the original, temporary entanglement network. In a stress relaxation experiment, this will be observed as a decrease of the modulus towards zero, as shown by the lower curve in Fig. 1. According to the reptation concept of de Gennes,[2] the relaxation time of the disentanglement process is approximately proportional to the third power of the molecular weight.

In a hypothetical experiment, increasing the molecular weight of a linear polymer by 1000 times would increase the relaxation time for the disentanglement process by about 10 orders of magnitude. Thus, such a polymer would appear to be a permanently crosslinked elastomer, since the disentanglement process would be experimentally inaccessible. The rate of relaxation could not be increased sufficiently by heating without causing degradation of the polymer. Although the rubber plateau modulus of such a polymer would appear to be an equilibrium modulus, it would not be the equilibrium in a true thermodynamic sense.

There is, however, another method for preventing disentanglement, namely the introduction of permanent chemical crosslinks. A single

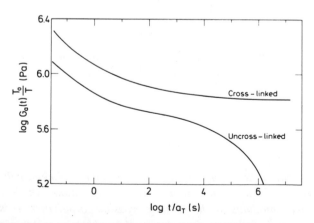

Fig. 1. Stress relaxation curves for polybutadiene with 88% vinyl structure and a weight-average molecular weight of 291 000. The reference temperature is 263 K. The crosslinked sample has about 60 crosslinks per chain. Reproduced from Ref. 1.

crosslink anywhere along a linear chain would prevent disentanglement by the reptation process. Each of the two ends would now have to disentangle by much slower processes, e.g. by the retracing process which was also suggested by de Gennes.[3] In a well-crosslinked elastomer, disentanglement will be prevented almost completely. The dangling chain ends will still be able to disentangle, but in well-crosslinked elastomers they amount to a very small fraction of the network only. The upper curve of Fig. 1 shows the stress relaxation modulus of a well-crosslinked polymer at 263 K, which is 8 K above the glass transition temperature. It can be seen that the equilibrium modulus in this case is well above the rubber plateau modulus of the uncrosslinked polymer. This would not be true at low degrees of crosslinking.

The first question that should be asked is whether chain entangling contributes appreciably to the equilibrium modulus of a well-crosslinked elastomer. If it does, the second question is by how much and whether the contribution is related to the rubber plateau modulus of the uncrosslinked polymer. So far, it has not even been possible to reach agreement on the first question, and the disagreement has become somewhat of a controversy. It should be pointed out that an equilibrium modulus which is unchanged by heating–cooling as well as by swelling–deswelling should be considered the true equilibrium modulus. It may turn out to be impossible to verify by experiment whether it is also the equilibrium in a true thermodynamic sense.

The hypothetical experiment described above suggests that the entanglement structure would give rise to an equilibrium modulus if the disentanglement process could be prevented. In the hypothetical experiment, disentanglement was prevented by the extremely high molecular weight. In a well-crosslinked elastomer, disentanglement is prevented by the presence of permanent crosslinks which effectively trap the entanglement structure. The Langley method[4] is based on the assumption that chemical crosslinks and the permanently trapped fraction of the original entanglement structure give additive contributions. Graessley and co-workers[5,6] have found excellent agreement with theoretical predictions. On the other hand, Flory[7] assumes that the elastic contribution from the trapped entanglement structure relaxes to zero at the true elastic equilibrium. Mark[8] and others have found remarkably good agreement with the predictions of Flory. However, the latter work is mostly based on end-linked networks which do not efficiently trap the entanglement structure, unless the end-linking process is nearly perfect.

There are enough theories of rubber elasticity to choose from. In order to resolve the controversy it therefore seems paramount to design an experiment which would require no theory of rubber elasticity and none of the usual assumptions, e.g. regarding the functionality of the crosslinks and the homogeneity of the crosslinking process. The present chapter describes such an experiment, which is a simplification of the so-called 'two-network method'.[9]

MATERIALS AND METHODS

Polymer Preparation and Characterization

High-vinyl polybutadiene was prepared by anionic polymerization, by use of high vacuum procedures outlined elsewhere.[10] The polymerization was performed in cyclohexane at 20°C with n-butyllithium as initiator, and tetramethylethylenediamine was used to control the polybutadiene microstructure. The reaction was terminated by addition of degassed methanol.

The weight-average molecular weight of the polymer was 2×10^6, as determined by low-angle laser light scattering in cyclohexane. Size exclusion chromatography indicated a narrow molecular weight distribution, $M_w/M_n < 1.1$. The polymer microstructure was determined by proton NMR analysis: 88% vinyl and 12% 1,4 units. The high vinyl content causes a convenient glass transition temperature of 255 K.

Sample Preparation and Stress Relaxation Measurements

A thin film was cast from a 2–3% solution of the polymer in hexane. After solvent evaporation, the film was kept in a vacuum oven for 1 month. Thin strips with dimensions $40 \times 3 \times 0.4$ in millimeters were cut with parallel knives and mounted in the stress relaxation apparatus. The stretch ratios used in the present study were about 1·4, and the stress was recorded in the time interval 1–50000 s, or until the sample broke. Using the time–temperature principle, stress relaxation curves measured at different temperatures were superimposed to give the behavior over a wide range of time.

Crosslinking in the Strained State and Equilibrium Measurements

The stress relaxation process can be stopped anywhere along the stress relaxation curve by quenching the sample to well below the glass transition temperature. Conformational rearrangements are now practically impossible and crosslinks may conveniently be introduced by high energy electrons.[11] In the present study, samples were stretched at 283 K and quenched to 25 K below T_g after 1000 s of relaxation. The extension ratio was about 1·6. Radiation doses were 50, 100 and 200 kGy (5, 10 and 20 Mrad) of 10 MeV electrons with a dose rate of about 100 kGy/h. The temperature was then raised to 263 K for 10 min to allow any unreacted free radicals to react, before heating the sample to 283 K for another ~ 10 min. The sample was then heated to 323 K where it was allowed to reach elastic equilibrium. At higher degrees of crosslinking, equilibrium was reached within a few minutes at this temperature. The sample was finally cooled to 283 K in order to measure the equilibrium force at the stress relaxation temperature. The entire experiment was conducted in a nitrogen atmosphere.

The three most important quantities determined in the present experiment are: f, the stretching force of the uncrosslinked sample after 1000 s of relaxation at 283 K; f_c, the force immediately after crosslinking and heating to 283 K; and f_e, the equilibrium force at 283 K. It should be emphasized that the stretched sample is held at constant length throughout the entire experiment. It means that the experiment allows a direct comparison of the three forces both at the same temperature and at constant length.

RESULTS AND DISCUSSION

The stress relaxation modulus of the uncrosslinked polymer at 283 K is shown in Fig. 2, while Table 1 gives the magnitudes of the required horizontal shifts. Minor vertical shifts were also required, probably due to inaccuracies in the determination of sample dimensions. The curve shows a very pronounced rubber plateau because of the extremely high molecular weight. It was, unfortunately, impossible to extend the experiment further into the terminal zone, since the sample invariably broke at this stage, probably due to disentanglement by reptation.

Figure 3 shows an experiment where crosslinks are introduced in the strained state after relaxation for 1000 s at 283 K, which is at the entrance to

TABLE 1
Horizontal Shift Factors Used to Construct Fig. 2

Temperature (K):	258	263	273	283	303
$\log a_T$	3·8	3·0	1·0	0	−1·5

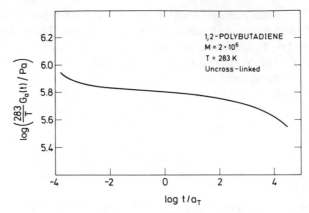

Fig. 2. Stress relaxation curve for uncrosslinked polybutadiene with 88% vinyl structure and a weight-average molecular weight of 2×10^6. The reference temperature is 283 K.

the terminal zone, i.e. disentanglement by reptation has just begun. The ordinate shows the directly measured forces, and the three forces f, f_c and f_e are indicated in the figure. Table 2 summarizes the results for three different doses of irradiation. The quantity $100(f - f_c)/f$ gives the percentage change in the stretching force caused by the crosslinking process, including the subsequent relaxation during heating to the stress relaxation temperature. The quantity $100(f - f_e)/f$ gives the relative difference between the stress relaxation force and the equilibrium force in per cent. The very low values of the quantity $100(f - f_e)/f$ irrespective of irradiation dose clearly show that the crosslinks do not contribute to the stretching force in the present experiment where the sample is held at constant length. The crosslinks merely serve the purpose of trapping the highly entangled structure. Table 2 also shows that the measured equilibrium forces are only slightly smaller

TABLE 2
Results from Crosslinking in the Strained State with Extension Ratios of about 1·6

	Sample 7	Sample 9	Sample 8
Dose (kGy)	50	100	200
$(f - f_c)/f \times 100$ (%)	3·0	2·9	3·6
$(f - f_e)/f \times 100$ (%)	8·9	7·7	5·4

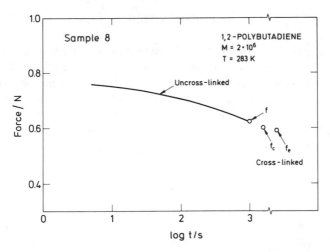

Fig. 3. Force at constant length for the same polymer as in Fig. 2, uncrosslinked (solid curve) and crosslinked below T_g in the strained state. The force f_c is obtained by subsequent heating of the sample to the stress relaxation temperature of 283 K. The equilibrium force at 283 K is obtained after heating to 323 K followed by cooling back to 283 K.

than the stress relaxation forces prior to crosslinking, as expressed by $100 \, (f-f_e)/f$. Since the crosslinks contribute nothing to the force in the present experiment, irrespective of degree of crosslinking, this shows that the trapped entanglement structure contributes substantially to the equilibrium modulus; and, furthermore, that the contribution from the trapped entanglement structure simply is equal to the rubber plateau modulus. This is of course true only because we have chosen an extremely high molecular weight and relatively high degrees of crosslinking. For lower molecular weight polymers, incomplete trapping should be expected in most cases, as given by the Langley method.[4]

Considerable chain scission would make it impossible to interpret the results of the present method. However, the small values of the quantity $100(f-f_e)/f$ directly prove that chain scission must be small or absent in the present system. It should also be pointed out that no theory is needed to make the above conclusions and that the functionality of the crosslinks need not be known. A distribution of crosslink functionalities as well as inhomogeneous crosslinking would not be serious problems in the present method, since the crosslinks merely serve the purpose of trapping the entanglement structure.

CONCLUSION

Crosslinking in the strained state is used to demonstrate the effect of chain entangling on the equilibrium modulus of well-crosslinked rubber networks with few dangling chain ends. The contribution of chain entangling in such networks is simply equal to the rubber plateau modulus of the uncrosslinked polymer. This conclusion may be reached without a theory of rubber elasticity and without any assumptions regarding chain scission, crosslink functionality, and efficiency and homogeneity of the crosslinking process.

ACKNOWLEDGEMENT

This work was supported by grants from the Danish Research Council for Natural Science and the Danish Council for Science and Industrial Research under the FTU program.

REFERENCES

1. Kramer, O., *Brit. Polym. J.*, 1985, **17**, 129.
2. de Gennes, P. G., *J. Chem. Phys.*, 1971, **55**, 572.
3. de Gennes, P. G., *J. Physique*, 1975, **36**, 1199.
4. Langley, N. R., *Macromolecules*, 1968, **1**, 348.
5. Dossin, L. M. and Graessley, W. W., *Macromolecules*, 1979, **12**, 123.
6. Pearson, D. S. and Graessley, W. W., *Macromolecules*, 1980, **13**, 1001.
7. Flory, P. J., *J. Chem. Phys.*, 1977, **66**, 5720.
8. Mark, J. E., *Acc. Chem. Res.*, 1985, **18**, 202.
9. Kramer, O., Carpenter, R. I., Ty, V. and Ferry, J. D., *Macromolecules*, 1974, **7**, 79.
10. Morton, M. and Fetters, L. J., *Rubber Chem. Techn.*, 1975, **48**, 360.
11. Batsberg, W. and Kramer, O., *J. Chem. Phys.*, 1981, **74**, 6507.

36

A SIMPLE MODEL OF RANDOM
TETRAFUNCTIONAL NETWORKS WITH DEFECTS

Andrzej Ziabicki

Polish Academy of Sciences,
Institute of Fundamental Technological Research,
Swietokrzyska 21, Warsaw, Poland

and

Janusz Walasek

Technical University, Radom, Poland

ABSTRACT

Real crosslinked systems include a wide variety of structures. Crosslinks can be connected in many ways, including connection by multiple chains, intramolecular loops, etc. A simple topological model of random tetrafunctional crosslinked systems has been described.

A system produced by crosslinking consists of uncrosslinked polymer chains and/or unused crosslinking agent, a low-crosslinked set of aggregates (sol) and (beyond some critical point) also a densely crosslinked network (gel). In a tetrafunctional system one can distinguish seven 'topological elements' connected to a crosslink: single- and multiple-connected chains, free-end chains, loops and voids (unsaturated functionalities). These elements can appear at a crosslink (junction) in 34 different combinations. First-order topological structure of the system is described by fractions of the 34 junction types. All macroscopic characteristics, such as gel point, modulus of elasticity, swelling equilibrium and sol–gel ratio, can be expressed as functions of the junction type distribution.

The distribution of junction types has been derived from the assumption of randomness, and expressed as an explicit function of the number of the seven topological elements. These, in turn, can be estimated from basic information about crosslinking kinetics, and from an additional model assumption related to conformational properties of the polymer chains. In this way, complex structure of a crosslinked system and its physical properties can be described in terms of a few parameters which can be determined experimentally.

INTRODUCTION

Topological structure of real crosslinked systems admits many defects. Not all potential functionalities of crosslinks (network junctions) are saturated, some polymer chains being unattached to any crosslinks, some remaining elastically ineffective as attached to the network with one end only. This fact, recognized early in the development of the rubber elasticity theory, led to several specific network models.[1−4] The presence of defects strongly affects the gel point, i.e. the total concentration of crosslinks required for gelation, sol and gel fractions, modulus of elasticity, swelling behaviour, etc. The existing models, however, are incomplete as they consider only some, rather arbitrarily selected, structures. Double-connected pairs of crosslinks ('doublets') are generally neglected, although the probability of their occurrence is comparable to that of widely recognized primary loops: the same can be said about higher multiplets, configurations of two loops at one junction, etc.

The classification of topological defects in crosslinked systems and their effect on physical properties has been discussed in two papers by the present authors.[5,6] Here we will show how the general rules outlined before can be applied to a tetrafunctional network (strictly speaking, a tetrafunctional crosslinked system). All possible first-order topological structures will be considered, and macroscopic properties of the system expressed as explicit functions of the corresponding distribution. Later, two model assumptions—uniformity and randomness of the distribution of defects, and quasi-equilibrium chain conformation—will reduce the original number of variables (34 fractions of topologically different crosslinks) to two, thus making possible their direct determination from the experimental data.

It should be emphasized that the model does not involve any specific assumptions about the way in which the system was crosslinked. The model is not intended for an *a priori* prediction of network structure. Its main goal is to provide a simple tool for the interpretation of the available experimental data in a more complete and correct way. The parameters required for numerical evaluation of structural characteristics and macroscopic behaviour can either be determined from physico-chemical measurements (modulus, swelling equilibrium, etc.) performed on the crosslinked system, and/or derived from information about the stoichiometry and kinetics of the process of crosslinking. Although the model can be used both for end-linking and vulcanization-type processes, we will discuss here only the first case, which yields the full set of topological structures. The junction distribution function for the other case (vulcanization) has been given in Ref. 6.

STRUCTURAL ELEMENTS AND JUNCTION DISTRIBUTION

Topologically different structural elements attached to tetrafunctional crosslinks are shown in Fig. 1. They include singlets, i.e. chains connecting a pair of junctions; free-end chains attached to a junction with one end only; loops, i.e. chains attached with both ends to the same junction; doublets, triplets and quadruplets, i.e. two, three or four chains connecting the same pair of junctions. To make the picture complete, we will also consider voids, or unsaturated functionalities, as structural elements of a junction, and unattached chains as elements of the system. The molecules of unreacted crosslinking agent are considered as 'junctions' with four voids (v^4). The structural elements shown in Fig. 1 saturate zero (unattached chains), one (singlet, free-end chain or void), two (loop, doublet), three (triplet) or four functionalities of the junction to which they are attached. There are 34

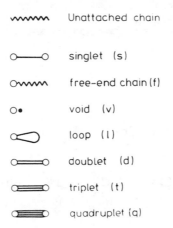

⌇⌇⌇⌇⌇	Unattached chain
O——O	singlet (s)
O⌇⌇⌇⌇	free-end chain (f)
O•	void (v)
⊂⊃	loop (l)
O══O	doublet (d)
⊂══O	triplet (t)
⊂══⊃	quadruplet (q)

Fig. 1. Topological elements appearing in a tetrafunctional crosslinked system.

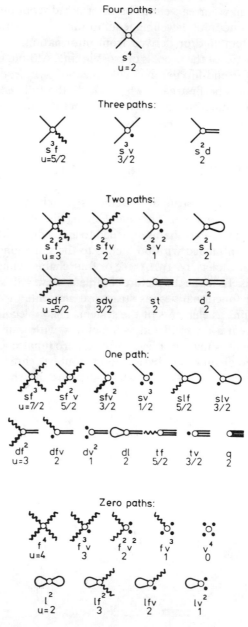

Fig. 2. Possible types of tetrafunctional junctions.

possible combinations of these elements which determine first-order structure of the crosslinked system (Fig. 2). We will use lower-case characters to denote fractions of functionalities in the system occupied by singlets (s), free-end chains (f), voids (v), loops (l), doublets (d), triplets (t) and quadruplets (q). These are normalised to unity:

$$s + f + v + l + d + t + q = 1 \qquad (1)$$

The same characters are used for labelling different junction configurations. Thus, s^4 denotes a junction with four singlets, ds^2 one with a doublet and two singlets, etc.

In Fig. 2 the junctions are segregated according to the number of independent paths leading to other junctions. There is only one type of junction (s^4) which provides four paths. Three types of junctions (s^3f, s^3v, s^2d) provide three paths. There are eight junction types with two paths, thirteen with one path and nine types without any connection to the gel. The numbers (u) shown in Fig. 2 below the junction codes denote the number of network chains per junction: voids contribute zero; free-end chains and loops contribute one chain; and singlets, doublets, triplets and quadruplets, each shared by two junctions, contribute 1/2, 1, 3/2 and 2 chains, respectively.

Thirty-four variables, fractions of different types of junctions, n_i, and the number of unattached chains, N_{un}, are required for a complete description of the primary structure of the system.

THE BALANCE OF CHAINS AND JUNCTIONS, CONTINUITY OF PATHS AND PROPERTIES OF THE SYSTEM

We will consider a system composed of N_0 linear chains with reacting groups A on their ends, and N_j molecules of a tetrafunctional crosslinking agent with groups B. According to the adopted convention, all these molecules are defined as 'junctions'. Crosslinking involves reaction between groups A and B. No intramolecular bonds, AA, or BB are admitted:

$$\begin{array}{ccc} & B & B \\ & | & | \\ A\!-\!A + B\!-\!R\!-\!B = A\!-\!AB\!-\!R\!-\!B \\ & | & | \\ & B & B \end{array} \qquad (2)$$

Continuity of paths emerging from an arbitrary junction and leading to other junctions provides a fundamental characteristic of a crosslinked system. Using the theory of cascade processes[7] with the probability-generating function based on the fractions $n^{(0)}$, $n^{(1)}$, ..., $n^{(4)}$ of junctions with 0, 1, ..., 4 emerging paths (cf. Ref. 4), one arrives at the following expression for the 'probability of extinction', e, i.e. the probability that a path starting from a randomly chosen junction will be terminated before reaching infinity:

$$e = (1/2)\{[(1 + 3n^{(3)}/4n^{(4)})^2 + n^{(1)}/n^{(4)}]^{1/2} - 1 - 3n^{(3)}/n^{(4)}\} \tag{3}$$

The probability of extinction, together with the distribution of junction types, yields several macroscopic characteristics of the crosslinked system.

Gel Point

The critical condition for gelation requires that the probability of extinction, e, be less than unity. From eqn (3):

$$(e < 1) \Rightarrow n^{(1)} < 8n^{(4)} + 3n^{(3)} \tag{4}$$

The above relation can be evaluated in terms of crosslink density when the fractions $n^{(k)}$ are known.

The Number of Junctions in the Gel

Junctions contained in the gel must have at least one path leading to infinity. Consequently

$$N_{j,gel}/N_j = \sum_{k=1}^{4} n^{(k)}[1 - e^k] \tag{5}$$

which, combined with the normalisation condition

$$n^{(0)} + n^{(1)} + n^{(2)} + n^{(3)} + n^{(4)} = 1 \tag{6}$$

yields

$$N_{j,gel}/N_j = 1 - n^{(0)} - e[n^{(1)} + en^{(2)} + e^2n^{(3)} + e^3n^{(4)}] \tag{7}$$

The Number of Elastically Effective Junctions

The junction is elastically effective if at least three paths emerging from it lead to infinity. In terms of the path distribution function $n^{(k)}$,

$$N_{j,el}/N_j = \sum_{k=3}^{4} \sum_{m=3}^{k} n^{(k)} \binom{k}{m} (1 - e)^m e^{k-m}$$

$$= (1 - e)^3 [n^{(3)} + n^{(4)}(1 + 3e)] \tag{8}$$

The fraction of elastically effective junctions determines the contraction ratio, A_c, appearing in the equation for the modulus

$$G = N_{ch,el} A_c kT \tag{9}$$

$$A_c = \langle h_{net}^2 \rangle / \langle h_0^2 \rangle \tag{10}$$

which, for Gaussian chains, reduces to [5,6]

$$A_c = (N_{j,el} - 1)/N_{ch,el} \simeq N_{j,el}/N_{ch,el} \tag{11}$$

$N_{ch,el}$ denotes the number of elastically effective network chains, and $N_{j,el}$ the number of elastically effective junctions in the system. Combination of eqns (8)–(11) yields the shear modulus, as a function of the fractions of junctions with one, three and four paths:

$$G = N_{j,el} kT = N_j kT (1 - e)^3 [n^{(3)} + n^{(4)}(1 + 3e)] \tag{12}$$

The number of elastically effective junctions, $N_{j,el}$, controls also the elastic contribution to the free energy of swelling.

The Distribution of Chains Between Sol and Gel

The total number of chains in the system, $N_{ch,0}$, includes the unreacted portion, N_{un}, not attached to any junctions, a portion attached to the gel, $N_{ch,gel}$, and one present in the soluble fraction (sol), $N_{ch,sol}$:

$$N_0 = N_{un} + N_{ch,gel} + N_{ch,sol} \tag{13}$$

The gel fraction, $N_{ch,gel}$, can be calculated from the condition (7) by summation of contributions introduced by various types of junctions

$$N_{ch,gel} = N_j \sum_{i=1}^{34} n_i u_i [1 - e^{k(i)}] \tag{14}$$

n_i represent fractions of junctions of the ith type and u_i the number of network chains shared by the junction (indicated in Fig. 2). $k(i)$ is the number of paths connecting the ith junction to other junctions. Similarly, the number of unattached chains, N_{un}, can be calculated as

$$N_{un} = N_0 - N_j \sum_{i=1}^{34} n_i u_i \tag{15}$$

The relation between the distribution of paths, $n^{(k)}$, and the distribution of junction configurations, n_i, is straightforward—$n^{(k)}$ represent sums of n_i

with the same number of paths:

$$n^{(k)} = \sum_{k \text{ paths}} n_i \tag{16}$$

The fractions n_i are normalized to unity, as were fractions $n^{(k)}$:

$$n_1 + n_2 + \cdots + n_{34} = 1 \tag{17}$$

It can be seen that the complete first-order structure of the network and its macroscopic properties are described by the total number of chains, N_0, the amount of the crosslinking agent (total number of junctions), N_j, and 33 independent fractions of junction types. So many variables cannot be determined from experimental measurements, and we will reduce their number by applying a few specific model assumptions.

RANDOM DISTRIBUTION OF STRUCTURAL ELEMENTS AMONG JUNCTIONS

The assumption that all structural elements, singlets, free-end chains, voids, loops, etc., are distributed randomly among the junctions and appear with frequencies dependent on their global concentration in the system (s, f, v, l, ..., q) seems acceptable as a first approximation. The distribution of tetrads of functionalities occupied by various structural elements has been obtained from a combinatorial analysis, the details of which will be published separately.[8] For a set of monofunctional structural units, such an analysis would yield a simple tetranomial distribution with fractions of structural elements (s, f, v, ...) appearing in the appropriate powers: in the case of mono-, bi-, tri- and tetrafunctional elements the result is less trivial. In addition to the fractions of functionalities occupied by various structural elements (not number fractions of elements themselves), there appear also sums of functionalities occupied by monofunctional and bifunctional elements:

$$\Sigma = s + f + v \tag{18}$$

$$\Delta = l + d \tag{19}$$

The resulting distribution can be presented in the form

$n(\alpha, \beta, \gamma, \delta, \varepsilon, \phi, \eta)$

$$= \left[\frac{2(4/3)^\phi (\alpha + \beta + \gamma)! \times (\Sigma - t/3)^{2(1-\phi-\eta)-\delta-\varepsilon}}{(2 - \delta - \varepsilon)! \alpha! \beta! \gamma! \delta! \varepsilon! (\Sigma + \Delta - t/3)^{1-\phi-\eta} \times \Sigma^{\alpha+\beta+\gamma}} \right] s^\alpha f^\beta v^\gamma d^\delta l^\varepsilon t^\phi q^\eta \tag{20}$$

α, β, γ, δ, ε, ϕ, η denote numbers of junction functionalities occupied, respectively, by singlets, free-end chains, voids, doublets, loops, triplets and quadruplets. Equation (20) reduces 33 independent variables (fractions of junction types, n_i) to six independent fractions of functionalities occupied by individual structural elements. This is still too much for the available experimental possibilities and additional reduction is required.

EQUILIBRIUM CONFORMATION OF NETWORK CHAINS

Additional relations reducing the number of independent variables will be obtained from the assumption that the distribution of end-to-end distances of network chains corresponds to thermodynamic equilibrium. This is equivalent to assuming that the formation of crosslinks was slow enough compared with intramolecular motions which control chain conformation. This assumption seems to be justified, at least for crosslinking in solutions.

The assumption of equilibrium conformation yields relations between the concentrations of individual structural elements. A similar approach was used by James and Guth,[9] and Jacobson and Stockmayer[10] for the calculation of the probability of intramolecular loops.

Given a system of volume V_0 with N_j randomly distributed junctions, one can fix one chain end in a chosen junction and ask what is the probability that the other end will contact one of the remaining $(N_j - 1)$ junctions uniformly distributed in space to produce a singlet, or, alternatively, that the chain will return to the original junction producing a loop. The ratio of these two probabilities is interpreted as (s/l). In a similar way, probability of a doublet, triplet and quadruplet can be compared with that of two, three or four singlets, yielding the ratios (d/s^2), (t/s^3) and (q/s^4). For a monodisperse system of network chains, these ratios result in the form[6]

$$1/s = W(h = 0)(V_0/N_j) \tag{21}$$

$$d/s^2 = \int W^2(h)\, d^3h(V_0/N_j) \tag{22}$$

$$t/s^3 = \int W^3(h)\, d^3h(V_0/N_j)^2 \tag{23}$$

$$q/s^4 = \int W^4(h)\, d^3h(V_0 N_j)^3 \tag{24}$$

$W(h)$ denotes normalised distribution of end-to-end vectors, h, and (N_j/V_0) is the junction density.

We will write down the above ratios for two extreme cases of molecular behaviour. For ideally flexible, freely jointed, chains with Gaussian end-to-end distribution, eqns (21)–(24) reduce to

$$l/s = 2^{3/2}Z \tag{21a}$$

$$d/s^2 = Z \tag{22a}$$

$$t/s^3 = (4/3)^{3/2}Z^2 \tag{23a}$$

$$q/s^4 = 2^{3/2}Z^3 \tag{24a}$$

The parameter involved

$$Z = (V_0/N_j)[3/4\pi\langle h_0^2\rangle]^{3/2} \tag{25a}$$

is inversely proportional to crosslink density and to the volume swept out by the macromolecular coil. In the case of ideally rigid rods with length L, the end-to-end distance is subject to delta-distribution, and eqns (21)–(24) reduce to

$$l/s = 0 \tag{21b}$$

$$d/s^2 = Q \tag{22b}$$

$$t/s^3 = Q^2 \tag{23b}$$

$$q/s^4 = Q^3 \tag{24b}$$

with the parameter

$$Q = (V_0/N_j)[1/4\pi L^2 R_o] \tag{25b}$$

R_o denotes molecular radius of the junction. Finally, the relative probabilities of individual defects have been calculated for worm-like chains. The end-to-end distribution $W(h)$ for worm-like chains was derived by Daniels,[11] and Gobush et al.,[12] and used for discussion of conformational properties by Yamakawa and Stockmayer.[13] 'Second-order Daniels distribution'[12,13] was used:

$$\begin{aligned}
W(h) = {} & [3/2\pi aL]^{3/2}\exp(-3h^2/2aL) \\
& \times [1 - x(5/8 - 2H^2 + 33H^4/40) \\
& - x^2(79/640 + 329H^2/640 - 6799H^4/1600 \\
& + 3441H^6/1400 - 1089H^8/3200) - x^3(\cdots) - \cdots] \tag{26}
\end{aligned}$$

i.e. one including second-order terms in the effective chain rigidity parameter $x = a/L$. L denotes the total contour length of the chain, and $2a$, a measure of the intrinsic stiffness of the chain, represents the so-called 'persistence length' of the chain. For freely jointed Gaussian chains ($x = 0$),

a reduces to the length of the Kuhn segment. $H^2 = h^2/aL$ is the dimensionless square end-to-end distance.

The probabilities of defects result in the form

$$l/s = 2^{3/2}(1 - 5x/8 - 79x^2/640 + \cdots)Z' \tag{21c}$$

$$d/s^2 = (1 + x/16 - 1\cdot755\,566x^2 + \cdots)Z' \tag{22c}$$

$$t/s^3 = (4/3)^{3/2}(1 - x/3 - 0\cdot897\,685x^2 + \cdots)Z'^2 \tag{23c}$$

$$q/s^4 = 2^{3/2}(1 - 27x/32 - 1\cdot384\,842x^2 + \cdots)Z'^3 \tag{24c}$$

The parameter Z' appearing in eqns (21c)–(24c) depends on chain rigidity, $x = a/L$:

$$Z' = (V_0/N_j)[3/4\pi aL]^{3/2} \tag{25c}$$

Depending on whether the comparison is made at constant length, L, or constant intrinsic stiffness, a, Z' is proportional to $x^{-3/2}$ or $x^{3/2}$.

The relative probability of all defects decreases with increasing chain stiffness and/or increasing molecular weight of network chains.

Significant difference in the behaviour of rigid vs. flexible chains is noted when the probability of loops is considered. For flexible chains, replacement of a singlet by a loop (probability ratio: l/s) is nearly as probable as the replacement of two singlets by a doublet (d/s^2). Ideally rigid rods do not form any loops. Both for flexible and rigid chains, the relative probabilities of doublets, triplets and quadruplets satisfy simple relations, namely

$$t/s^3 = \text{const. } (d/s^2)^2 \tag{27}$$

$$q/s^4 = \text{const.' } (d/s^2)^3 \tag{28}$$

The relative probability of loops and multiplets increases with increasing parameters Z, Z' and Q, i.e. decreases with increasing volume concentration of crosslinks and decreases with increasing molecular weight of network chains, M. For flexible Gaussian chains the concentration of loops and multiplets increases with the corresponding powers of the parameters Z and Q. Introducing volume fraction of polymer in the system

$$v_2 = \text{const.} \times N_0 M/V_0 \tag{29}$$

one finds that for Gaussian chains the probabilities of defects are controlled by

$$Z \propto (V_0/N_j)M^{-3/2} \propto v_2^{-1}M^{-1/2}$$

while in the case of ideally rigid rods (with molecular weight proportional to the length, L) by

$$Q \propto (V_0/N_j)M^{-2} \propto v_2^{-1}M^{-1}$$

In both cases the relative probability of triplets and quadruplets is proportional to the square and cube of the relative probability of doublets. In the range of small values of Z, or Q, i.e. high volume fractions of the polymer and/or high molecular weight between crosslinks, M, triplets and quadruplets can be neglected. On the other hand, the appearance of loops and doublets in a system of flexible chains is equally probable.

The relative probabilities of loops, doublets, triplets and quadruplets are plotted vs. parameters Z and Q in Fig. 3.

Figure 4 presents the ratios (ls/d), (ts/d^2) and (qs^2/d^3) as functions of the effective chain flexibility, $x = a/L$. It is evident that for chains with low rigidity $(x < 0.03)$ all ratios are practically constant, equal to those predicted for Gaussian chains (eqns (21a)–(24a)). In the range of higher values of x, the probabilities of higher multiplets and loops decrease, the loops reaching asymptotically (rods: $x = \infty$) zero concentration, triplets and quadruplets dropping to the level $(ts/d^2 = qs^2/d^3 = 1)$ determined by eqns (21b)–(24b). The end-to-end distribution which led to eqns (21c)–(24c) and the characteristics presented in Fig. 4 are meaningful in the range of rather small x, not exceeding 0.1.

Equations (21)–(24) reduce the number of independent parameters describing junction distribution, n_i, to two: the fraction of singlets, s, and that of free-end chains, f. The fractions of loops, doublets, triplets and quadruplets can be expressed through s, and the fraction of voids, v, results

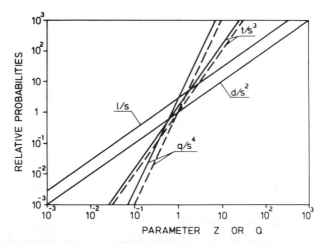

Fig. 3. Relative probabilities of loops (l/s), doublets (d/s^2), triplets (t/s^3) and quadruplets (q/s^4) plotted vs. parameter Z or Q. Solid lines: flexible chains, abscissa $= Z$. Dotted lines: rigid rods, abscissa $= Q$.

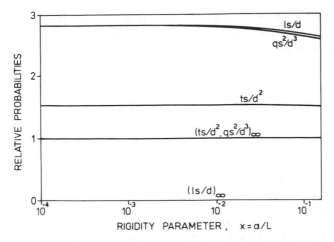

Fig. 4. Reduced probabilities of loops (ls/d), triplets (ts/d^2) and quadruplets (qs^2/d^3) vs. chain rigidity, $x = a/L$. Horizontal line represents asymptotic level of the ratios (ts/d^2) and (qs^2/d^3) at $x = \infty$ (ideally rigid rods).

from the normalization equation (1). Information about the volume concentration of network chains, N_0/V_0 and their molecular weight, M, as well as the total concentration of the crosslinking agent, N_j/V_0, is given by the conditions of crosslinking. When the chains are short and/or stiff, the concentration of loops and multiplets should be calculated from the appropriate end-to-end distribution function $W(h)$.

CLOSING REMARKS

The simple model presented in this paper describes first-order topological structure of a crosslinked system formed by end-linking of linear chains with a tetrafunctional agent. The structure allows for 34 different configurations of seven structural elements (single chains connecting a pair of junctions, free-end chains, void functionalities and multiplets). The macroscopic characteristics of the system (gel point, gel/sol ratio, shear modulus, contraction factor, swelling behaviour) are expressed through 34 variables: fractions of individual types of junctions. The assumption of random distribution of structural elements among the crosslinks reduces the number of independent variables to seven. Further reduction, based on the assumption of equilibrium chain conformation, leaves two independent structural parameters: the fraction of single chains connecting a pair of junctions, s, and the fraction of free-end chains, f. These can easily be determined from experimental data.

REFERENCES

1. Flory, P. J., *Principles of Polymer Chemistry*, Cornell University Press, Ithaca, New York, 1953.
2. Tobolsky, A. V., Metz, D. J. and Mesrobian, R. B., *J. Am. Chem. Soc.*, 1950, **72**, 1942.
3. Scanlan, J., *J. Polymer Sci.*, 1960, **43**, 501; Mullins, L. and Thomas, A. G., ibid., 1960, **43**, 13; Case, L. C., ibid., 1960, **45**, 397.
4. Ziabicki, A. and Klonowski, W., *Rheol. Acta*, 1975, **14**, 105, 113; Ziabicki, A., *Colloid & Polymer Sci.*, 1974, **252**, 767.
5. Ziabicki, A. and Walasek, J., *Macromolecules*, 1978, **11**, 471.
6. Ziabicki, A., *Polymer*, 1979, **20**, 1373.
7. Rosenblatt, M., *Random Processes*, Oxford University Press, Oxford, 1962.
8. Walasek, J. and Ziabicki, A., *Colloid & Polymer Sci.*, 1988, **266**, in press.
9. James, H. and Guth, E., *J. Chem. Phys.*, 1947, **15**, 669.
10. Jacobson, H. and Stockmayer, W. H., *J. Chem. Phys.*, 1950, **18**, 1600.
11. Daniels, H. E., *Proc. Roy. Soc. (Edinburgh)*, 1952, **63A**, 290.
12. Gobush, W., Yamakawa, H., Stockmayer, W. H. and Magee, W. S., *J. Chem. Phys.*, 1972, **57**, 2839.
13. Yamakawa, H. and Stockmayer, W. H., 1972, **57**, 2843.

CONTRIBUTING AUTHOR INDEX

531

SUBJECT INDEX